Fundamental Electrical and Electronic Principles

Fundamental Electrical and Electronic Principles covers the essential principles that form the foundations for electrical and electronic engineering courses. This new edition is extensively updated with a greater focus on electronic principles, evenly balanced with electrical principles. Fuller coverage is given to active electronics, with the additional topics of diodes and transistors, and core topics such as oscilloscopes now reflect state-of-the-art technology.

Each chapter starts with learning outcomes tied to the syllabus. All theory is explained in detail and backed up with numerous worked examples. Students can test their understanding with end-of-chapter assignment questions for which answers are provided. The book also includes suggested practical assignments and handy summaries of equations.

The book forms an excellent core work for beginning further education students with some mathematics background preparing for careers as technicians, and an introductory text for first-year undergraduate students in all engineering disciplines.

Jo Verhaevert is an associate professor and programme leader of the Electronic Engineering curriculum at Ghent University in Belgium, and a senior researcher in Electromagnetics at IMEC, Belgium.

Fundamental Electrical and Electronic Principles

Fourth Edition

Jo Verhaevert

Routledge
Taylor & Francis Group
LONDON AND NEW YORK

Cover image: Shutterstock ©

First published 2024
by Routledge
4 Park Square, Milton Park, Abingdon, Oxon OX14 4RN

and by Routledge
605 Third Avenue, New York, NY 10158

Routledge is an imprint of the Taylor & Francis Group, an informa business

British Library Cataloguing-in-Publication Data
A catalogue record for this book is available from the British Library

Library of Congress Cataloging-in-Publication Data
Names: Verhaevert, Jo, author.
Title: Fundamental electrical and electronic principles / Jo Verhaevert.
Description: Fourth edition. | Abingdon, Oxon; New York, NY: Routledge, 2023. | Revised edition of: Fundamental electrical and electronic principles / Christopher R. Robertson, 2nd ed. Amsterdam; Boston: Elsevier/Newnes, 2001, 2004 reprinted.
Identifiers: LCCN 2023024353 | ISBN 9781032311487 (hbk) | ISBN 9781032311470 (pbk) | ISBN 9781003308294 (ebk)
Subjects: LCSH: Electronics–Textbooks. | Electric circuits–Textbooks.
Classification: LCC TK7816 .R59 2023 | DDC 621.381–dc23/eng/20230729
LC record available at https://lccn.loc.gov/2023024353

ISBN: 978-1-032-31148-7 (hbk)
ISBN: 978-1-032-31147-0 (pbk)
ISBN: 978-1-003-30829-4 (ebk)

DOI: 10.1201/9781003308294

Typeset in Sabon
by Deanta Global Publishing Services, Chennai, India

Contents

Preface

This book is the successor to the book *Fundamental Electrical and Electronic Principles*, third edition, by the author Christopher Robertson. I expressly thank previous author Christopher Robertson for allowing me to completely rework and update his version. This has resulted in the fourth edition of this book, which contains over 240 figures, over 120 worked examples and over 190 additional assignment questions with 13 of them practical in approach.

The main objective of this book is to introduce the reader to the fascinating world of electricity on the one hand and electronics on the other, while providing a meaningful description and analysis of various electrical and electronic systems. Furthermore, it also provides insight and basic knowledge related to recent developments in the field of electricity and electronics. To gain the necessary understanding, this book first covers fundamentals such as resistors and D.C. circuits. Afterwards, capacitors and inductors are covered, and electrical, magnetic and electromagnetic fields are discussed. The second part of this book focuses on semiconductor technology with the basic components of diodes and transistors. Also alternating quantities, D.C. machines and D.C. transients are discussed.

Prof. Jo Verhaevert
Ghent, Belgium

Chapter 1

Fundamentals

1.1 UNITS

Wherever measurements are performed there is a need for a coherent and practical system of units. In science and engineering the International System of units (SI units) forms the basis of all units used. There are seven 'base' units, shown in Table 1.1.

All other units are derived from these 'base' units. A few examples of derived units are shown in Table 1.2, and it is worth noting that different symbols are used to represent the quantity and its associated unit in each case.

For a more comprehensive list of SI units see Appendix A at the back of the book.

1.2 STANDARD FORM NOTATION

Standard form is a method of writing large and small numbers in a form that is more convenient than writing a large number of trailing or leading zeros. For example the speed of light is approximately 300 000 000 m/s. When written in standard form this figure would appear as

$$\mathbf{3.0 \times 10^8\ m/s,\ where\ 10^8\ represents\ 100\ 000\ 000}$$

Similarly, if the wavelength of 'red' light is approximately 0.000 000 767 m, it is more convenient to write it in standard form as

$$\mathbf{7.67 \times 10^{-7}\ m,\ where\ 10^{-7} = 1/10\ 000\ 000}$$

It should be noted that whenever a 'multiplying' factor is required, the base 10 is raised to a positive power. When a 'dividing' factor is required, a negative power is used. This is illustrated below:

10	$=10^1$	1/10	$=0.1$	$=10^{-1}$
100	$=10^2$	1/100	$=0.01$	$=10^{-2}$
1000	$=10^3$	1/1000	$=0.001$	$=10^{-3}$
	etc.		etc.	

One restriction that is applied when using standard form is that only the first non-zero digit must appear before the decimal point. Thus, 46 500 is written as

4.65×10^4 and *not* as 46.5×10^3

DOI: 10.1201/9781003308294-1

Table 1.1 The SI base units

Quantity	*Unit*	*Unit symbol*
Mass	kilogram	kg
Length	metre	m
Time	second	s
Electric current	ampere	A
Temperature	kelvin	K
Luminous intensity	candela	cd
Amount of substance	mole	mol

Table 1.2 Some SI derived units

Quantity		*Unit*	
Name	*Symbol*	*Name*	*Symbol*
Force	*F*	Newton	N
Power	*P*	Watt	W
Energy	W	Joule	J
Resistance	*R*	Ohm	Ω

Similarly, 0.002 69 is written as

$$2.69\times10^{-3} \text{ and } \textit{not} \text{ as } 26.9\times10^{-4} \text{ or } 269\times10^{-5}$$

1.3 'SCIENTIFIC' NOTATION

This notation has the advantage of using the base 10 raised to a power but it is not restricted to the placement of the decimal point. It has the added advantage that the base 10 raised to certain powers has unique symbols assigned. For example if a resistor has a resistance value R = 500 000 Ω. In standard form this would be written as

$$R = 5.0\times10^5\ \Omega.$$

Using scientific notation it would appear as

$$R = 500\ \text{k}\Omega\ (500\ \text{kiloOhm})$$

where the 'k' in front of the Ω represents 10^3.

Not only is the latter notation much neater but it gives a better 'feel' to the meaning and relevance of the quantity.

See Table 1.3 for the symbols (prefixes) used to represent the various powers of 10. It should be noted that these prefixes are arranged in multiples of 10^3. It is also a general rule that the positive powers of 10 are represented by capital letters, with the negative powers being represented by lower-case (small) letters. The exception to this rule is the 'k' used for kilo.

Table 1.3 Unit prefixes used in 'scientific' notation

Multiplying factor	*Prefix name*	*Symbol*
10^{15}	peta	P
10^{12}	tera	T
10^{9}	giga	G
10^{6}	mega	M
10^{3}	kilo	k
10^{-3}	milli	m
10^{-6}	micro	μ
10^{-9}	nano	n
10^{-12}	pico	p

WORKED EXAMPLE 1.1

Q Write the following quantities in a concise form using (a) standard form and (b) scientific notation: (i) 0.000 018 A, (ii) 15 000 V, (iii) 250 000 000 W.

(a) (i) $0.000\,018\text{A} = 1.8 \times 10^{-5}$ A

(ii) 15000 V $= 1.5 \times 10^{4}$ V

(iii) $250\,000\,000$ W $= 2.5 \times 10^{8}$ W

(b) (i) $0.000\,018$ A $= 18\ \mu$A

(ii) 15000 V $= 15$ kV

(iii) $250\,000\,000$ W $= 250$ MW

The above example illustrates the neatness and convenience of the scientific or engineering notation.

WORKED EXAMPLE 1.2

Q Write the following quantities in scientific (engineering) notation: (a) 25×10^{-5} A, (b) 3×10^{4} W, (c) 850 000 J, (d) 0.0016 V.

(a) 25×10^{-5} A $= 250 \times 10^{-6}$ A

and since 10^{-6} is represented by μ (micro)

then 25×10^{-5} A $= 250\ \mu$A

Alternatively, 25×10^{-5} A $= 0.25 \times 10^{-3} \times 10^{-2}$ A

so 25×10^{-5} A $= 0.25$ mA

(b) 3×10^{-4} W $= 0.3 \times 10^{-3}$ W or 300×10^{-6} W

so 3×10^{-4} W $= 0.3$ mW or $300\ \mu$W

(c) $850\,000$ J $= 850 \times 10^{3}$ J or 0.85×10^{6} J

so $850\,000$ J $= 850$ kJ or 0.85 MJ

(d) 0.0016 V $= 1.6 \times 10^{-3}$ V

so 0.0016 V $= 1.6$ mV

1.4 ELECTRIC VERSUS ELECTRONIC

Although the electronics domain was not invented until 1883 through the Edison effect, electrical appliances had existed for 100 years before then. The first electric batteries were invented by Alessandro Volta around 1800. His contribution was so important that the unit volt was named after him. In addition, the electric telegraph was invented and commercialised in the 1830s by Samuel Morse. This led to the first transatlantic telegraph cable in 1866 that used real-time communication between Europe and the United States. Other examples of electrical appliances still in use today are lamps, vacuum cleaners and toasters.

Thomas Edison (1847–1931) was an American inventor and founder of the General Electric Company, who made his fortune buying inventions from others and patenting them by himself. If these proved successful, he perfected them and put them into production. Edison was the long-time record holder for the largest number of patents granted to a person (approximately 1400). The light bulb (1879) and the phonograph (1877) are two of his best-known products.

Samuel Morse (1791–1872) was an American inventor and painter. Morse became famous for his portraits and his paintings of historical events. At a later age he became interested in quick communication over long distances. He designed the Morse code for communication in 1835, for use in a telegraph device combining electricity and magnetism.

So what is the difference between electrical versus electronic appliances? The answer is given by how those devices manipulate the electricity to do their job. Electrical appliances absorb energy from the electrical current and simply transform it into another form of energy, such as light, heat or movement. In a toaster, the heating elements convert the electrical energy in heat to toast your sandwich. In a vacuum cleaner, the electrical energy is converted into a motor movement to suck up your toast crumbs from the floor.

Electronic devices, on the other hand, do more. They are designed to protect the electrical current to do interesting and useful things. In the first electronic device by Thomas Edison in 1883, the electric current through a lamp was manipulated in such a way that the device could actually monitor the voltage being delivered. Monitoring also allowed it to automatically increase the voltage if it was too low and reduce it if it was too high. More advanced examples of this manipulation are audio signals, in which sound information has been added to the electrical current. In the case of video signals, image information is also added.

As with any difference, there are always doubts. Simple electrical appliances are sometimes extended with electronic components. Let's just think back to the toaster where an electronic thermostat controls the temperature to make perfect toast. On the other hand, even the most complex electronic appliances sometimes contain simple electrical parts. The remote control of your television set is quite a complex electronic device, but it also contains electrical parts such as batteries.

A list of the possibilities of electronic devices is given below.

- Sound: Whether it's noise or music, electronic devices can handle audio signals very well. Human speech or musical instruments are recorded by a microphone that converts the sound into an audio signal. The sound changes somehow give rise to changes in the audio signal. The succeeding amplifier receives the small audio signal (with limited amplitude) to a large audio signal (with a large amplitude). Finally, the loudspeakers convert the electrical current back into audible sound.

- Light: The simplest electronic devices are LEDs or light-emitting diodes which are the electronic equivalent of light bulbs. More complex devices build complete images, based on video signals. Note that visible light is not always used. The remote control of a television set, for example, transmits infrared light.
- Communication: Communication allows all forms of information to be forwarded. This can be done with and without cables. For audio information, for example, this is possible with a speaker cable or a radio set (where the antenna is sometimes hidden or integrated in the set itself).
- Computers: Last but not least: computers. They have evolved in the last 50 years from simple calculators to devices that sometimes transcend human capabilities. Computers are the most advanced form of electronics today. They are based on digital electronics, which manipulate the information in the binary language of zeros and ones very efficiently. It is certain that in a few years new features will be added.

Etymologically, the word 'electricity' is derived from the Ancient Greek word for amber: electron! Amber is a special substance. It is actually not a stone, but resin from trees that has hardened. Traditionally, amber has been known for its ability to become electrically charged (static) when rubbed against an animal's fur, for example. Scientists were so impressed by these properties that they called these static possibilities electron force. So the original meaning of electricity is amber power!

1.5 BASIC ELECTRICAL CONCEPTS

1.5.1 Bohr Model

All matter is made up of atoms, and there are a number of 'models' used to explain physical effects that have been both predicted and subsequently observed. One of the oldest and simplest of these is the Bohr model. This describes the atom as consisting of a central nucleus containing minute particles called protons and neutrons. Surrounding the nucleus, there are a number of electrons in various orbits. This model is illustrated in Figure 1.1. The possible

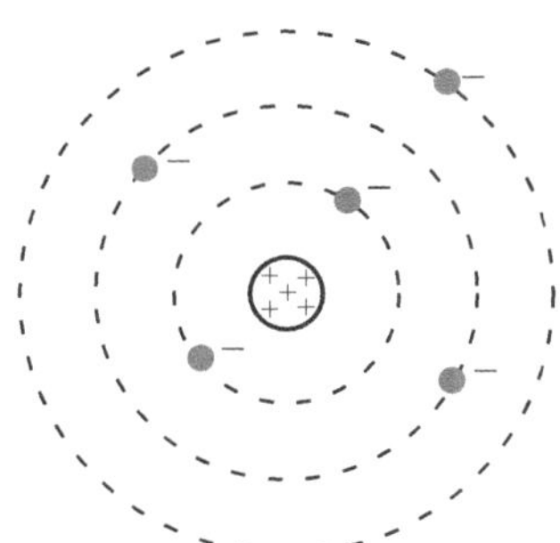

Figure 1.1 The Bohr model

presence of neutrons in the nucleus has been ignored, since these particles play no part in the electrical concepts to be described. It should be noted that this atomic model is greatly over-simplified. It is this very simplicity that makes it ideal for the beginner to achieve an understanding of what electricity is and how electrical devices operate.

Niels Bohr (1885–1962) was a Danish theoretical physicist and theoretical chemist. In 1922 he was awarded the Nobel Prize in Physics, as being one of the founders of atomic physics. He formulated a theoretical basis for a new atomic model, based on quantum mechanics. He described how the orbits of electrons around the atomic nucleus could only have certain values (quantified) and have energies with fixed distinct energy levels.

The model shown in Figure 1.1 is not drawn to scale since a proton is approximately 2000 times more massive than an electron. Due to this *relatively* large mass the proton does not play an active part in electrical current flow. It is the behaviour of the electrons that is more important. However, protons and electrons do share one thing in common; they both possess a property known as electric charge. The unit of charge is called the coulomb (C). Since charge is considered as the *quantity* of electricity it is given the symbol Q. An electron and proton have exactly the same amount of charge. The electron has a negative charge, whereas the proton has a positive charge. Any atom in its 'normal' state is electrically neutral (has no net charge). So, in this state the atom must possess as many orbiting electrons as there are protons in its nucleus. If one or more of the orbiting electrons can somehow be persuaded to leave the parent atom then this charge balance is upset. In this case the atom acquires a net positive charge, and is then known as a positive ion. On the other hand, if 'extra' electrons can be made to orbit the nucleus then the atom acquires a net negative charge. It then becomes a negative ion.

Charles-Augustin de Coulomb (1736–1806) was a French physicist who studied electricity and magnetism, and after whom the unit of electric charge coulomb and Coulomb's law are named. He discovered that the force that two charged particles have on each other is inversely proportional to the square of the distance between those two particles. He demonstrated that with a torsion balance, allowing very small electrical charges to be measured.

An analogy is a technique where the behaviour of one system is compared to the behaviour of another system. The system chosen for this comparison will be one that is more familiar and so more easily understood. However, it must be borne in mind that an analogy should not be extended too far. Since the two systems are usually very different physically there will come a point where comparisons are no longer valid.

You may now be wondering why the electrons remain in orbit around the nucleus anyway. This can best be explained by considering an analogy. Thus, an electron orbiting the nucleus may be compared to a satellite orbiting the Earth. The satellite remains in orbit due to a balance of forces. The gravitational force of attraction towards the Earth is balanced by the centrifugal force on the satellite due to its high velocity. This high velocity means that the satellite has high kinetic energy. If the satellite is required to move into a higher orbit, then its motor must be fired to speed it up. This will increase its energy. Indeed, if its velocity is increased sufficiently, it can be made to leave Earth's orbit and travel out into space.

In the case of the electron, there is also a balance of forces involved. Since both electrons and protons have mass, there will be a gravitational force of attraction between them. However, the masses involved are so minute that the gravitational force is negligible. So, what force of attraction does apply here? Remember that electrons and protons are oppositely charged particles, and oppositely charged bodies experience a force of attraction. Compare this to two simple magnets, whereby opposite polarities attract and like (the same) polarities repel each other. The same rule applies to charged bodies. Thus it is the balance between this force of electrostatic attraction and the kinetic energy of the electron that maintains the orbit. It may now occur to you to wonder why the nucleus remains intact, since the protons within it are all positively charged particles! It is beyond the scope of this book to give a comprehensive answer. Suffice to say that there is a force within the nucleus far stronger than the electrostatic repulsion between the protons that binds the nucleus together.

All materials may be classified into one of three major groups: conductors, insulators and semiconductors. In simple terms, the group into which a material falls depends on how many 'free' electrons it has. The term 'free' refers to those electrons that have acquired sufficient energy to leave their orbits around their parent atoms. In general we can say that conductors have many free electrons which will be drifting in a *random* manner within the material. Insulators have very few free electrons (ideally none), and semiconductors fall somewhere between these two extremes.

1.5.2 Electric Current

This is the rate at which free electrons can be made to drift through a material *in a particular direction*. In other words, it is the rate at which charge is moved around a circuit. Since charge is measured in coulombs and time in seconds then logically the unit for electric current would be the coulomb/second or abbreviated to C/s. In fact, the amount of current flowing through a circuit may be calculated by dividing the amount of charge passing a given point by the time taken. The unit, however, is given a special name, the ampere (often abbreviated to A). This is fairly common practice with SI units, whereby the names chosen are those of famous scientists whose pioneering work is thus commemorated. The relationship between current, charge and time can be expressed as a mathematical equation as follows:

$$I = \frac{Q}{t}, \text{ or } Q = It \tag{1.1}$$

André-Marie Ampère (1775–1836) was a French physicist and mathematician widely regarded as one of the discoverers of electromagnetism, thanks to his mathematical description of the relationship between magnetism and electricity. Using experiments, he discovered that two parallel currents attract each other if the currents flow in the same direction, but repel each other if they flow in opposite directions. He discovered that the force between the wires was proportional to the two currents and inversely proportional to the square of the distance between the wires. This formula was later called Ampère's law.

WORKED EXAMPLE 1.3

Q A charge of 35 mC is transferred between two points in a circuit in a time of 20 ms. Calculate the value of current flowing.

$Q = 35 \times 10^{-3}\,C;\ t = 20 \times 10^{-3}\ \text{s}$

$$I = \frac{Q}{t} = \frac{35 \times 10^{-3}}{20 \times 10^{-3}}$$

$I = 1.75\ \text{A}$

WORKED EXAMPLE 1.4

Q If a current of 120 μA flows for a time of 15 s, determine the amount of charge transferred.

$I = 120 \times 10^{-6}\ \text{A};\ t = 15\ \text{s}$

$Q = It = 120 \times 10^{-6} \times 15$

$Q = 1.8\ \text{mC}$

WORKED EXAMPLE 1.5

Q 80 Coulombs of charge was transferred by a current of 0.5 A. Calculate the time for which the current flowed.

$Q = 80\ \text{C};\ I = 0.5\ \text{A}$

$$t = \frac{Q}{I} = \frac{80}{0.5}$$

$t = 160\ \text{s}$

1.5.3 Electromotive Force (emf)

The *random* movement of electrons within a material does not constitute an electrical current. This is because it does not result in a drift in one particular direction. In order to cause the 'free' electrons to drift in a given direction an electromotive force or, in short, emf must be applied. Thus the emf is the 'driving' force in an electrical circuit. The symbol for emf is E and the unit of measurement is the volt (V). Typical sources of emf are cells, batteries and generators.

Alessandro Volta (1745–1827) was an Italian physicist known for his discovery of the electric battery or the voltaic cell (voltaic pile). It was composed of a number of bowls, filled with a salt solution, which were connected with copper and zinc wires. He showed that a current ran through a closed circuit; he had built a device that could deliver an uninterrupted electrical current. Later he made a simpler design consisting of plates of copper and zinc separated by a slice of cardboard soaked in a salt solution.

The amount of current that will flow through a circuit is directly proportional to the size of the emf applied to it. The circuit diagram symbols for a cell and a battery are shown in Figures 1.2(a) and (b) respectively. Note that the positively charged plate (the long line) usually does not have a plus sign written alongside it. Neither does the negative plate normally have a minus sign written by it. These signs have been included here merely to indicate (for the first time) the symbol used for each plate.

1.5.4 Resistance (R)

Although the amount of electrical current that will flow through a circuit is directly proportional to the applied emf, the other property of the circuit (or material) that determines the resulting current is the opposition offered to the flow. This opposition is known as the electrical resistance, which is measured in ohms (Ω). Thus conductors, which have many 'free' electrons available for current carrying, have a low value of resistance. Conductive materials such as copper or aluminium allow the current to flow freely. In fact, all materials have a certain resistance; even the best conductors have a small, but sometimes measurable resistance. Some values can be found in Table 1.4, where the resistivity is expressed in Ωm. On the other hand, since insulators have very few 'free' charge carriers, they have a very high resistance. Insulating materials such as plastic or glass completely prevent the current from flowing. When a conductive material and an insulating material are mixed together, it results in a composition that conducts current, but certainly not optimally. We say that this component has a resistance. Pure semiconductors tend to behave more like insulators in this respect. However, in practice, semiconductors tend to be used in an impure form, where the added impurities improve the conductivity of the material. An electrical device that is designed to have a specified value of resistance is called a resistor. Resistors are actually the brakes of the electric current. Just like the brakes on a bike or in a car, resistors work on the current through the electrical equivalent of friction. The corresponding frictional energy is largely dissipated in heat. This heat dissipation is sometimes a disadvantage (as a loss of energy), but sometimes it is an advantage such as with the rear window heating of a car. The voltage applied then, to different resistance wires, results in slowing down the current and largely converting it into heat. The circuit diagram symbol for a resistor is shown in Figure 1.3.

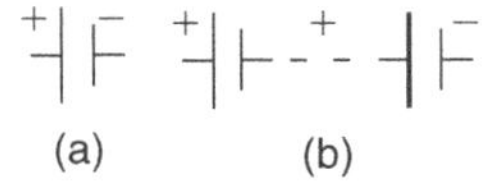

Figure 1.2 The circuit diagram symbols for a cell and a battery

Table 1.4 Resistivity values for different materials

Material	*Resistivity in Ωm*
Silver	1.59×10^{-8}
Copper	1.68×10^{-8}
Aluminium	2.65×10^{-8}
Graphite	From 3×10^{-5} to 60×10^{-5}
Glass	1×10^{9}
Plastics	From 1×10^{10} to 1×10^{19}

Figure 1.3 The circuit diagram symbol for a resistor

Georg Ohm (1789–1854) was a German mathematician and physicist. He became known for Ohm's law, named after him, in which the relationship between electrical voltage, electrical current and resistance is expressed. The unit of electrical resistance – the ohm – is also named after him. He ran current through wires of different lengths and discovered that the current decreased as the length increased: the longer a wire is, the harder it is for the electricity to work its way through the wire, so more voltage is needed to travel that way.

Note: You can assume that the resistance value of an ordinary copper wire is zero. However, copper has a very limited resistance (as can be seen in Table 1.4). For most electrical and electronic circuits, the copper wires are much limited in length. Insulators and open circuits are supposed to have infinite resistance, which is also not quite consistent with reality. There will always be an albeit very small current through the air, because the resistance value of air is certainly not infinite.

Graphite is a crystalline form of the element carbon and is one of the basic materials of a pencil, where it is mixed with and encased in a wooden shaft. When drawing lines on a sheet of paper, the graphite forms a dark line and results in a resistor. The corresponding resistor value is higher than when copper is used and lower in the case of plastics. It is very unpractical due to the paper, but this resistor can be used as an electrical component. Nowadays, most resistors are carbon film resistors. They are made from carbon, placed on a plastic foil or film and covered with plastic.

The resistance of a sample of material depends upon four factors:

- Its length
- Its cross-sectional area (csa)
- The actual material used
- Its temperature

Simple experiments can show that the resistance is directly proportional to the length and inversely proportional to the csa. Combining these two statements we can write:

$$R \alpha \frac{\ell}{A} \text{ where } \ell = \text{length(in metres) and } A = \text{csa(in square metres)}$$

The constant of proportionality in this case concerns the third factor listed above, and is known as the resistivity of the material. This is defined as the resistance that exists between the opposite faces of a 1 m cube of that material, measured at a defined temperature. The symbol for resistivity is ρ . The unit of measurement is the ohm-metre (Ωm). Thus the equation for resistance using the above factor is

$$R = \rho \frac{\ell}{A} \tag{1.2}$$

WORKED EXAMPLE 1.6

Q A coil of copper wire 200 m long and of csa 0.8 mm^2 has a resistivity of 0.02 μΩm at normal working temperature. Calculate the resistance of the coil.

$\ell = 200$ m; $\rho = 2 \times 10^{-8}$ Ωm; $A = 8 \times 10^{-7}$ m^2

$$R = \frac{\rho \ell}{A} = \frac{2 \times 10^{-8} \times 200}{8 \times 10^{-7}}$$

$R = 5\ \Omega$

WORKED EXAMPLE 1.7

Q A wire-wound resistor is made from a 250 m length of copper wire having a circular cross-section of diameter 0.5 mm. Given that the wire has a resistivity of 0.018 μΩm, calculate its resistance value.

$\ell = 250$ m; $d = 5 \times 10^{-4}$ m; $\rho = 1.8 \times 10^{-8}$ Ωm

$R = \frac{\rho \ell}{A}$, where cross-sectional area, $A = \frac{\pi d^2}{4}$ m^2

$$\text{hence, } A = \frac{\pi \times \left(5 \times 10^{-4}\right)^2}{4} = 1.9635 \times 10^{-7} \text{m}^2$$

$$\text{hence, } R = \frac{1.8 \times 10^{-8} \times 250}{1.9635 \times 10^{-7}} = 22.92\ \Omega$$

The resistance of a material also depends on its temperature and has a property known as its temperature coefficient of resistance. The resistance of all pure metals increases with the increase of temperature. The resistance of carbon, insulators, semiconductors and electrolytes decreases with the increase of temperature. For these reasons, conductors (metals) are said to have a positive temperature coefficient of resistance. Insulators etc. are said to have a negative temperature coefficient of resistance. Apart from this there is another major difference. Over a moderate range of temperature, the change of resistance for conductors is relatively small and is a very close approximation to a straight line. Semiconductors on the other hand tend to have very much larger changes of resistance over the same range of temperatures, and follow an exponential law. These differences are illustrated in Figure 1.4.

Temperature coefficient of resistance is defined as the ratio of the change of resistance per degree change of temperature, to the resistance at some specified temperature. The quantity symbol is α and the unit of measurement is per degree, e.g./°C. The reference temperature usually quoted is 0°C, and the resistance at this temperature is referred to as R_0. Thus the resistance at some other temperature θ_1°C can be obtained from:

$$R_1 = R_0\left(1 + \alpha\theta_1\right) \qquad (1.3)$$

In general, if a material having a resistance R_0 at 0°C has a resistance R_1 at θ_1°C and R_2 at θ_2°C, and if α is the temperature coefficient at 0°C, then

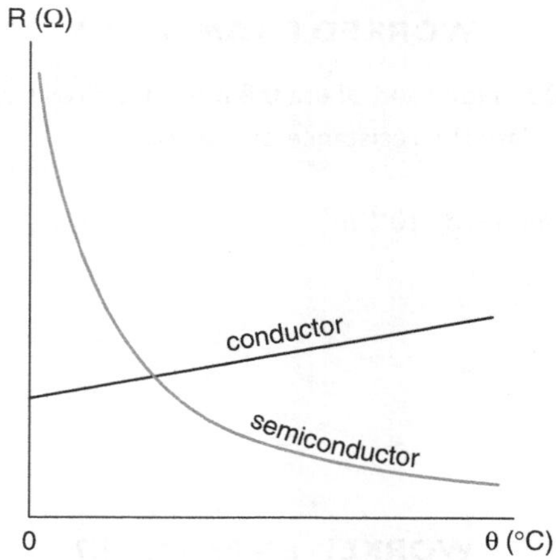

Figure 1.4 The resistance of a material as function of its temperature

$$R_1 = R_0(1+\alpha\theta_1) \text{ and } R_2 = R_0(1+\alpha\theta_2)$$

$$\text{so } \frac{R_1}{R_2} = \frac{1+\alpha\theta_1}{1+\alpha\theta_2} \quad (1.4)$$

WORKED EXAMPLE 1.8

Q The field coil of an electric motor has a resistance of 250 Ω at 15°C. Calculate the resistance if the motor attains a temperature of 45°C when running. Assume that $\alpha = 0.00428/°C$ referred to 0°C.

$R_1 = 250\ \Omega$; $\theta_1 = 15°C$; $\theta_2 = 45°C$; $\alpha = 4.28 \times 10^{-3}\ /°C$

Using Equation (1.4):

$$\frac{250}{R_2} = \frac{1+(4.28\times10^{-3}\times15)}{1+(4.28\times10^{-3}\times45)}$$

$$\frac{250}{R_2} = 0.8923$$

$$R_2 = 280.2\ \Omega$$

WORKED EXAMPLE 1.9

Q A coil of wire has a resistance value of 350 Ω when its temperature is 0°C. Given that the temperature coefficient of resistance of the wire is 4.26×10^{-3}/°C referred to 0°C, calculate its resistance at a temperature of 60°C.

$R_0 \quad = 350\ \Omega;\ \alpha = 4.26\times10^{-3}\ /^\circ C;\ \theta_1 = 60^\circ C$

$R_1 \quad = R_0(1+\alpha\theta_1)$; where R_1 is the resistance at $60^\circ C = 350\left[1+\left(4.26\times10^{-3}\times60\right)\right]$

$= 350[1+0.22556] = 350\times1.2556$

so, $R_1 \quad = 439.6\ \Omega$

WORKED EXAMPLE 1.10

Q A carbon resistor has a resistance value of 120 Ω at a room temperature of 16°C. When it is connected as part of a circuit, with current flowing through it, its temperature rises to 32°C. If the temperature coefficient of resistance of carbon is −0.00048/°C referred to 0°C, calculate its resistance under these operating conditions.

$\theta_1 = 16^\circ C;\ \theta_2 = 32^\circ C;\ R_1 = 120\ \Omega;\ \alpha = -0.00048\ /^\circ C$

$$\frac{R_1}{R_2} = \frac{1+\alpha\theta_1}{1+\alpha\theta_2}$$

$$\frac{120}{R_2} = \frac{1+(-0.00048\times16)}{1+(-0.00048\times32)}$$

$$\frac{120}{R_2} = 1.0078$$

$$R_2 = \frac{120}{1.0078}$$

$$\text{so, } R_2 = 119.1\Omega$$

Usually a resistor is packed as a small component and designed in such a way to obtain a certain amount of resistance in a circuit. The most common resistors are carbon film resistors, because they are made of a layer of carbon together with an insulating material and packed in a small cylinder. The ratio of carbon and the insulating material determines the resistance value. On the outside, the sequence and colour of the painted bands always show the resistance value in combination with their tolerance. The tolerance indicates how close the value is to the indicated resistance value. Usually there are four coloured bands, the first three of which indicate the resistance value and the last its tolerance. Of these first three coloured bands, the first two indicate the value and the third the exponent. Sometimes there are five bands of which the last one also indicates the tolerance on the resistance value. Then the first three indicate the value and the fourth the exponent. Since you can also reverse the resistor itself (and thus change the order), the first coloured band is always placed closest to the edge and the last one not at all. The tolerance indicates a percentage by which the real resistance value can differ from the resistance value indicated on the component itself. A 470 Ω resistor with a 10% tolerance has a true resistance value between 423 Ω and 517 Ω. The smaller the tolerance, the more expensive it is to purchase the component. Usually resistors with a tolerance of 5% or 10% will suffice. Each colour corresponds to a specific value, as can also be found in Table 1.5. You can easily check that the colour combination Red–Red–Orange corresponds to the resistance value 22 kΩ (or digit 2–digit 2–exponent 1 k). Note that a 20% tolerance is sometimes also indicated without a dash (instead of the black dash).

Table 1.5 Conversion from colour code to resistance value

Colour	*Digit*	*Exponent*	*Tolerance*
Black	0	1	20%
Brown	1	10	1%
Red	2	100	2%
Orange	3	1k	3%
Yellow	4	10k	4%
Green	5	100k	
Blue	6	1M	
Violet	7	10M	
Grey	8	100M	
White	9	1000M	
Gold		0.1	5%
Silver		0.01	10%
Nothing			20%

The first two (or three) coloured bands can theoretically represent 100 different combinations with the digits. In practice there are only a limited number of colour combinations. The greater the tolerance, the wider the range of the value. So it only makes sense to fabricate values whose tolerance ranges do not overlap. The E-series has been developed for this, based on an approximate geometric sequence. The ratio between two successive values is approximately constant. If the number of steps in the sequence is n, then the ratio is approximately $10^{1/n}$. A decade is a range of which the largest is greater than the smallest by a factor of 10. That decade contains three values in the E3 series and therefore only has 10, 22 and 47 (with mutual ratio $10^{1/3} = \pm 2.15$). The most famous series is the E12 series with 12 values in each decade: 10, 12, 15, 18, 22, 27, 33, 39, 47, 56, 68 and 82 with mutual ratio $10^{1/12} = \pm 1.21$ (each subsequent step is thus about more than 20% greater, which means that a tolerance greater than 10% is meaningless).

A special resistor is the thermistor. The thermistor R_{PTC} has a positive temperature coefficient (or PTC), hence its name. It means that as the temperature increases, its electrical resistance also increases (albeit non-linear). If the temperature rises due to, for example, excessive power consumption, its resistance will increase abruptly and greatly limit the current flowing through it. However, a maintenance current will continue to flow to keep the resistor warm. The thermistor can be used as temperature protection in electrical appliances. It acts as a fuse that can be reset: by switching off the voltage, the maintenance current is lost and the thermistor can cool down again. The counterpart is the R_{NTC} with a negative temperature coefficient (or NTC): as the temperature rises, the electrical resistance will decrease.

Another special resistor is the photoresistor or Light-Dependent Resistor (LDR). This component acts as an electronic switch that is operated by the incidence of light. The resistor value is influenced by the amount of light. The more light, the lower the resistance value. It ranges from several MΩ when it is dark to a few hundred ohms in the light.

1.5.5 Potential Difference (p.d.)

Whenever current flows through a resistor there will be a potential difference (shortened to p.d.) developed across it. The p.d. is measured in volts, and is quite literally the difference in voltage levels between two points in a circuit. Although both p.d. and emf are measured in volts they are not the same quantity. Essentially, emf (being the driving force) causes current to flow; whilst a p.d. is the result of current flowing through a resistor. Thus emf is a *cause* and p.d. is an *effect*. It is a general rule that the symbol for a quantity is different to the symbol used for the unit in which it is measured. One of the few exceptions to this rule is that the quantity symbol for p.d. happens to be the same as its unit symbol, namely *V*. In order to explain the difference between emf and p.d. we shall consider another analogy.

Figure 1.5 represents a simple hydraulic system consisting of a pump, the connecting pipework and two restrictors in the pipe. The latter will have the effect of limiting the rate at which the water flows around the circuit. Also included is a tap that can be used to interrupt the flow completely. Figure 1.6 shows the equivalent electrical circuit, comprising a battery, the connecting conductors (cables or leads) and two resistors. The latter will limit the amount of current flow. Also included is a switch that can be used to 'break' the circuit and so prevent any current flow. As far as each of the two systems is concerned we are going to make some assumptions.

For the water system we will assume that the connecting pipework has no slowing down effect on the flow, and so will not cause any pressure drop. Provided that the pipework is

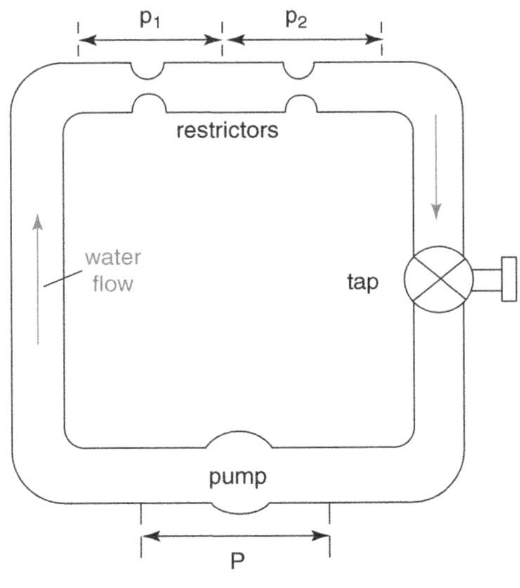

Figure 1.5 A simple hydraulic system consisting of a pump, the connecting pipework and two restrictors in the pipe

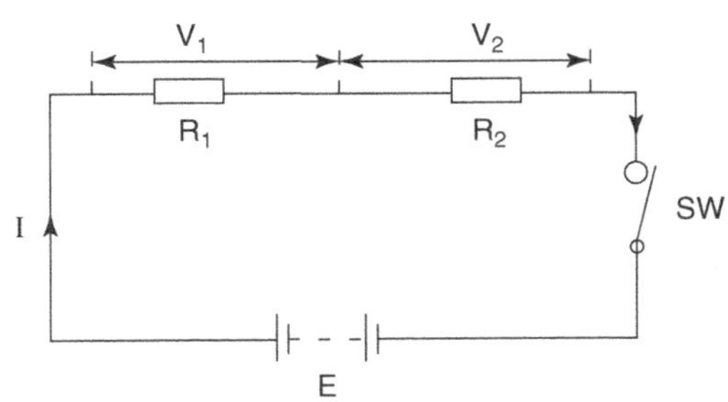

Figure 1.6 An equivalent electrical circuit, comprising a battery, the connecting conductors (cables or leads) and two resistors

relatively short then this is a reasonable assumption. A similar assumption in the electrical circuit is that the connecting wires have such a low resistance that they will cause no p.d. If anything, this is probably a more legitimate assumption to make. Considering the water system, the pump will provide the total system pressure (P) that circulates the water through it. Using some form of pressure measuring device it would be possible to measure this pressure together with the pressure drops (p_1 and p_2) that would occur across the two restrictors. Having noted these pressure readings it would be found that the total system pressure is equal to the sum of the two pressure drops. Using a similar technique for the electrical circuit, it would be found that the sum of the two p.d.s (V_1 and V_2) is equal to the total applied emf E volts. These relationships may be expressed in mathematical form as:

$$P = p_1 + p_2 \text{ pascal}$$

and

$$E = V_1 + V_2 \tag{1.5}$$

When the potential at some point in a circuit is quoted as having a particular value (say 10 V) then this implies that it is 10 V above some reference level or datum. Compare this with altitudes. If a mountain is said to be 5000 m high it does not necessarily mean that it rises 5000 m from its base to its peak. The figure of 5000 m refers to the height of its peak above mean sea level. Thus, the mean sea level is the reference point or datum from which altitudes are measured. In the case of electrical potentials the datum is taken to be the potential of the Earth which is 0 V. Similarly, –10 V means 10 V below or less than 0 V.

1.5.6 Conventional Current and Electron Flow

You will notice in Figure 1.6 that the arrows used to show the direction of current flow indicate that this is from the positive plate of the battery, through the circuit, returning to the negative battery plate. This is called conventional current flow. However, since electrons are negatively charged particles, then these must be moving in the opposite direction. The latter is called electron flow. Now, this poses the problem of which one to use. It so happens that before science was sufficiently advanced to have knowledge of the electron, it was assumed that the positive plate represented the 'high' potential and the negative the 'low' potential. So the convention was adopted that the current flowed around the circuit from the high potential to the low potential. This compares with water which can naturally only flow from a high level to a lower level. Thus the concept of conventional current flow was adopted. All the subsequent 'rules' and conventions were based on this direction of current flow. On the discovery of the nature of the electron, it was decided to retain the concept of conventional current flow. Had this not been the case then all the other rules and conventions would have needed to be changed! Hence, true electron flow is used only when it is necessary to explain certain effects (as in semiconductor devices such as diodes and transistors). Whenever we are considering basic electrical circuits and devices we shall use *conventional current flow*, i.e. current flowing around the circuit *from* the *positive* terminal of the source of emf *to* the *negative* terminal.

1.5.7 Ohm's Law

This states that the p.d. developed between the two ends of a resistor is directly proportional to the value of current flowing through it, provided that all other factors (e.g. temperature) remain constant. Writing this in mathematical form we have:

$V \alpha I$

However, this expression is of limited use since we need an equation. This can only be achieved by introducing a constant of proportionality; in this case the resistance value of the resistor.

$$\text{Thus } V = IR \tag{1.6}$$

$$\text{or } I = \frac{V}{R} \tag{1.7}$$

$$\text{and } R = \frac{V}{I} \tag{1.8}$$

WORKED EXAMPLE 1.11

Q A current of 5.5 mA flows through a 33 kΩ resistor. Calculate the p.d. thus developed across it.

$I = 5.5 \times 10^{-3}$ A; $R = 33 \times 10^{3}$ Ω

$V = IR = 5.5 \times 10^{-3} \times 33 \times 10^{3}$

$V = 181.5$ V

WORKED EXAMPLE 1.12

Q If a p.d. of 24 V exists across a 15 Ω resistor then what must be the current flowing through it?

$V = 24$ V; $R = 15$ Ω

$$I = \frac{V}{R} = \frac{24}{15}$$

$I = 1.6$ A

1.5.8 Internal Resistance (r)

So far we have considered that the emf E volts of a source are available at its terminals when supplying current to a circuit. If this were so then we would have an ideal source of emf. Unfortunately, this is not the case in practice. This is due to the internal resistance of the source. As an example consider a typical 12 V car battery. This consists of a number of oppositely charged plates, appropriately interconnected to the terminals, immersed in an **electrolyte**. The plates themselves, the internal connections and the electrolyte all combine to produce a small but **finite** resistance, and it is this that forms the battery's internal resistance.

An **electrolyte** is the chemical 'cocktail' in which the plates are immersed. In the case of a car battery, this is an acid/water mixture. In this context, **finite** simply means measurable.

Figure 1.7 shows such a battery with its terminals on open circuit (no external circuit connected). Since the circuit is incomplete no current can flow. Thus there will be no p.d. developed across the battery's internal resistance *r*. Since the term p.d. quite literally means a difference in potential between the two ends of *r*, then the terminal A must be at a potential of 12 V, and terminal B must be at a potential of 0 V. Hence, under these conditions, the full emf 12 V is available at the battery terminals.

Figure 1.8 shows an external circuit, in the form of a 2 Ω resistor, connected across the terminals. Since we now have a complete circuit then current *I* will flow as shown. The value of this current will be 5.71 A (the method of calculating this current will be dealt with early in the next chapter). This current will cause a p.d. across *r* and also a p.d. across *R*. These calculations and the consequences for the complete circuit now follow:

$$\text{p.d. across } r = Ir \text{ (Ohm's law applied)} = 5.71\ A \times 0.1\ \Omega = 0.571\ \text{V}$$

$$\text{p.d. across } R = IR = 5.71 \times 2 = 11.42\ \text{V}$$

Note that 0.571 V + 11.42 V = 11.991 V but this figure *should* be 12 V. The very small difference is simply due to 'rounding' the figures obtained from the calculator.

The p.d. across *R* is the battery terminal p.d. *V*. Thus it may be seen that when a source is supplying current, the terminal p.d. will always be less than its emf. To emphasise this point let us assume that the external resistor is changed to one of 1.5 Ω resistance. The current now drawn from the battery will be 7.5 A. Hence:

$$\text{p.d. across } r \quad = 7.5 \times 0.1 = 0.75\ \text{V}$$
$$\text{and p.d. across } R \quad = 7.5 \times 1.5 = 11.25\ \text{V}$$

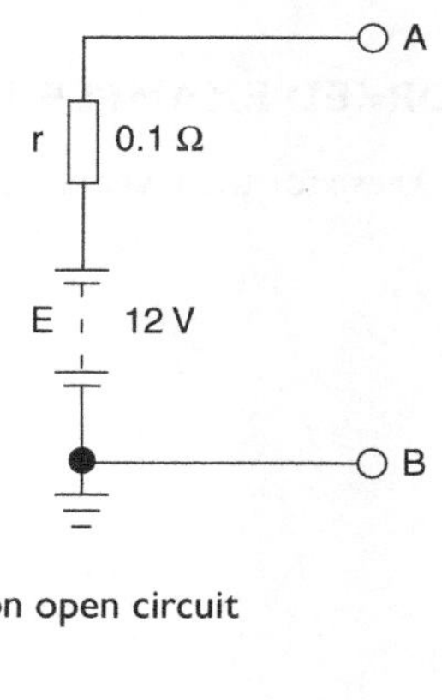

Figure 1.7 A battery with its terminals on open circuit

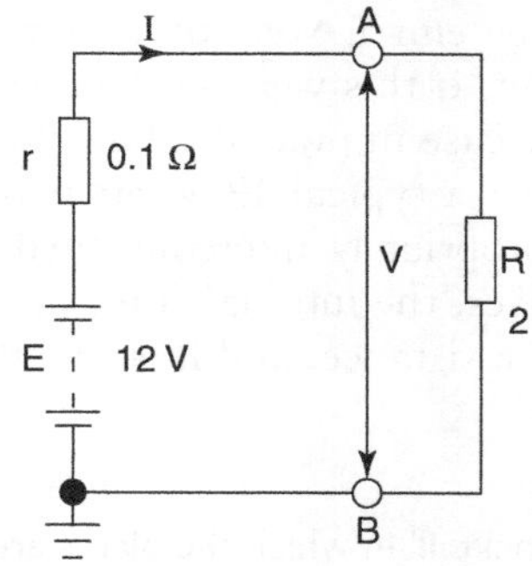

Figure 1.8 An external circuit connected across the battery terminals

Note that 11.25 + 0.75 = 12 V (rounding error not involved). Hence, the battery terminal p.d. has fallen still further as the current drawn has increased. This example brings out the following points.

1 Assuming that the battery's charge is maintained, then its emf remains constant. But its terminal p.d. varies as the current drawn is varied, such that

$$V = E - Ir \tag{1.9}$$

2 Rather than having to write the words 'p.d. across R' it is more convenient to write this as V_{AB} which, translated, means the potential difference between points A and B.
3 In future, if no mention is made of the internal resistance of a source, then for calculation purposes you may assume that it is zero, i.e. an ideal source.

WORKED EXAMPLE 1.13

Q A battery of emf 6 V has an internal resistance of 0.15 Ω. Calculate its terminal p.d. when delivering a current of (a) 0.5 A, (b) 2 A and (c) 10 A.

$E = 6\text{ V};\ r = 0.15\ \Omega$

(a) $V = E - Ir = 6\text{ V} - (0.5 \times 0.15) = 6\text{ V} - 0.175\text{ V} = 5.925\text{ V}$

(b) $V = 6\text{ V} - (2 \times 0.15) = 6\text{ V} - 0.3\text{ V} = 5.7\text{ V}$

(c) $V = 6\text{ V} - (10 \times 0.15) = 6\text{ V} - 1.5\text{ V} = 4.5\text{ V}$

Note: This example verifies that the terminal p.d. of a source of emf decreases as the load on it (the current drawn from it) is increased.

WORKED EXAMPLE 1.14

Q A battery of emf 12 V supplies a circuit with a current of 5 A. If, under these conditions, the terminal p.d. is 11.5 V, determine (a) the battery internal resistance, (b) the resistance of the external circuit.

$E = 12\text{ V};\ I = 5\text{ A};\ V = 11.5\text{ V}$

As with the vast majority of electrical problems, a simple sketch of the circuit diagram will help you to visualise the problem. For the above problem the circuit diagram would be as shown in Figure 1.9.

(a) $E = V + Ir\ E - V = Ir$ so, $r = \dfrac{E - V}{I} = \dfrac{12 - 11.5}{5} = 0.1\ \Omega$

(b) $R = \dfrac{V}{I} = \dfrac{11.5}{5} = 2.3\ \Omega$

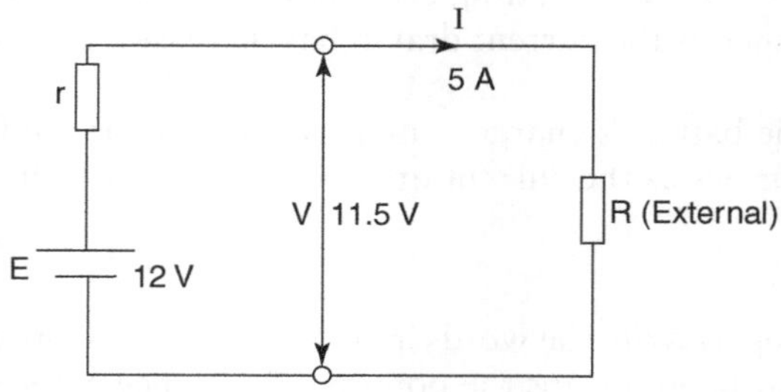

Figure 1.9 The circuit diagram for Worked Example 1.14

1.5.9 Energy (*W*) and Power (*P*)

Energy (*W*) is the property of a system that enables it to do work. Whenever work is done energy is transferred from that system to another one. The most common form into which energy is transformed is heat. The heat produced (or energy dissipated) is expressed in Joule (shortened to J) and is given by the equation

$$W = I^2Rt \tag{1.10}$$

and applying Ohm's law as shown in Equations (1.6) to (1.8)

$$W = \frac{V^2t}{R} \tag{1.11}$$

or

$$W = VIt \tag{1.12}$$

James Joule (1818–1889) was an English physicist. He carried out an investigation on one of the effects of an electric current: the production of heat (e.g. an electric kettle). He reached the conclusion that the amount of heat so produced was proportional to the value of the square of the current flowing and the time for which it flowed. Once more a constant of proportionality is required, and again it is the resistance of the circuit that is used.

WORKED EXAMPLE 1.15

Q A current of 200 mA flows through a resistance of 750 Ω for a time of 5 minutes. Calculate (a) the p.d. developed and (b) the energy dissipated.

$I = 200$ mA $= 0.2$ A; $t = 5 \times 60 = 300$ s; $R = 750$ Ω

(a) $V = IR = 0.2 \times 750 = 150$ V

(b) $W = I^2Rt = 0.2 \times 0.2 \times 750 \times 300$ W $= 9000$ J or 9 kJ

Note: It would have been possible to use either Equation (1.14) or (1.15) to calculate *W*. However, this would have involved using the calculated value for *V*. If this value had been miscalculated, then the answers to both parts of the question would have been incorrect. So, whenever possible, make use of data that are given in the question in preference to values that you have calculated. Please also note that the time has been converted to its basic unit, the second.

Power (P) is the rate at which work is done, or at which energy is dissipated. The unit in which power is measured is the watt (W).

Warning: Do not confuse this unit symbol with the quantity symbol for energy. In general terms we can say that power is energy divided by time.

i.e. $P = \dfrac{W}{t}$

Thus, by dividing each of Equations (1.10), (1.11), and (1.12), in turn, by t, the following equations for power result:

$$P = I^2R \tag{1.13}$$

$$P = \frac{V^2}{R} \tag{1.14}$$

$$\text{or } P = VI \tag{1.15}$$

The production of heat can be advantageous, can be disadvantageous, or can be both. In an electric kettle, the purpose is to heat the water, whereas in a battery the produced heat is a loss of energy. In an incandescent bulb, electrical energy is converted into both light and heat, where the heat is lost energy.

James Watt (1736–1819) was a Scottish engineer, who is regarded as the inventor of the modern steam engine. He is also the inventor of the first copier, for which he received a patent in 1781. He introduced horsepower as a unit of power for classifying the steam engines. His version of the unit is equivalent to 550 foot-pounds per second (735.5 Watts).

WORKED EXAMPLE 1.16

Q A resistor of 680 Ω, when connected in a circuit, dissipates a power of 85 mW. Calculate (a) the p.d. developed across it and (b) the current flowing through it.

$R = 680\ \Omega;\ P = 85 \times 10^{-3}\ \text{W}$

(a) $P = \dfrac{V^2}{R}$ so, $V^2 = PR$ and $V = \sqrt{PR} = \sqrt{85 \times 10^{-3} \times 680} = \sqrt{57.8} = 7.6\ \text{V}$

(b) $P = I^2R$ so, $I^2 = \dfrac{P}{R}$ and $I = \sqrt{\dfrac{P}{R}} = \sqrt{\dfrac{85 \times 10^{-3}}{680}} = \sqrt{1.25 \times 10^{-4}} = 11.18\ \text{mA}$

Note: Since $P = VI$, the calculations may be checked as follows

$$P = 7.6\ V \times 11.18 \times 10^{-3}\ A$$

so, P = 84.97 mW, which when rounded up to one decimal place gives 85.0 mW – the value given in the question.

WORKED EXAMPLE 1.17

Q A current of 1.4 A when flowing through a circuit for 15 minutes dissipates 200 kJ of energy. Calculate (a) the p.d., (b) power dissipated and (c) the resistance of the circuit.

$I = 1.4$ A; $t = 15 \times 60 = 900$ s; $W = 2 \times 10^5$ J

(a) $W = VIt$ so $V = \dfrac{W}{It} = \dfrac{2 \times 10^5}{1.4 \times 900} = 158.7$ V

(b) $P = VI = 158.7 \times 1.4 = 222.2$ W

(c) $R = \dfrac{V}{I} = \dfrac{158.7}{1.4} = 113.4\ \Omega$

Although the joule is the SI unit of energy, it is too small a unit for some practical uses, e.g. where large amounts of power are used over long periods of time. The electricity meter in your home actually measures the energy consumption. So, if a 3 kW heater was in use for 12 hours the amount of energy used would be 129.6 MJ. In order to record this, the meter would require at least ten dials to indicate this very large number. Hence the commercial unit of energy is the kilowatt-hour (kWh). Kilowatt-hours are the 'units' that appear on electricity bills. The number of units consumed can be calculated by multiplying the power (in kW) by the time interval (in hours). So, for the heater mentioned above, the number of 'units' consumed would be written as 36 kWh.

WORKED EXAMPLE 1.18

Q Calculate the cost of operating a 12.5 kW machine continuously for a period of 8.5 h if the cost per unit is 7.902p.

$W = 12.5 \times 8.5 = 106.25$ kWh

and cost $= 106.25 \times 7.902 = £8.40$

Note: When calculating energy in kWh the power must be expressed in kW, and the time in hours respectively, rather than in their basic units of watts and seconds respectively.

WORKED EXAMPLE 1.19

Q An electricity bill totalled £78.75, which included a standing charge of £15.00. The number of units charged for was 750. Calculate (a) the charge per unit and (b) the total bill if the charge/unit had been 9p, and the standing charge remained unchanged.

Total bill = £78.75; standing charge = £15.00; units used = 750 = 750 kWh

(a) Cost of the energy (units) used = total − standing charge

$$= 78.75 - 15.00 = 63.75$$

$$\text{Cost/unit} = \frac{63.75}{750} = 0.085 = 8.5\text{p}$$

(b) If the cost/unit is raised to 9p, then cost of energy used = 0.09 × 750 = £67.50

total bill = cost of units used + standing charge = 67.50 + 15.00 = £82.50

1.5.10 Alternating and Direct Quantities

The sources of emf and resulting current flow so far considered are called d.c. quantities. This is because a battery or cell once connected to a circuit is capable of driving current around the circuit in one direction only. If it is required to reverse the current it is necessary to reverse the battery connections. The term d.c., strictly speaking, means 'direct current'. However, it is also used to describe unidirectional voltages. Thus a d.c. voltage refers to a unidirectional voltage that may only be reversed as stated above.

However, the other commonly used form of electrical supply is that obtained from the electrical mains. This is the supply that is generated and distributed by the power companies. This is an alternating or a.c. supply in which the current flows alternately in opposite directions around a circuit. Again, the term strictly means 'alternating current', but the emfs and p.d.s associated with this system are referred to as a.c. voltages. Thus, an a.c. generator (or alternator) produces an alternating voltage. Most a.c. supplies provide a sinusoidal waveform (a sinewave shape). Both d.c. and a.c. waveforms are illustrated in Figure 1.10. The treatment of a.c. quantities and circuits is dealt with in Chapter 8, and need not concern you any further at this stage.

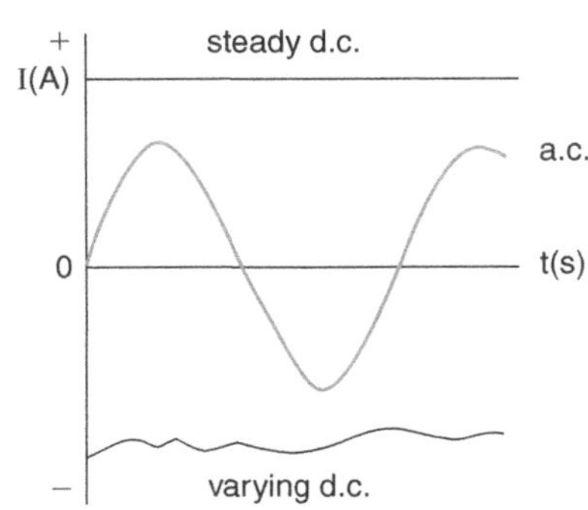

Figure 1.10 Illustration of d.c. and a.c. waveforms

1.5.11 Use of Meters

The measurement of electrical quantities is an essential part of engineering, so you need to be proficient in the use of the various types of measuring instrument. In this chapter we will consider only the use of the basic current and voltage measuring instruments, namely the ammeter and voltmeter respectively.

An ammeter is a current measuring instrument. It has to be connected into the circuit in such a way that the current to be measured is forced to flow through it. If you need to measure the current flowing in a section of a circuit that is already connected together, you will need to 'break' the circuit at the appropriate point and connect the ammeter in the 'break'. If you are connecting a circuit (as you will frequently have to do when carrying out practical assignments), then insert the ammeter as the circuit connections are being made. Most ammeters will have their terminals colour coded: red for the positive and black for the negative. Please note that these polarities refer to *conventional current flow*, so the current should enter the meter at the red terminal and leave via the black terminal. The ammeter circuit symbol is shown in Figure 1.11.

As you would expect, a voltmeter is used for measuring voltages; in particular, p.d.s. Since a p.d. is a voltage between two points in a circuit, this meter is NOT connected into the circuit in the same way as an ammeter. In this sense it is a simpler instrument to use, since it need only be connected across (between the two ends of) the component whose p.d. is to be measured. The terminals will usually be colour coded in the same way as an ammeter, so the red terminal should be connected to the more positive end of the component, i.e. follow the same principle as with the ammeter. The voltmeter symbol is shown in Figure 1.12.

It is most probable that you will have to make use of meters that are capable of combining the functions of an ammeter, a voltmeter and an ohmmeter. These instruments are known as multimeters, where the 'readings' are in the form of a numerical display, using either light-emitting diodes or a liquid crystal, as on calculator displays. They have switches, either rotary or pushbutton, that are used to select between a.c. or d.c. measurements. There is also a facility for selecting a number of current and voltage ranges.

All measuring instruments are quite fragile, not only mechanically but even more so electrically. So whenever you use them please observe the following rules:

1 Do not switch on (or connect) the power supply to a circuit until your connections have been double checked.
2 Starting with all meters switched to the OFF position, select the highest possible range, and then carefully select lower ranges until a suitable figure is displayed.
3 When taking a series of readings try to select a range that will accommodate the whole series. This is not always possible. However, if the range(s) *are* changed and the results are used to plot a graph, then a sudden unexpected gap or 'jump' in the plotted curve may well occur.

Figure 1.11 The ammeter circuit symbol

Figure 1.12 The voltmeter circuit symbol

4 When finished, turn off and disconnect all power supplies, and turn all meters to their OFF position.

'Warning – High Voltage' is a warning you have probably seen several times. All forms of energy – including electrical energy – can be dangerous. But it is not just the voltage that is harmful. When you touch your car during the winter months, for example, the voltage shock of static electricity can be hundreds to thousands of times greater than the voltage that can harm you. After all, it is the electrical current that flows through the body that can cause an injury. Then why the warning 'Warning – High Voltage'? Most electrical sources produce a constant voltage. It is actually easier to measure a voltage than a current. The warning therefore warns for what is easiest to measure.

The danger of electric shock or electrocution is not in burns, but in temporary or permanent damage of the nerve system. Nerves use electrochemical signals and an electrical current can interrupt these signals. If the current flows only through skeletal muscles, it can lead to temporary paralysis or involuntary contractions. Fortunately, these are not life-threatening. If the current runs along nerves and muscles that supply oxygen to the brain, the problem is much greater. Temporary paralysis can lead to oxygen deprivation and unnatural contractions can interrupt the signals that determine the heart's rhythm, resulting in death after a few minutes. Table 1.6 shows the physiological reactions at different current levels. They are approximations because they are obtained from accident analyses and not from human experiments. Obviously, any good electrical and electronic design will limit current to a few mA or less in all circumstances.

Electricity/electronics and water are big enemies. Pure water is actually an insulator. But it is usually contaminated with very good conductors, so that water becomes dangerous. Because there are applications with high humidity, for example in wet areas (think of cellars, garages and bathrooms) or outside in difficult-to-predict weather conditions, an IP code has been drawn up. IP stands for Ingress Protection Rating (or ingress protection rating) and is internationally established. It indicates the degree of protection against touching and ingress of objects on the one hand and moisture on the other. So this classification is better and more detailed than a vague marketing term like 'waterproof'. After the letters IP there are usually two numbers, the higher the more protection.

1.6 SIMULATION PACKAGES

There are many simulation packages on the market that fully simulate analogue and/or digital electrical and electronic circuits. In this way you can visualise the electrical and electronic behaviour of the different currents and voltages and make adjustments where necessary. In addition to the paid software, there also exist simulation packages available for free. One only does analogue, the other only digital, and yet another only allows you to integrate electrical and electronic components from the manufacturer itself.

Table 1.6 Relationship between current and physiological reaction

Current	*Physiological response*
3–5 mA	Barely perceptible
35–50 mA	Severe pain
50–70 mA	Muscle paralysis
500 mA	Cardiac arrest

A lot of that software falls back on SPICE (short for Simulation Program with Integrated Circuit Emphasis). This is a widely used analogue circuit simulator. It includes models for most common electronic components and can handle complex, non-linear circuits. This simulator was developed at the University of California at Berkeley and was first launched in 1972. Today we are already on the third version: SPICE3. The program was originally in the public domain, but there are many companies that have released their own commercial version (sometimes paid, sometimes not), improving ease of use and simulation convergence.

LTspice (website: http://www.linear.com/designtools/software/) is a freely available, high-performance simulation software, together with an extensive manual. You can enter your scheme graphically and visualise the analysis of the different voltages and currents graphically. The big advantage is that it is not limited to the number of knots and that models of components from other manufacturers can also be integrated. In addition, so-called netlists can be generated that can serve as a basis for, for example, the design of printed circuit boards or pcbs.

Another example of publicly available software is Circuit Lab (website: http://www.circuitlab.com). This software is completely web-based and requires no installation on your PC. Because it is not based on SPICE, it is a lot slower in calculation time. The underlying models to simulate the various components are usually enormously simplified and commercialised. However, it gives us the possibility to simulate an electrical or electronic circuit quickly and easily – and yet sufficiently accurately.

Some more electric/electronic simulation packages are PartQuest (website: http://www.partquest.com) and TinkerCad (website: http://www.tinkercad.com).

SUMMARY OF EQUATIONS

Charge: $Q = It$

Resistance: $R = \dfrac{\rho\ell}{A}$

Resistance at specified temp.: $R_1 = R_0(1+\alpha\theta)$ or $\dfrac{R_1}{R_2} = \dfrac{1+\alpha\theta_1}{1+\alpha\theta_2}$

Ohm's law: $V = IR$

Terminal p.d.: $V = E - Ir$

Energy: $W = VIt = I^2Rt = \dfrac{V^2t}{R}$

Power: $P = \dfrac{W}{t} = VI = I^2R = \dfrac{V^2}{R}$

ASSIGNMENT QUESTIONS

1 Convert the following into standard form.
   (a) 456.3 (b) 902 344 (c) 0.000 285 (d) 8000 (e) 0.047 12 (f) 180 μ*A* (g) 38 mV (h) 80 GN (i) 2000 μF
2 Write the following quantities in scientific notation.
   (a) 1500 Ω (b) 0.0033 Ω (c) 0.000 025 A (d) 750 V (e) 800 000 V (f) 0.000 000 047 F
3 Calculate the charge transferred in 25 minutes by a current of 500 mA.

4 A current of 3.6 A transfers a charge of 375 mC. How long would this take?

5 Determine the value of charging current required to transfer a charge of 0.55 mC in a time of 600 μs.

6 Calculate the p.d. developed across a 750 Ω resistor when the current flowing through it is (a) 3 A, (b) 25 mA.

7 An emf of 50 V is applied in turn to the following resistors: (a) 22 Ω, (b) 820 Ω, (c) 2.7 MΩ, (d) 330 kΩ. Calculate the current flow in each case.

8 The current flowing through a 470 Ω resistance is 4 A. Determine the energy dissipated in a time of 2 h. Express your answer in both joules and in commercial units.

9 A small business operates three pieces of equipment for nine hours continuously per day for six days a week. If the three pieces of equipment consume 10 kW, 2.5 kW and 600 W respectively, calculate the weekly cost if the charge per unit is 7.9 pence.

10 A charge of 500 μC is passed through a 560 Ω resistor in a time of 1 ms. Under these conditions determine (a) the current flowing, (b) the p.d. developed, (c) the power dissipated and (d) the energy consumed in 5 minutes.

11 A battery of emf 50 V and internal resistance 0.2 Ω supplies a current of 1.8 A to an external load. Under these conditions determine (a) the terminal p.d. and (b) the resistance of the external load.

12 The terminal p.d. of a d.c. source is 22.5 V when supplying a load current of 10 A. If the emf is 24 V calculate (a) the internal resistance and (b) the resistance of the external load.

13 For the circuit arrangement specified in Q12 above, determine the power and energy dissipated by the external load resistor in 5 minutes.

14 A circuit of resistance 4 Ω dissipates a power of 16 W. Calculate (a) the current flowing through it, (b) the p.d. developed across it and (c) the charge displaced in a time of 20 minutes.

15 In a test the velocity of a body was measured over a period of time, yielding the results shown in the table below. Plot the corresponding graph and use it to determine the acceleration of the body at times $t = 0$, $t = 5$ s and $t = 9$ s. You may assume that the graph consists of a series of straight lines.

v(m/s)	0.0	3.0	6.0	10.0	14.0	15.0	16.0
t(s)	0.0	1.5	3.0	4.5	6.0	8.0	10.0

16 The insulation resistance between a conductor and earth is 30 MΩ. Calculate the leakage current if the supply voltage is 240 V.

17 A 3 kW immersion heater is designed to operate from a 240 V supply. Determine its resistance and the current drawn from the supply.

18 A 110 V d.c. generator supplies a lighting load of forty 100 W bulbs, a heating load of 10 kW and other loads which consume a current of 15 A. Calculate the power output of the generator under these conditions.

19 The field winding of a d.c. motor is connected to a 110 V supply. At a temperature of 18°C, the current drawn is 0.575 A. After running the machine for some time the current has fallen to 0.475 A, the voltage remaining unchanged. Calculate the temperature of the winding under the new conditions, assuming that the temperature coefficient of resistance of copper is 0.004 26/°C at 0°C.

20 A coil consists of 1500 turns of aluminium wire having a cross-sectional area of 0.75 mm^2. The mean length per turn is 60 cm and the resistivity of aluminium at the working temperature is 0.028 μΩm. Calculate the resistance of the coil.

Chapter 2

D.C. Circuits

LEARNING OUTCOMES

This chapter explains how to apply circuit theory to the solution of simple circuits and networks by the application of Ohm's law and Kirchhoff's laws, and the concepts of potential and current dividers.

This means that on completion of this chapter you should be able to:

1. Calculate current flows, potential differences, power and energy dissipations for circuit components and simple circuits, by applying Ohm's law.
2. Carry out the above calculations for more complex networks using Kirchhoff's laws.
3. Calculate circuit p.d.s using the potential divider technique, and branch currents using the current divider technique.
4. Understand the principles and use of a slidewire potentiometer.

2.1 RESISTORS IN SERIES

When resistors are connected 'end-to-end' so that the same current flows through them all they are said to be cascaded or connected in series. Such a circuit is shown in Figure 2.1. Note that, for the sake of simplicity, an ideal source of emf has been used (no internal resistance).

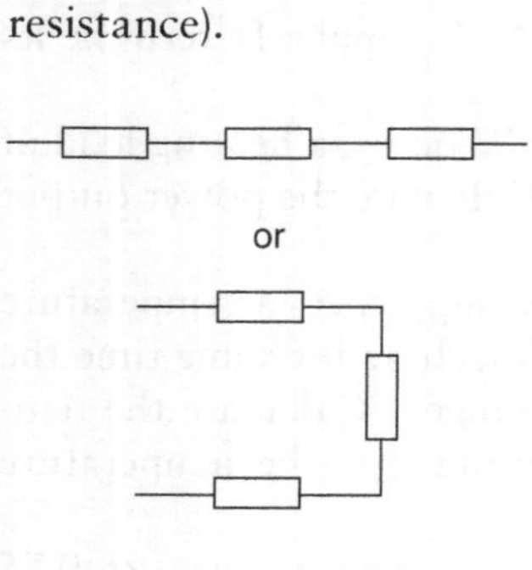

From the previous chapter we know that the current flowing through the resistors will result in p.d.s being developed across them. We also know that the sum of these p.d.s must equal the value of the applied emf. Thus

$$V_1 = IR_1$$

 DOI: 10.1201/9781003308294-2

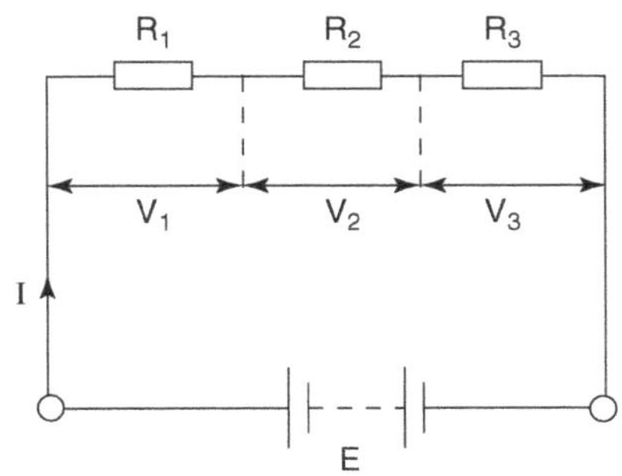

Figure 2.1 Resistors cascaded or connected in series

$$V_2 = IR_2$$

$$V_3 = IR_3$$

However, the circuit current I depends ultimately on the applied emf E and the total resistance R offered by the circuit. Hence

$$E = IR$$
$$E = V_1 + V_2 + V_3$$

and substituting for E, V_1, V_2 and V_3 in this last equation, we have

$$IR = IR_1 + IR_2 + IR_3$$

and dividing this last equation by the common factor I

$$R = R_1 + R_2 + R_3 \qquad (2.1)$$

where R is the total circuit resistance. From this result it may be seen that when resistors are connected in series or cascaded, the total resistance is found simply by adding together the resistor values.

WORKED EXAMPLE 2.1

Q For the circuit shown in Figure 2.2 calculate (a) the circuit resistance, (b) the circuit current, (c) the p.d. developed across each resistor and (d) the power dissipated by the complete circuit.

$E = 24$ V; $R_1 = 330\ \Omega$; $R_2 = 1500\ \Omega$; $R_3 = 470\ \Omega$

(a) $R = R_1 + R_2 + R_3 = 330 + 150 + 470 = 2300\ \Omega$ or $2.3\ \text{k}\Omega$

(b) $I = \dfrac{E}{R} = \dfrac{24}{2300} = 10.43$ mA

(c) $V_1 = IR_1 = 10.43 \times 10^{-3} \times 330 = 3.44$ V

$V_2 = IR_2 = 10.43 \times 10^{-3} \times 1500 = 15.65$ V

$V_3 = IR_3 = 10.43 \times 10^{-3} \times 470 = 4.90$ V

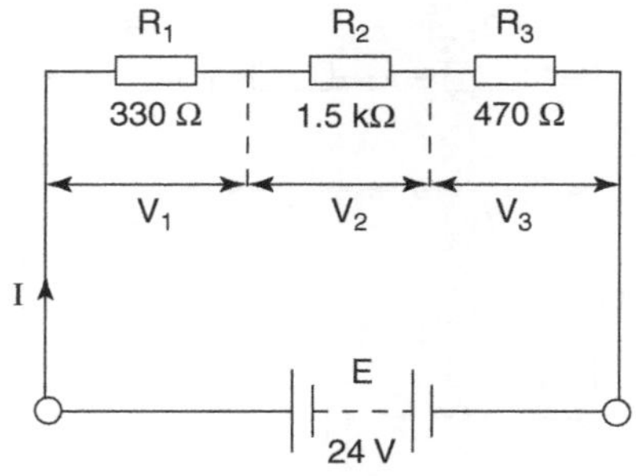

Figure 2.2 The circuit diagram for Worked Example 2.1

Note: The sum of the above p.d.s is 23.99 V instead of 24 V due to the rounding errors in the calculation. It should also be noted that the value quoted for the current was 10.43 mA whereas the calculator answer is 10.4347 mA. This latter value was then stored in the calculator memory and used in the calculations for part (c), thus reducing the rounding errors to an acceptable minimum.

(d) $P = EI = 24 \times 10.43 \times 10^{-3} = 0.25$ W or 250 mW

It should be noted that the power is dissipated by the three resistors in the circuit. Hence, the circuit power could have been determined by calculating the power dissipated by each of these and adding these values to give the total. This is shown below, and serves as a check for the last answer.

$$P_1 = I^2R_1 = \left(10.43 \times 10^{-3}\right)^2 \times 330 = 35.93 \text{ mW}$$

$$P_2 = \left(10.43 \times 10^{-3}\right)^2 \times 1500 = 163.33 \text{ mW}$$

$$P_3 = \left(10.43 \times 10^{-3}\right)^2 \times 470 = 51.18 \text{ mW}$$

total power: $P = P_1 + P_2 + P_3 = 250.44$ mW

(Note the worsening effect of the continuous rounding error.)

WORKED EXAMPLE 2.2

Q Two resistors are connected in series across a battery of emf 12 V. If one of the resistors has a value of 16 Ω and it dissipates a power of 4 W, then calculate (a) the circuit current and (b) the value of the other resistor.

Since the only two pieces of data that are directly related to each other concern the 16 Ω resistor and the power that it dissipates, this information must form the starting point for the solution of the problem. Using these data we can determine either the current through or the p.d. across the 16 Ω resistor (and it is not important which of these is calculated first). To illustrate this point both methods will be demonstrated. The appropriate circuit diagram, which forms an integral part of the solution, is shown in Figure 2.3.

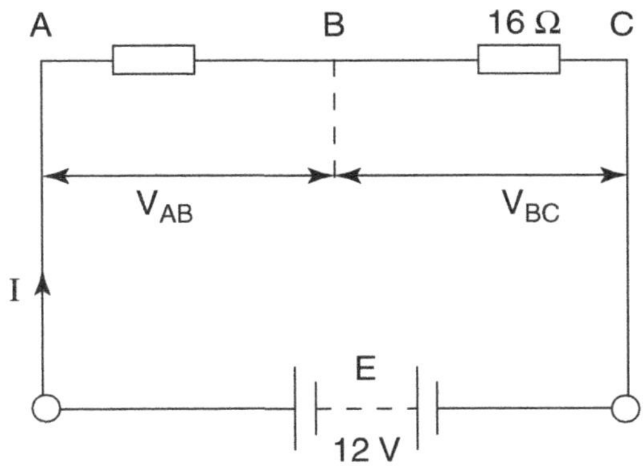

Figure 2.3 The circuit diagram for Worked Example 2.2

$E = 12$ V; $R_{BC} = 16\ \Omega$; $P_{BC} = 4$ W

(a) $I^2 R_{BC} = P_{BC}$

$$I^2 = \frac{P_{BC}}{R_{BC}} = \frac{4}{16} = 0.25 \text{ so } I = 0.5 \text{ A}$$

(b) total resistance, $R = \frac{E}{I} = \frac{12}{0.5} = 24\ \Omega$

$$R_{AB} = R - R_{BC} = 24 - 16 = 8\ \Omega$$

Alternatively, the problem can be solved thus:

(a) $\frac{V_{BC}^2}{R_{BC}} = P_{BC}$

$$V_{BC}^2 = P_{BC} \times R_{BC} = 4 \times 16 = 64 \text{ V so } V_{BC} = 8 \text{ V}$$

$$I = \frac{V_{BC}}{R_{BC}} = \frac{8}{16} = 0.5 \text{ A}$$

(b) $V_{AB} = E - V_{BC} = 12 - 8$ $V_{AB} = 4$ V

$$R_{AB} = \frac{V_{AB}}{I} = \frac{4}{0.5} = 8\ \Omega$$

As can be easily noticed, the p.d. developed across each resistor is in direct proportion to its resistance value. This is a useful fact to bear in mind, since it means it is possible to calculate the p.d.s without first having to determine the circuit current. Consider two resistors connected across a 50 V supply as shown in Figure 2.4. In order to demonstrate the potential divider effect we will in this case firstly calculate circuit current and hence the two p.d.s by applying Ohm's law:

$$R = R_1 + R_2 = 75\ \Omega + 25\ \Omega = 100\ \Omega$$

$$I = \frac{E}{R} = \frac{50}{100} = 0.5 \text{ A}$$

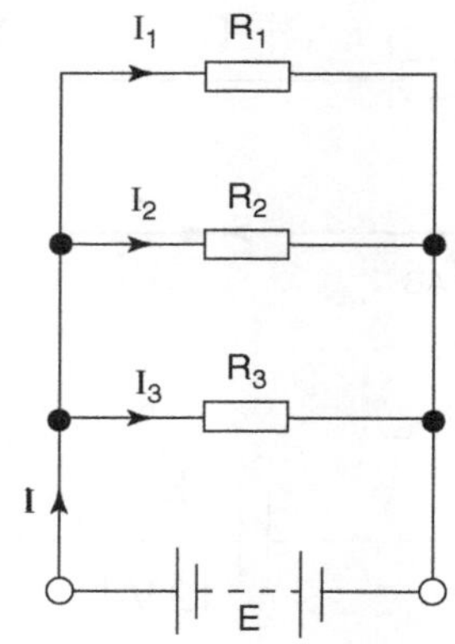

Figure 2.4 Two resistors in series across a 50 V supply

$$V_1 = IR_1 = 0.5 \times 75 = 37.5 \text{ V}$$
$$V_2 = IR_2 = 0.5 \times 25 = 12.5 \text{ V}$$

Applying the potential divider technique, the two p.d.s may be obtained by using the fact that the p.d. across a resistor is given by the ratio of its resistance value to the total resistance of the circuit, expressed as a proportion of the applied voltage. Although this sounds complicated it is very simple to put into practice. Expressed in the form of an equation it means

$$V_1 = \frac{R_1}{R_1 + R_2} \times E \tag{2.2}$$

and

$$V_2 = \frac{R_2}{R_1 + R_2} \times E \tag{2.3}$$

and using the above equations the p.d.s can more simply be calculated as follows:

$$V_1 = \frac{75}{100} \times 50 = 37.5 \text{ V}$$
$$V_2 = \frac{25}{100} \times 50 = 12.5 \text{ V}$$

This technique is not restricted to only two resistors in series, but may be applied to any number. For example, if there were three resistors in series, then the p.d. across each may be found from

$$V_1 = \frac{R_1}{R_1 + R_2 + R_3} \times E$$
$$V_2 = \frac{R_2}{R_1 + R_2 + R_3} \times E$$
$$V_3 = \frac{R_3}{R_1 + R_2 + R_3} \times E$$

Christmas lights are used to decorate trees or public buildings around December every year. Small electric light bulbs in the shape of a candle flame are used as a replacement for the fire danger of candles. Instead of incandescent light bulbs, light-emitting diodes (LEDs) are also used, being more energy efficient. Although many variations exist, those bulbs are cascaded or connected in series, ranging from 12 to 24 bulbs up to 200 miniature lights. It reduces the length of the swinging electric cable in the tree, used for connecting all bulbs. Older or cheaper light sets go completely dark when a single bulb in the series connection fails. Troubleshooting can be done by a one-by-one replacement with a known working bulb, or by using a multimeter to find out where the voltage gets interrupted. Recently, shunts have been used to maintain the connection when a bulb fails.

2.2 RESISTORS IN PARALLEL

When resistors are joined 'side-by-side' so that their corresponding ends are connected together, they are said to be connected in parallel. Using this form of connection means that there will be a number of paths through which the current can flow. Such a circuit consisting of three resistors is shown in Figure 2.5, and the circuit may be analysed as follows:

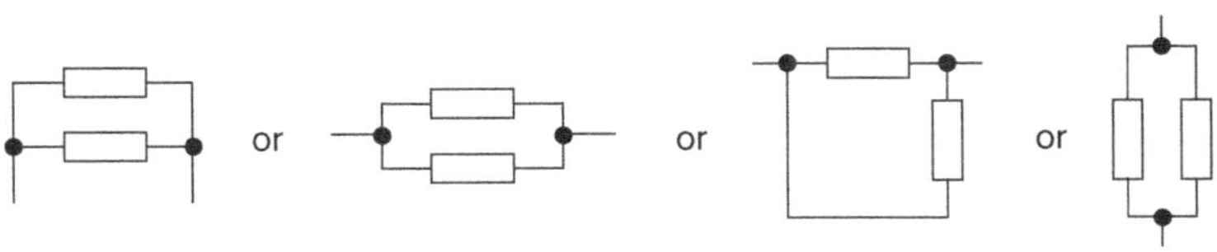

Since all three resistors are connected directly across the battery terminals, they all have the same voltage developed across them. In other words the voltage is the common factor in this arrangement of resistors. Now, each resistor will allow a certain value of current to flow through it, depending upon its resistance value. Thus

$$I_1 = \frac{E}{R_1}$$

$$I_2 = \frac{E}{R_2}$$

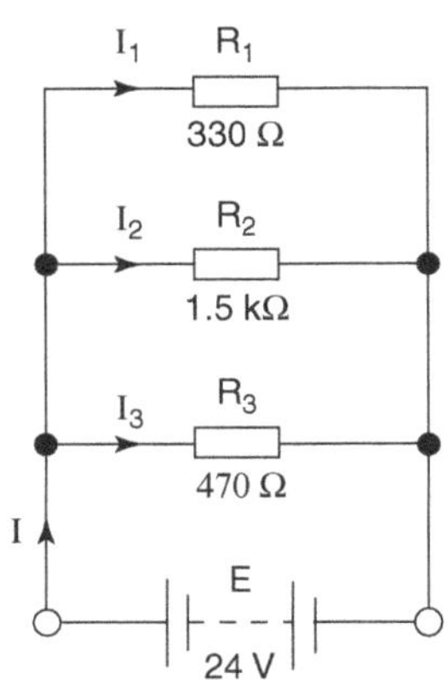

Figure 2.5 Resistors in parallel

$$I_3 = \frac{E}{R_3}$$

The total circuit current I is determined by the applied emf and the total circuit resistance R,

$$I = \frac{E}{R}$$

Also, since all three branch currents originate from the battery, the total circuit current must be the sum of the three branch currents

$$I = I_1 + I_2 + I_3$$

and substituting the above expression for the currents:

$$\frac{E}{R} = \frac{E}{R_1} + \frac{E}{R_2} + \frac{E}{R_3}$$

and dividing the above equation by the common factor E:

$$\frac{1}{R} = \frac{1}{R_1} + \frac{1}{R_2} + \frac{1}{R_3} \quad (2.4)$$

Note: The above equation does *not* give the total resistance of the circuit, but does give the total circuit conductance (G), which is measured in Siemens (S). Thus, conductance is the reciprocal of resistance, so to obtain the circuit resistance you must then take the reciprocal of the answer obtained from an equation of the form of Equation (2.4). The inverse of a resistor R (expressed in Ω) is the conductance G (expressed in S) and vice versa.

Conductance is a measure of the 'willingness' of a material or circuit to allow current to flow through it.

Ernst Werner von Siemens (1816–1892) was a German industrialist and inventor. He contributed to the invention and development of electrical machines. The unit of conductance Siemens is the reciprocal of the unit of resistance ohm; electrical engineers refer to the conductance as mho (ohm, reversed), which was proposed by William Thomson (1824–1907), a British physicist. *Note*: Siemens is abbreviated to S (capitalised), not to be distinguished from second, abbreviated to s (lower case).

The inverse of a resistor R (expressed in Ω) is the conductance G (expressed in S) and vice versa. That is

$$\frac{1}{R} = G \text{ and } \frac{1}{G} = R \quad (2.5)$$

However, when only two resistors are in parallel, the combined resistance may be obtained directly by using the following equation:

$$R = \frac{R_1 \times R_2}{R_1 + R_2} \quad (2.6)$$

If there are 'x' identical resistors in parallel the total resistance is simply R/x ohms.

In this context, the word 'identical' means having the same value of resistance.

WORKED EXAMPLE 2.3

Q Considering the circuit of Figure 2.6, calculate (a) the total resistance of the circuit, (b) the three branch current and (c) the current drawn from the battery.

$E = 24\ \text{V};\ R_1 = 330\ \Omega;\ R_2 = 1500\ \Omega;\ R_3 = 470\ \Omega$

(a) $$\frac{1}{R} = \frac{1}{R_1} + \frac{1}{R_2} + \frac{1}{R_3} = \frac{1}{330} + \frac{1}{1500} + \frac{1}{470} = 0.00303 + 0.000667 + 0.00213 = 0.005825\ \text{S}$$

$$R = 171.68\ \Omega\left(\text{inverse of } 0.005825\ \text{S}\right)$$

(b) $$I_1 = \frac{E}{R_1} = \frac{24}{330} = 72.73\ \text{mA}$$

$$I_2 = \frac{E}{R_2} = \frac{24}{1500} = 16\ \text{mA}$$

$$I_3 = \frac{E}{R_3} = \frac{24}{470} = 51.06\ \text{mA}$$

(c) $I = I_1 + I_2 + I_3 = 72.73 + 16 + 51.06 = 139.8\ \text{mA}$

Alternatively, the circuit current could have been determined by using the values for E and R as follows

$$I = \frac{E}{R} = \frac{24}{171.68} = 139.8\ \text{mA}$$

Compare this example with Worked Example 2.1 (the same values for the resistors and the emf have been used). From this it should be obvious that when resistors are connected in parallel the total resistance of the circuit is reduced. This results in a corresponding increase of current drawn from the source. This is simply because the parallel arrangement provides more paths for current flow.

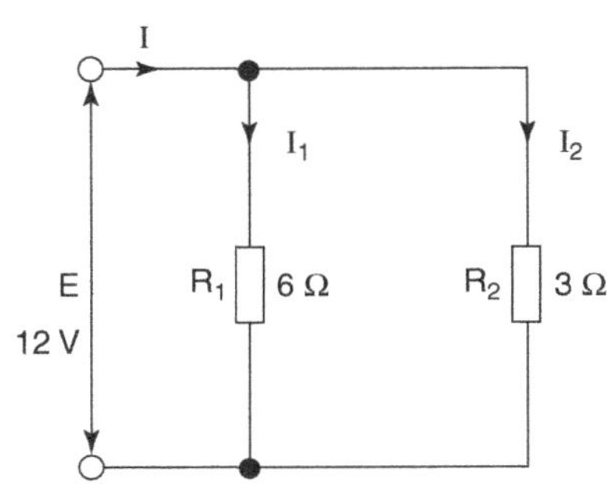

Figure 2.6 The circuit diagram for Worked Example 2.3

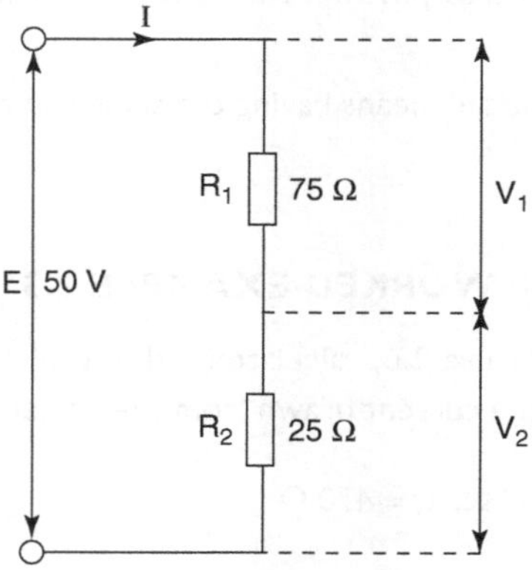

Figure 2.7 The circuit diagram for Worked Example 2.4

WORKED EXAMPLE 2.4

Q Two resistors, one of 6 Ω and the other of 3 Ω resistance, are connected in parallel across a source of emf 12 V. Determine (a) the effective resistance of the combination, (b) the current drawn from the source and (c) the current through each resistor.

The corresponding circuit diagram, suitably labelled, is shown in Figure 2.7.

$E = 12\ \text{V};\ R_1 = 6\ \Omega;\ R_2 = 3\ \Omega$

(a) $R = \dfrac{R_1 R_2}{R_1 + R_2} = \dfrac{6 \times 3}{6 + 3} = 2\ \Omega$

(b) $I = \dfrac{E}{R} = \dfrac{12}{2} = 6\ \text{A}$

(c) $I_1 = \dfrac{E}{R_1} = \dfrac{12}{6} = 2\ \text{A}$

$I_2 = \dfrac{E}{R_2} = \dfrac{12}{3} = 4\ \text{A}$

WORKED EXAMPLE 2.5

Q A 10 Ω resistor, a 20 Ω resistor and a 30 Ω resistor are connected (a) in series, and then (b) in parallel with each other. Calculate the total resistance for each of the two connections.

(a) $R_1 = 10\ \Omega;\ R_2 = 20\ \Omega;\ R_3 = 30\ \Omega$

$R = R_1 + R_2 + R_3 = 10 + 20 + 30 = 60\ \Omega$

(b) $\dfrac{1}{R} = \dfrac{1}{R_1} + \dfrac{1}{R_2} + \dfrac{1}{R_3} = \dfrac{1}{10} + \dfrac{1}{20} + \dfrac{1}{30} = 0.1 + 0.05 + 0.033$

$R = \dfrac{1}{0.183\ \text{S}} = 5.46\ \Omega$

Alternatively, $\frac{1}{R} = \frac{1}{10} + \frac{1}{20} + \frac{1}{30} = \frac{6+3+2}{60} = \frac{11}{60}\,\text{S}$

$$R = \frac{60}{11} = 5.46\ \Omega$$

It has been shown that when resistors are connected in parallel the total circuit current divides between the alternative paths available. So far we have determined the branch currents by calculating the common p.d. across a parallel branch and dividing this by the respective resistance values. However, these currents can be found directly, without the need to calculate the branch p.d., by using the current divider theory. Consider two resistors connected in parallel across a source of emf 48 V as shown in Figure 2.8. Using the p.d. method we can calculate the two currents as follows:

$$I_1 = \frac{E}{R_1} = \frac{48}{12} = 4\ \text{A} \quad \text{and} \quad I_2 = \frac{E}{R_2} = \frac{48}{24} = 2\ \text{A}$$

It is now worth noting the values of the resistors and the corresponding currents. It is clear that R_1 is half the value of R_2. So, from the calculation we obtain the quite logical result that I_1 is twice the value of I_2. That is, a ratio of 2:1 applies in each case. Thus, the smaller resistor carries the greater proportion of the total current. By stating the ratio as 2:1 we can say that the current is split into three equal 'parts'. Two 'parts' are flowing through one resistor and the remaining 'part' through the other resistor.

Thus $\frac{2}{3} \times I$ flows through R_1

and $\frac{1}{3} \times I$ flows through R_2

Since $I = 6$ A then

$$I_1 = \frac{2}{3} \times 6 = 4\ \text{A}$$

$$I_2 = \frac{1}{3} \times 6 = 2\ \text{A}$$

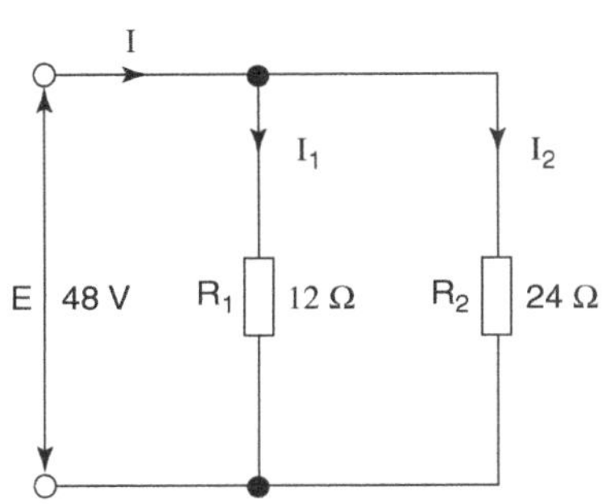

Figure 2.8 Two resistors in parallel across an emf of 48 V

In general, we can say that

$$I_1 = \frac{R_2}{R_1 + R_2} \times I \tag{2.7}$$

and

$$I_2 = \frac{R_1}{R_1 + R_2} \times I \tag{2.8}$$

Note: This is *not* the same ratio as for the potential divider. If you compare (2.2) with (2.7) you will find that the numerator in (2.2) is R_1 whereas in (2.7) the numerator is R_2. There is a similar 'cross-over' when (2.3) and (2.8) are compared.

Again, the current divider theory is not limited to only two resistors in parallel. Any number can be accommodated. However, with three or more parallel resistors the current division method can be cumbersome to use, and it is much easier for mistakes to be made. For this reason it is recommended that where more than two resistors exist in parallel the 'p.d. method' is used. This will be illustrated in the next section, but for completeness the application to three resistors is shown below.

Consider the arrangement shown in Figure 2.9:

$$\frac{1}{R} = \frac{1}{R_1} + \frac{1}{R_2} + \frac{1}{R_3} = \frac{1}{3\,\Omega} + \frac{1}{4\,\Omega} + \frac{1}{6\,\Omega} = \frac{4+3+2}{12}\,\text{S}$$

and examining the numerator, we have 4 + 3 + 2 = 9 'parts'.

Thus, the current ratios will be 4/9, 3/9 and 2/9 respectively for the three resistors.

$$I_1 = \frac{4}{9} \times 18 = 8\ \text{A}$$

$$I_2 = \frac{3}{9} \times 18 = 6\ \text{A}$$

$$I_3 = \frac{2}{9} \times 18 = 4\ \text{A}$$

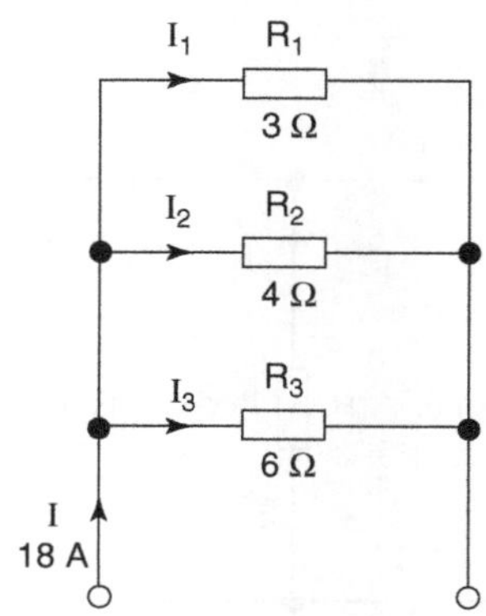

Figure 2.9 Three resistors in parallel

The rear window heating of a car is constructed as a parallel connection of several constantan wires. Constantan is a copper-nickel alloy with a resistivity of 4.9×10^{-7} Ω ·m. In total 13 wires of a length of 1.1 m each are used in parallel. The cross-section of each wire is 4.2×10^{-2} mm^2, resulting in the resistance of one single wire of 4.9×10^{-7} Ω ·m × 1.1 m/4.2×10^{-8} m^2 = 12.83 Ω, resulting in a total resistance of 12.83 Ω/13 = 0.99 Ω, because there are 13 wires in parallel. If you know that the voltage between the poles of the battery is 12.8 V, the total current can by calculated by using Ohm's law: 12.8 V/0.99 Ω = 12.97 A. The power dissipated by the rear window heating equals 12.8 V × 12.97 A = 165.97 W, assuming that the resistance of the cables connecting the rear window defogger to the battery is neglected. It is left to the reader to calculate what happens if one wire is broken.

2.3 SERIES/PARALLEL COMBINATIONS

Most practical circuits consist of resistors which are interconnected in both series and parallel forms. The simplest method of solving such a circuit is to reduce the parallel branches to their equivalent resistance values and hence reduce the circuit to a simple series arrangement. This is best illustrated by means of a worked example.

WORKED EXAMPLE 2.6

Q For the circuit shown in Figure 2.10, calculate (a) the current drawn from the supply, (b) the current through the 6 Ω resistor and (c) the power dissipated by the 5.6 Ω resistor.

The first step in the solution is to sketch and label the circuit diagram, clearly showing all currents flowing and identifying each part of the circuit as shown in Figure 2.11. Also note that since there is no mention of internal resistance it may be assumed that the source of emf is ideal.

(a) To determine the current I drawn from the battery we need to know the total resistance R_{AC} of the circuit.

$$R_{BC} = \frac{6\,\Omega \times 4\,\Omega}{6\,\Omega + 4\,\Omega}\left(\text{using } \frac{\text{product}}{\text{sum}} \text{ for two resistors in parallel}\right) = 2.4\,\Omega$$

The original circuit may now be redrawn as in Figure 2.12.

$$R_{AC} = R_{AB} + R_{BC} \text{ (resistors in series)} = 5.6 + 2.4 = 8\,\Omega$$

$$I = \frac{E}{R_{AC}} = \frac{64}{8} = 8\text{ A}$$

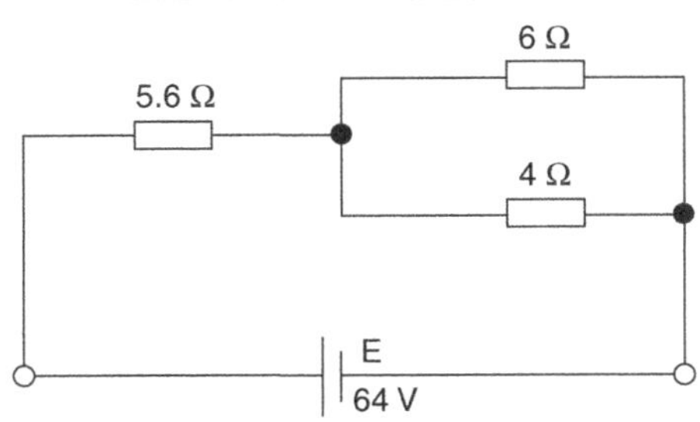

Figure 2.10 The circuit diagram for Worked Example 2.6

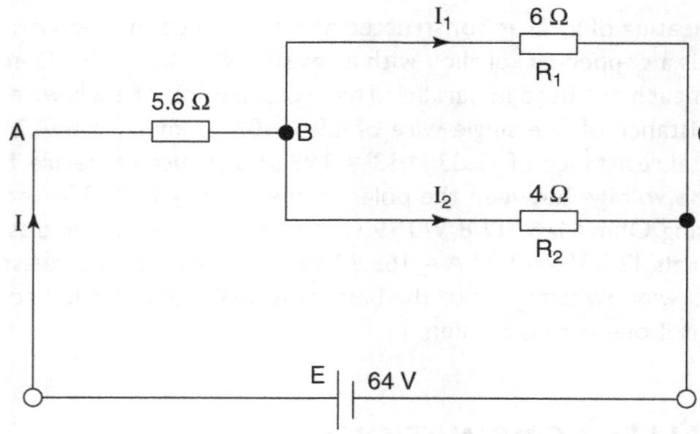

Figure 2.11 The circuit diagram for Worked Example 2.6, with labelling

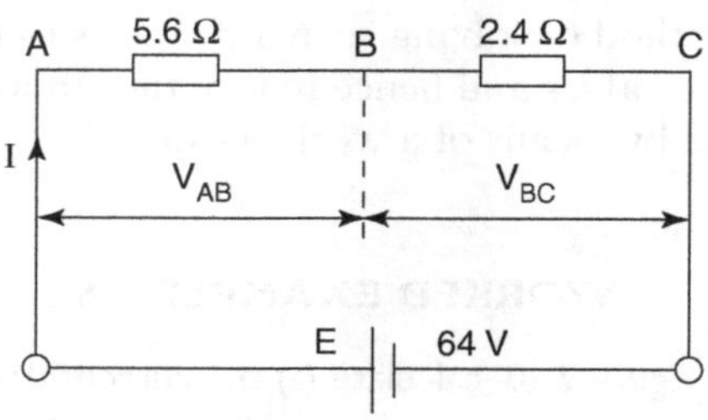

Figure 2.12 The circuit diagram for Worked Example 2.6, redrawn

(b) To find the current I_1 through the 6 Ω resistor we may use either of two methods. Both of these are now demonstrated.

p.d. method:

$$V_{BC} = IR_{BC} \text{ (Fig. 2.12)} = 8 \times 2.4 = 19.2 \text{ V}$$

$$I_1 = \frac{V_{BC}}{R_1} \text{ (Fig. 2.11)} = \frac{19.2}{6} = 3.2 \text{ A}$$

This answer may be checked as follows:

$$I_1 = \frac{V_{BC}}{R_2} = \frac{19.2}{4} = 4.8 \text{ A}$$

and since $I = I_1 + I_2 = 3.2 + 4.8 = 8$ A

which agrees with the value found in (a).

current division method:

Considering Figure 2.11, the current I splits into the components I_1 and I_2 according to the ratio of the resistor values. However, you must bear in mind that the *larger* resistor carries the *smaller* proportion of the total current.

$$I_1 = \frac{R_2}{R_1 + R_2} \times I = \frac{4}{6+4} \times 8 = 3.2 \text{ A}$$

(c) $P_{AB} = I^2 R_{AB} = 8 \text{ A} \times 8 \text{ A} \times 5.6\ \Omega = 358.4 \text{ W}$

Alternatively, $P_{AB} = V_{AB} I$

where $V_{AB} = E - V_{BC} = 64 - 19.2 = 44.8 \text{ V}$

$P_{AB} = 44.8 \times 8 = 358.4 \text{ W}$

WORKED EXAMPLE 2.7

Q For the circuit of Figure 2.13 calculate (a) the current drawn from the source, (b) the p.d. across each resistor, (c) the current through each resistor and (d) the power dissipated by the 5 Ω resistor.

The first step in the solution is to label the diagram clearly with letters at the junctions and identifying p.d.s and branch currents. This is shown in Figure 2.14.

$$R_{AB} = \frac{R_1 R_2}{R_1 + R_2} = \frac{4 \times 6}{4 + 6} = 2.4\ \Omega$$

$$R_{BC} = 5\ \Omega$$

(a)

$$\frac{1}{R_{CD}} = \frac{1}{R_4} + \frac{1}{R_5} + \frac{1}{R_6} = \frac{1}{3} + \frac{1}{6} + \frac{1}{8} = \frac{8+4+3}{24} = \frac{15}{24} \text{ S}$$

$$R_{CD} = \frac{24}{15} = 1.6\ \Omega$$

$$R = R_{AB} + R_{BC} + R_{CD} = 2.4 + 5 + 1.6 = 9\ \Omega$$

$$I = \frac{E}{R} = \frac{18}{9} = 2 \text{ A}$$

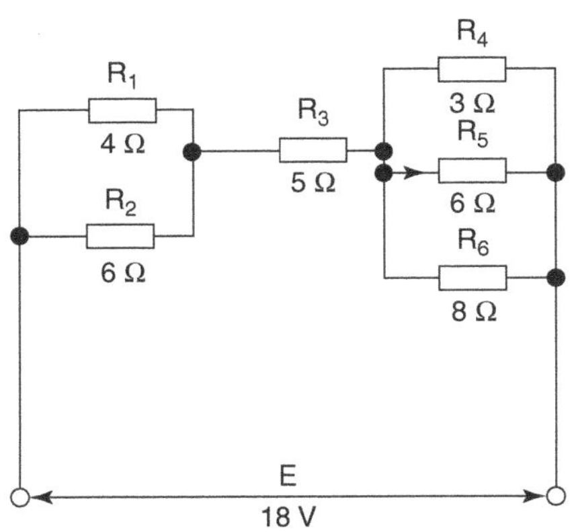

Figure 2.13 The circuit diagram for Worked Example 2.7

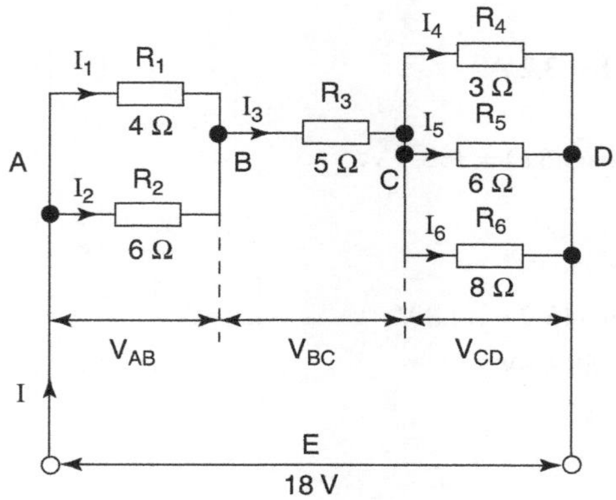

Figure 2.14 The circuit diagram for Worked Example 2.7, with labelling

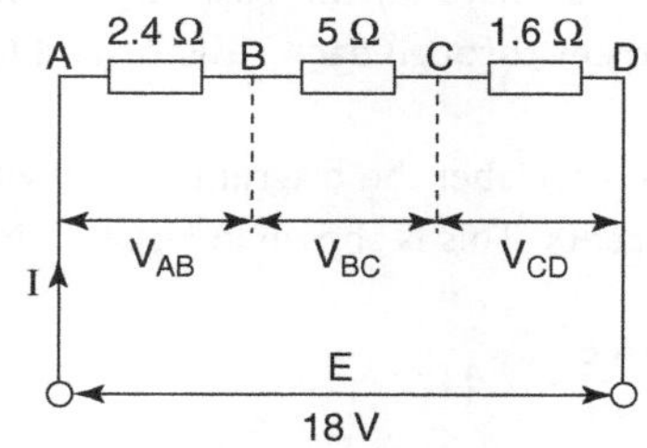

Figure 2.15 The circuit diagram for Worked Example 2.7, redrawn

(b) The circuit has been reduced to its series equivalent as shown in Figure 2.15. Using this equivalent circuit it is now a simple matter to calculate the p.d. across each section of the circuit.

$V_{AB} = IR_{AB} = 2 \times 2.4 = 4.8\text{ V}$

(this p.d. is common to both R_1 and R_2)

$V_{BC} = IR_{BC} = 2 \times 5 = 10\text{ V}$

$V_{CD} = IR_{CD} = 2 \times 1.6 = 3.2\text{ V}$

(this p.d. is common to R_4, R_5 and R_6)

(c)

$$I_1 = \frac{V_{AB}}{R_1} = \frac{4.8}{4} = 1.2\text{ A}$$

$$I_1 = \frac{R_2}{R_1 + R_2} \times I = \frac{6}{10} \times 2 = 1.2\text{ A}$$

$$I_2 = \frac{R_1}{R_1 + R_2} \times I = \frac{4}{10} \times 2 = 0.8\text{ A}$$

$$I_2 = \frac{V_{AB}}{R_2} = \frac{4.8}{6} = 0.8\text{ A}$$

$I_3 = I = 2\text{ A}$

$$\frac{1}{R_{CD}} = \frac{1}{R_4} + \frac{1}{R_5} + \frac{1}{R_6} = \frac{1}{3} + \frac{1}{6} + \frac{1}{8} = \frac{8+4+3}{24} = \frac{15}{24}\text{ S}$$

$$I_4 = \frac{V_{CD}}{R_4} = \frac{3.2}{3} = 1.067 \text{ A} \qquad I_4 = \frac{8}{15} \times 2 = 1.067 \text{ A}$$

$$I_5 = \frac{4}{15} \times 2 = 0.533 \text{ A}$$

$$I_6 = \frac{3}{15} \times 2 = 0.4 \text{ A}$$

$$I_5 = \frac{V_{CD}}{R_5} = \frac{3.2}{6} = 0.533 \text{ A}$$

$$I_6 = \frac{V_{CD}}{R_6} = \frac{3.2}{8} = 0.4 \text{ A}$$

Notice that the p.d. method is an easier and less cumbersome one than current division when more than two resistors are connected in parallel.

(d) $P_3 = I_3^2 R_3$ or $V_{BC} I_3$ or $\frac{V_{BC}^2}{R_3}$ and using the first of these alternative equations:

$$P_3 = 2 \times 2 \times 5 = 20 \text{ W}$$

It is left to the reader to confirm that the other two power equations above yield the same answer.

There are several websites where you can easily calculate series/parallel combinations of several resistors:

- https://www.digikey.com/en/resources/conversion-calculators/conversion-calculator-parallel-and-series-resistor (series and parallel calculation)
- https://www.allaboutcircuits.com/tools/parallel-resistance-calculator/ (parallel calculation)
- https://www.amplifiedparts.com/tech-articles/resistor-parallel-series-calculator (series and parallel calculation and combinations thereof)

The electrical behaviour of the human body is quite complex. In order to predict and control electrical phenomena, simplified models are used in which mathematical equations can approximate the relationship between voltage and current. The human body is actually a conductor of the current, so it can be modelled with resistors. Figure 2.16 shows a simplified model of series/parallel resistor combination. The arms, legs, trunk and head all have their own characteristic resistance. The resistor values not only depend on the length and diameter of that human body part, but also on the balance between muscles and blood. The resistance can be measured with a multimeter, by applying a small voltage and measuring the corresponding current. The ratio of both numbers then indicates the resistance.

Obviously, the current that runs through the torso also passes through the heart…

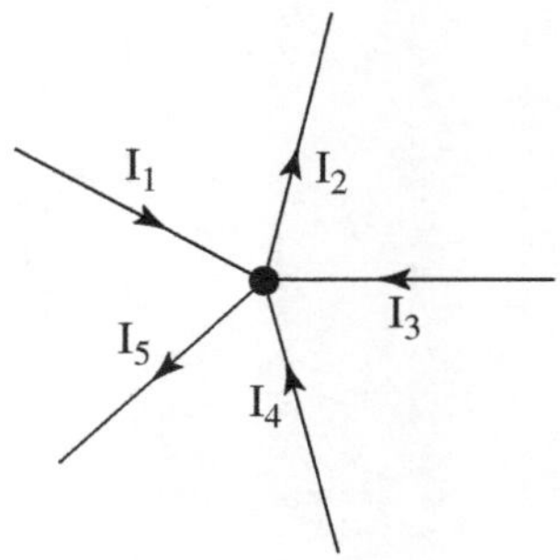

Figure 2.16 The human body modelled with resistors

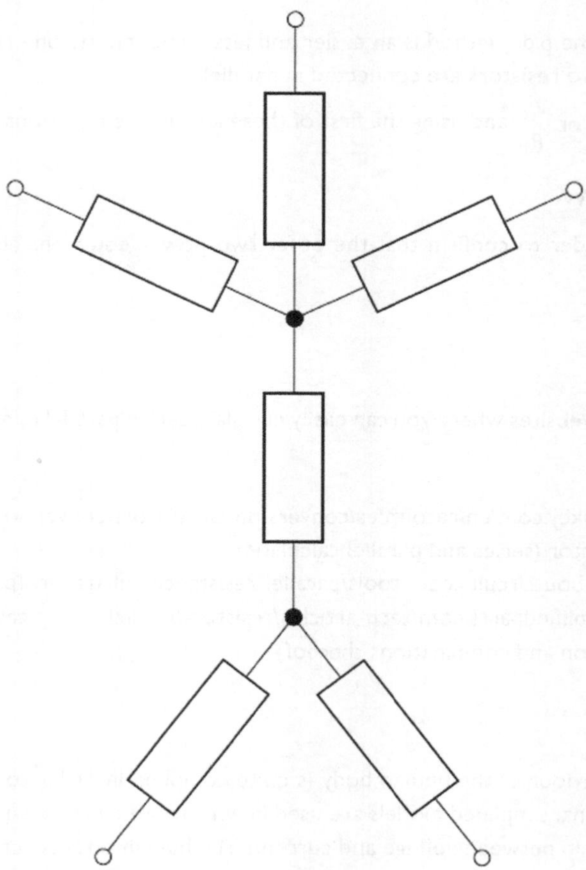

Figure 2.17 Kirchhoff's current law

2.4 KIRCHHOFF'S CURRENT LAW

We have already put this law into practice, though without stating it explicitly. The law states that the algebraic sum of the currents at any junction of a circuit is zero. Another, and perhaps simpler, way of stating this is to say that the sum of the currents arriving at a junction is equal to the sum of the currents leaving that junction. Thus we have applied the

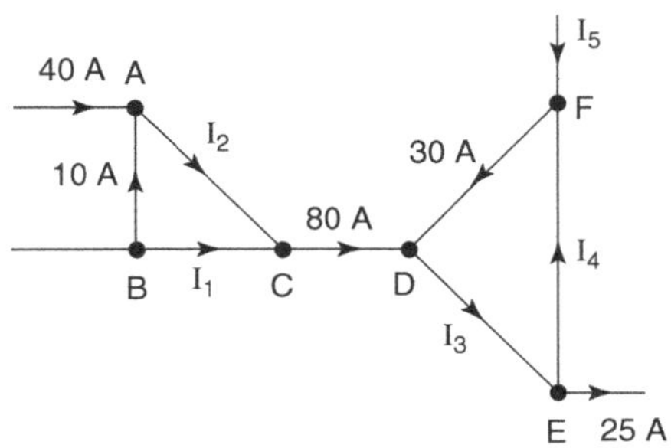

Figure 2.18 The circuit diagram for Worked Example 2.8

law with parallel circuits, where the assumption has been made that the sum of the branch currents equals the current drawn from the source. Expressing the law in the form of an equation we have:

$$\sum I = 0 \tag{2.9}$$

where the symbol $\sum$ means 'the sum of'.

Figure 2.17 illustrates a junction within a circuit with a number of currents arriving and leaving the junction. Applying Kirchhoff's current law yields:

$$I_1 - I_2 + I_3 + I_4 - I_5 = 0$$

where '+' signs have been used to denote currents arriving and '–' signs for currents leaving the junction. This equation can be transposed to comply with the alternative statement for the law, thus:

$$I_1 + I_3 + I_4 = I_2 + I_5$$

Gustav Robert Kirchhoff (1824–1887) was a German physicist. Besides his laws in the field of electrical engineering, he also performed research in the domain of spectroscopy and radiation of black bodies under heating. Together with the German chemist Robert Bunsen (1811–1899), he co-discovered the chemical elements cesium and rubidium.

WORKED EXAMPLE 2.8

Q For the network shown in Figure 2.18 calculate the values of the marked currents.

Junction A: $I_2 = 40 + 10 = 50$ A

Junction C: $I_1 + I_2 = 80$ A

$I_1 + 50$ A $= 80$ A

$I_1 = 30$ A

Junction D: $I_3 = 80 + 30 = 110$ A

Junction E: $I_4 + 25 = I_3$

$I_4 = 110 - 25 = 85$ A

Junction F: $I_5 + I_4 = 30$ A

$I_5 + 85$ A $= 30$ A

$I_5 = 30 - 85 = -55$ A

Note: The minus sign in the last answer tells us that the current I_5 is actually flowing away from the junction rather than towards it as shown.

2.5 KIRCHHOFF'S VOLTAGE LAW

This law also has already been used – in the explanation of p.d. and in the series and series/parallel circuits. This law states that in any closed network the algebraic sum of the emfs is equal to the algebraic sum of the p.d.s taken in order about the network. Once again, the law sounds very complicated, but it is really only common sense, and is simple to apply. So far, it has been applied only to very simple circuits, such as resistors connected in series across a source of emf. In this case we have said that the sum of the p.d.s is equal to the applied emf (e.g. $V_1 + V_2 = E$). However, these simple circuits have had only one source of emf, and could be solved using simple Ohm's law techniques. When more than one source of emf is involved, or the network is more complex, then a network analysis method must be used. Kirchhoff's is one of these methods.

Expressing the law in mathematical form:

$$\Sigma E = \Sigma IR \qquad (2.10)$$

A generalised circuit requiring the application of Kirchhoff's laws is shown in Figure 2.19. Note the following:

1 The circuit has been labelled with letters so that it is easy to refer to a particular loop and the direction around the loop that is being considered. Thus, if the left-hand loop is considered, and you wish to trace a path around it in a clockwise direction, this would be referred to as ABEFA. If a counterclockwise path was required, it would be referred to as FEBAF or AFEBA.

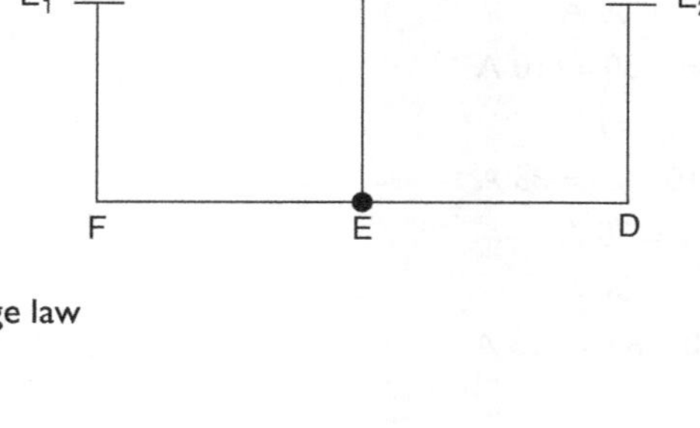

Figure 2.19 Kirchhoff's voltage law

2 Current directions have been assumed and marked on the diagram. As was found in the previous worked example (2.8), it may well turn out that one or more of these currents actually flows in the opposite direction to that marked. This result would be indicated by a negative value obtained from the calculation. However, to ensure consistency, make the initial assumption that all sources of emf are discharging current into the circuit; i.e. current leaves the positive terminal of each battery and enters at its negative terminal. The current law is also applied at this stage, which is why the current flowing through R_3 is marked as $(I_1 + I_2)$ and not as I_3. This is an important point since the solution involves the use of simultaneous equations, and the fewer the number of 'unknowns' the simpler the solution. Thus marking the third branch current in this way means that there are only two 'unknowns' to find, namely I_1 and I_2. The value for the third branch current, I_3, is then simply found by using the values obtained for I_1 and I_2.

3 If a negative value is obtained for a current then the minus sign *must* be retained in any subsequent calculations. However, when you quote the answer for such a current, make a note to the effect that it is flowing in the opposite direction to that marked, e.g. from C to D.

4 When tracing the path around a loop, concentrate solely on that loop and ignore the remainder of the circuit. Also note that if you are following the marked direction of current then the resulting p.d.(s) are assigned positive values. If the direction of 'travel' is opposite to the current arrow then the p.d. is assigned a negative value.

Let us now apply these techniques to the circuit of Figure 2.19.

Consider first the left-hand loop in a clockwise direction. Tracing around the loop it can be seen that there is only one source of emf within it (namely E_1). Thus the sum of the emfs is simply E_1 volt. Also, within the loop there are only two resistors (R_1 and R_2) which will result in two p.d.s, I_1R_1 and $(I_1 + I_2)R_3$ volt. The resulting loop equation will therefore be:

$$\text{ABEFA}: E_1 = I_1R_1 + (I_1 + I_2)R_3 \quad [1]$$

Now taking the right-hand loop in a counterclockwise direction it can be seen that again there is only one source of emf and two resistors. This results in the following loop equation:

$$\text{CBEDC}: E_2 = I_2R_2 + (I_1 + I_2)R_3 \quad [2]$$

Finally, let us consider the loop around the edges of the diagram in a clockwise direction. This follows the 'normal' direction for E_1 but is opposite to that for E_2, so the sum of the emfs is $E_1 - E_2$ volt. The loop equation is therefore

$$\text{ABCDEFA}: E_1 - E_2 = I_1R_1 - I_2R_2 \quad [3]$$

Since there are only two unknowns, only two simultaneous equations are required, and three have been written. However, it is a useful practice to do this as the 'extra' equation may contain more convenient numerical values for the coefficients of the 'unknown' currents.

The complete technique for the applications of Kirchhoff's laws becomes clearer by the consideration of a worked example containing numerical values.

WORKED EXAMPLE 2.9

Q For the circuit of Figure 2.20 determine the value and direction of the current in each branch, and the p.d. across the 10 Ω resistor.

The circuit is first labelled and current flows identified and marked by applying the current law. This is shown in Figure 2.21.

ABEFA:

$$10-4 = 3\,\Omega \times I_1 - 2\,\Omega \times I_2$$

$$6\text{ V} = 3I_1 - 2I_2 \qquad [1]$$

ABCDEFA:

$$10\text{ V} = 3\,\Omega \times I_2 + 10\,\Omega \times (I_1 + I_2) = 3I_2 + 10I_1 + 10I_2 = 13I_1 + 10I_2 \qquad [2]$$

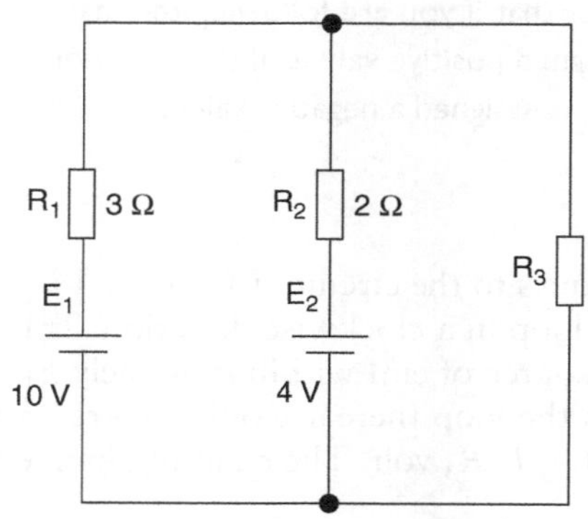

Figure 2.20 The circuit diagram for Worked Example 2.9

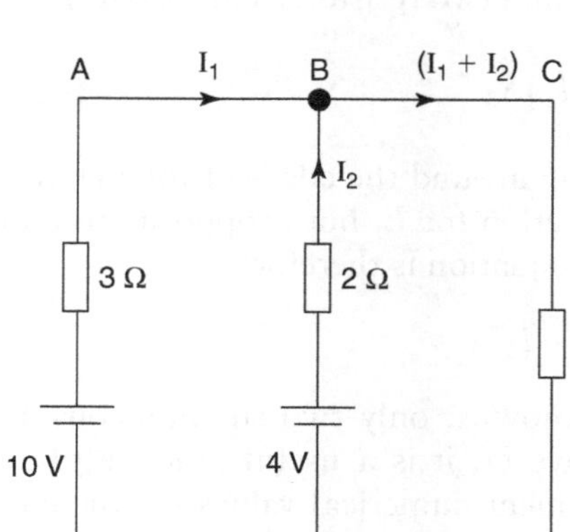

Figure 2.21 The circuit diagram for Worked Example 2.9, with labelling

BCDEB:

$$4\text{ V} = 2\ \Omega \times I_2 + 10\ \Omega \times (I_1 + I_2) = 2I_2 + 10I_1 + 10I_2 = 10I_1 + 12I_2 \qquad [3]$$

Inspection of Equations [1] and [2] shows that if Equation [1] is multiplied by 5, then the coefficient of I_2 will be the same in both equations. Thus, if the two are now added, the term containing I_2 will be eliminated, and hence a value can be obtained for I_1.

$$30 = 15I_1 - 10I_2 \cdots\cdots\cdots\cdots [1] \times 5$$
$$\underline{10} = \underline{13I_1 + 10I_2} \cdots\cdots\cdots\cdots [2]$$
$$40 = 28I_1$$
$$I_1 = \frac{40}{28} = 1.43\text{ A}$$

Substituting this value for I_1 into Equation [3] yields:

$$4 = 14.29 + 12I_2$$
$$12I_2 = 4 - 14.29$$
$$I_2 = \frac{-10.29}{12} = -0.86\text{ A}$$

$$(I_1 + I_2) = 1.43 - 0.86 = 0.57\text{ A}$$
$$V_{CD} = (I_2 + I_2) \cdot R_{CD} = 0.57 \times 10 = 5.70\text{ V}$$

WORKED EXAMPLE 2.10

Q For the circuit shown in Figure 2.22, use Kirchhoff's laws to calculate (a) the current flowing in each branch of the circuit and (b) the p.d. across the 5 Ω resistor.

Firstly the circuit is sketched and labelled, and currents identified using Kirchhoff's current law. This is shown in Figure 2.23.

(a) We can now consider three loops in the circuit and write down the corresponding equations using Kirchhoff's voltage law:

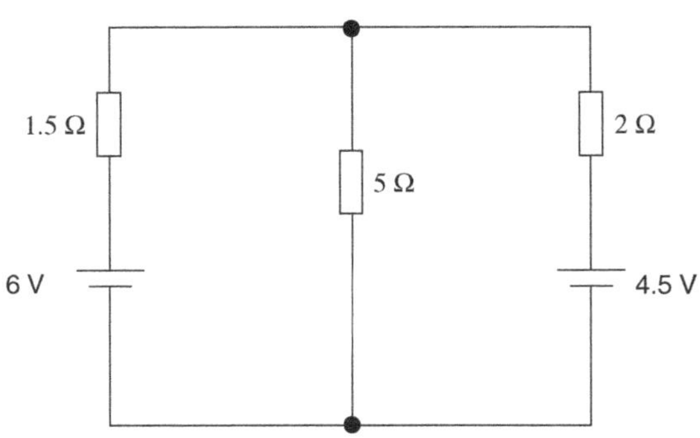

Figure 2.22 The circuit diagram for Worked Example 2.10

ABEFA:

$$E_1 = I_1R_1 + (I_1 + I_2)R_3 \qquad [1]$$

$$6\ \text{V} = 1.5 \times I_1 + 5 \times (I_1 + I_2) = 1.5I_1 + 5I_1 + 5I_2 = 6.5I_1 + 5I_2$$

CBEDC:

$$E_2 = I_2R_2 + (I_1 + I_2)R_3 \qquad [2]$$

$$4.5\ \text{V} = 2 \times I_2 + 5 \times (I_1 + I_2) = 2I_2 + 5I_1 + 5I_2 = 5I_1 + 7I_2$$

ABCDEFA:

$$E_1 - E_2 = I_1R_1 - I_2R_2$$

$$6 - 4.5 = 1.5 \times I_1 - 2 \times I_2 \qquad [3]$$

$$1.5\ \text{V} = 1.5I_1 - 2I_2$$

Now, any pair of these three equations may be used to solve the problem, using the technique of simultaneous equations. We shall use Equations [1] and [3] to eliminate the unknown current I_2, and hence obtain a value for current I_1. To do this we can multiply [1] by 2 and [3] by 5, and then add the two modified equations together, thus:

$$12 = 13I_1 + 10I_2 \ldots\ldots\ldots\ldots [1] \times 2$$

$$7.5 = 7.5I_1 - 10I_2 \ldots\ldots\ldots [3] \times 5$$

$$19.5 = 20.5I_1$$

$$\text{hence, } I_1 = \frac{19.5}{20.5} = 0.951\ \text{A}$$

Substituting this value for I_1 into Equation [3] gives:

$$1.5 = (1.5 \times 0.951) - 2I_2$$

$$1.5 = 1.427 - 2I_2$$

$$\text{hence, } 2I_2 = 1.427 - 1.5 = -0.073$$

$$\text{and } I_2 = -0.037\ \text{A}$$

Note: The minus sign in the answer for I_2 indicates that this current is actually flowing in the opposite direction to that marked in Figure 2.23. This means that battery E_1 is both supplying current to the 5 Ω resistor *and charging* battery E_2.

Current through 5 Ω resistor = $I_1 + I_2$ = 0.951 A + (−0.037 A) = 0.914 A

(b) To obtain the p.d. across the 5 Ω resistor we can either subtract the p.d. (voltage drop) across R_1 from the emf E_1 or *add* the p.d. across R_2 to emf E_2, because E_2 is being *charged*. A third alternative is to multiply R_3 by the current flowing through it. All three methods will be shown here, and, provided that the same answer is obtained each time, the correctness of the answers obtained in part (a) will be confirmed.

$$V_{BE} = E_1 - I_1R_1 = 6 - (0.951 \times 1.5) = 6 - 1.427 = 4.573\ \text{V}$$

or:

$$V_{BE} = E_2 + I_2R_2 = 4.5 + (0.037 \times 2) = 4.5 + 0.074 = 4.574\ \text{V}$$

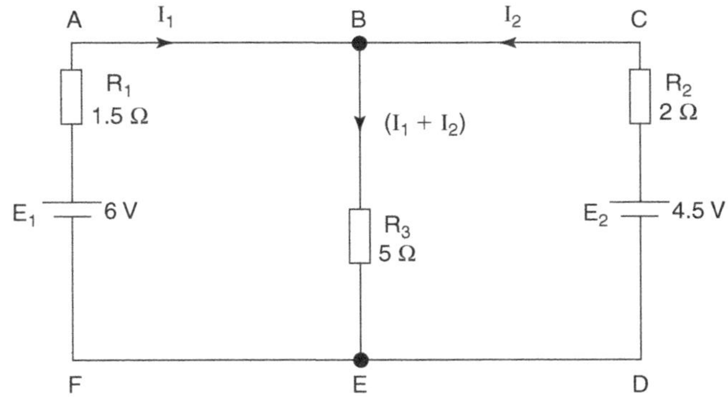

Figure 2.23 The circuit diagram for Worked Example 2.10, with labelling

or:

$$V_{BE} = (I_1 + I_2)R_3 = 0.914 \times 5 = 4.57 \text{ V}$$

The very small differences between these three answers are due simply to rounding errors, and so the answers to part (a) are verified as correct.

2.6 POTENTIOMETER

In addition to the ordinary resistors described above, there are also variable resistors. They are designed so that their resistance can be easily changed by the position of a mechanically movable component. This three-terminal resistor with a sliding or rotating contact forms an adjustable voltage divider and is called a potentiometer. There are slide potentiometers (adjusted by sliding the wiper lift or right), thumbwheel potentiometers (adjusted by means of a small thumbwheel), trimmer potentiometers (to be adjusted once or infrequently for fine-tuning) and even multiturn potentiometers (with a spiral element). Each time a contact point is moved across a fixed resistance, both that point of contact and both ends have an electrical contact, hence the name three-terminal. These three form a ratio of resistances, as in the voltage divider described above. Potentiometers often appear as a volume knob on a television or radio speaker. There the logarithmic type of potentiometer is used, because the amplitude sensitivity of the human ear is also logarithmic.

A slidewire potentiometer is used for the accurate measurement of small voltages. In its simplest form it comprises a metre length of wire held between two brass or copper blocks on a base board, with a graduated metre scale beneath the wire. Connected to one end of the wire is a contact, the other end of which can be placed at any point along the wire. A 2 V cell causes current to flow along the wire. This arrangement, including a voltmeter, is shown in Figure 2.24. The wire between the blocks A and B must be of uniform cross-section and resistivity throughout its length, so that each millimetre of its length has the same resistance as the next. Thus it may be considered as a number of equal resistors connected in series between points A and B. In other words it is a continuous potential divider.

Let us now conduct an imaginary experiment. If the movable contact is placed at point A then both terminals of the voltmeter will be at the same potential, and it will indicate zero volts. If the contact is now moved to point B then the voltmeter will indicate 2 V. Consider

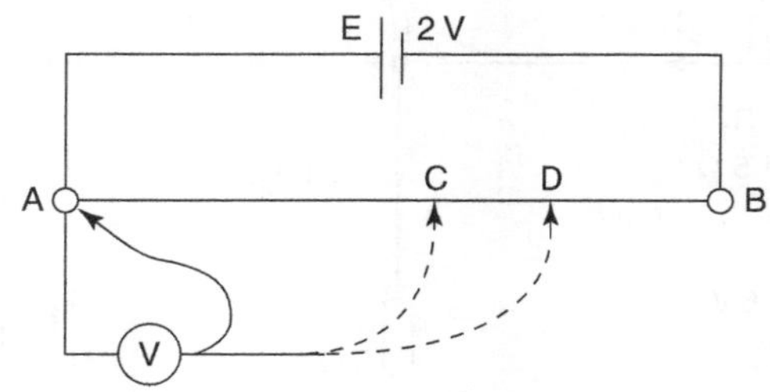

Figure 2.24 A slidewire potentiometer

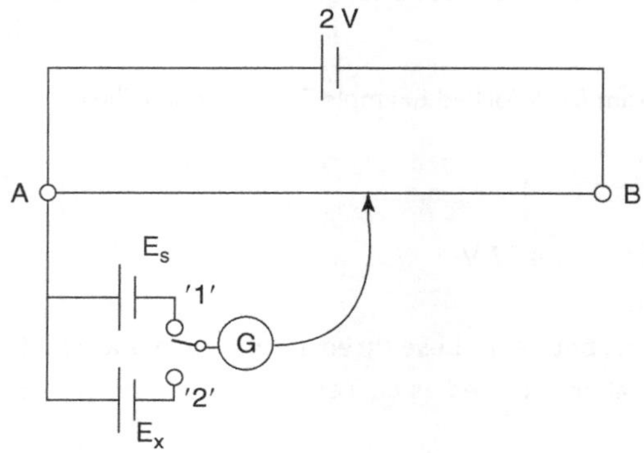

Figure 2.25 A slidewire potentiometer measuring instrument

now the contact placed at point C which is midway between A and B. In this case it is exactly halfway along our 'potential divider', so it will indicate 1 V. Finally, placing the contact at a point D (say 70 cm from A), the voltmeter will indicate 1.4 V. These results can be summarised by the statement that there is a uniform potential gradient along the wire. Therefore, the p.d. 'tapped off' by the moving contact is in direct proportion to the distance travelled along the wire from point A. Since the source has an emf of 2 V and the wire is of 1 metre length, the potential gradient must be 2 V/m. In general we can say that

$$V_{AC} = \frac{AC}{AB}E \tag{2.11}$$

where AC = distance travelled along wire
AB = total length of the wire
E = the source voltage

Utilising these facts the simple circuit can be modified to become a measuring instrument, as shown in Figure 2.25. In this case the voltmeter has been replaced by a galvo, a device to measure electrical current. The movable contact can be connected either to the cell to be measured or the standard cell, via a switch. Using this system the procedure would be as follows:

1 The switch is moved to position '1' and the slider moved along the wire until the galvo indicates zero current. The position of the slider on the scale beneath the wire is then noted. This distance from A represents the emf E_s of the standard cell.

2 With the switch in position '2', the above procedure is repeated, whereby distance along the scale represents the emf E_x of the cell to be measured.
3 The value of E_x may now be calculated from

$$E_x = \frac{\text{AD}}{\text{AC}} \times E_s$$

where AC represents the scale reading obtained for the standard cell and AD the scale reading for the unknown cell.

It should be noted that this instrument will measure the true emf of the cell since the readings are taken when the galvo carries zero current (i.e. no current is being drawn from the cell under test), hence there will be no p.d. due to its internal resistance.

WORKED EXAMPLE 2.11

Q A slidewire potentiometer when used to measure the emfs of two cells provided balance conditions at scale settings of (a) 600 mm and (b) 745 mm. If the standard cell has an emf of 1.0186 V and a scale reading of 509.3 mm then determine the values for the two cell emfs.

Let E_s, ℓ_1 and ℓ_2 represent the scale readings for the standard cell and cells 1 and 2 respectively. Hence:

$\ell_s = 509.3$ mm; $\ell_1 = 600$ mm; $\ell_2 = 745$ mm; $E_s = 1.0186$ V

$$E_1 = \frac{\ell_1}{\ell_s} \times E_s = \frac{600}{509.3} \times 1.0186 = 1.2 \text{ V}$$

$$E_2 = \frac{\ell_2}{\ell_s} \times E_s = \frac{745}{509.3} \times 1.0186 = 1.49 \text{ V}$$

It is obviously inconvenient to have an instrument that needs to be one metre in length and requires the measurements of lengths along a scale. In the commercial version of the instrument the long wire is replaced by a series of precision resistors plus a small section of wire with a movable contact. The standard cell and galvo would also be built-in features. Also, to avoid the necessity for separate calculations, there would be provision for standardising the potentiometer. This means that the emf values can be read directly from dials on the front of the instrument.

SUMMARY OF EQUATIONS

Resistors in series: $R = R_1 + R_2 + R_3 + \cdots$

Resistors in parallel: $\frac{1}{R} = \frac{1}{R_1} + \frac{1}{R_2} + \frac{1}{R_3} + \cdots$

and for *only* two resistors in parallel, $R = \frac{R_1 R_2}{R_1 + R_2} \left(\frac{\text{product}}{\text{sum}} \right)$

Potential divider: $V_1 = \dfrac{R_1}{R_1 + R_2} \times E$

Current divider: $I_1 = \dfrac{R_2}{R_1 + R_2} \times I$

Kirchhoff's laws: $\sum I = 0$ (sum of the currents at a junction = 0)

$\sum E = \sum IR$ (sum of the emfs = sum of the p.d.s, in order)

Slidewire potentiometer: $V_{AC} = \dfrac{AC}{AB} \times E$

ASSIGNMENT QUESTIONS

1 Two 560 Ω resistors are placed in series across a 400 V supply. Calculate the current drawn.

2 When four identical hotplates on a cooker are all in use, the current drawn from a 240 V supply is 33 A. Calculate (a) the resistance of each hotplate, (b) the current drawn when only three plates are switched on. The hotplates are connected in parallel.

3 Calculate the total current when six 120 Ω torch bulbs are connected in parallel across a 9 V supply.

4 Two 20 Ω resistors are connected in parallel and this group is connected in series with a 4 Ω resistor. What is the total resistance of the circuit?

5 A 12 Ω resistor is connected in parallel with a 15 Ω resistor and the combination is connected in series with a 9 Ω resistor. If this circuit is supplied at 12 V, calculate (a) the total resistance, (b) the current through the 9 Ω resistor and (c) the current through the 12 Ω resistor.

6 For the circuit shown in Figure 2.26 calculate the values for (a) the current through each resistor, (b) the p.d. across each resistor and (c) the power dissipated by the 20 Ω resistor.

7 Determine the p.d. between terminals E and F of the circuit in Figure 2.27.

8 For the circuit of Figure 2.28 calculate (a) the p.d. across the 8 Ω resistor, (b) the current through the 10 Ω resistor and (c) the current through the 12 Ω resistor.

9 Three resistors of 5 Ω, 6 Ω and 7 Ω respectively are connected in parallel. This combination is connected in series with another parallel combination of 3 Ω and 4 Ω. If the complete circuit is supplied from a 20 V source, calculate (a) the total resistance, (b) the total current, (c) the p.d. across the 3 Ω resistor and (d) the current through the 4 Ω resistor.

10 Two resistors of 18 Ω and 12 Ω are connected in parallel and this combination is connected in series with an unknown resistor R_x. Determine the value of R_x if the complete circuit draws a current of 0.6 A from a 12 V supply.

11 Three loads of 24 A, 8 A, and 12 A are supplied from a 200 V source. If a motor of resistance 2.4 Ω is also connected across the supply, calculate (a) the total resistance and (b) the total current drawn from the supply.

12 Two resistors of 15 Ω and 5 Ω are connected in series with a resistor R_x and the combination is supplied from a 240 V source. If the p.d. across the 5 Ω resistor is 20 V calculate the value of R_x.

13 A 200 V, 0.5 A lamp is to be connected in series with a resistor across a 240 V supply. Determine the resistor value required for the lamp to operate at its correct voltage.

14 A 12 Ω and a 6 Ω resistor are connected in parallel across the terminals of a battery of emf 6 V and internal resistance 0.6 Ω. Sketch the circuit diagram and calculate (a) the current drawn from the battery, (b) the terminal p.d. and (c) the current through the 6 Ω resistor.

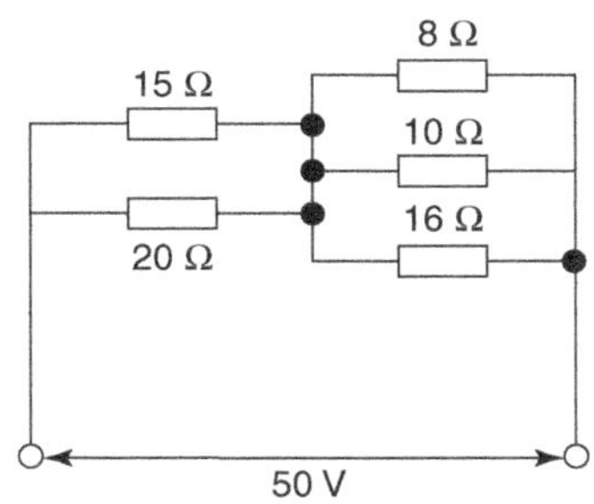

Figure 2.26 The circuit diagram for Assignment Question 6

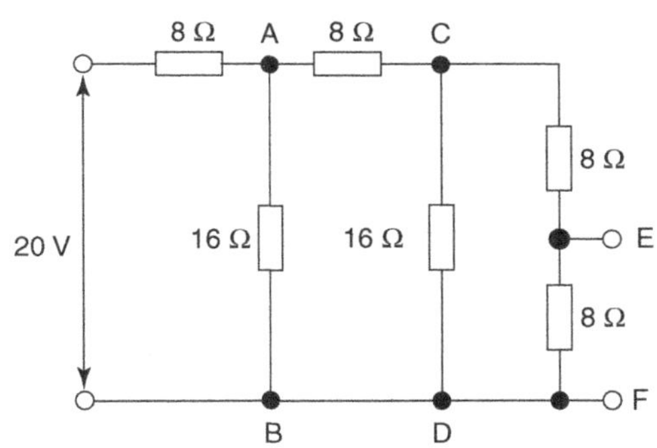

Figure 2.27 The circuit diagram for Assignment Question 7

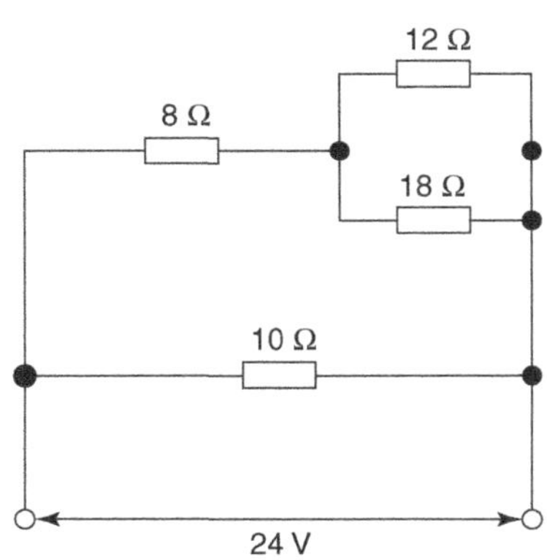

Figure 2.28 The circuit diagram for Assignment Question 8

15 An electric cooker element consists of two parts, each having a resistance of 18 Ω, which can be connected (a) in series, (b) in parallel or (c) using one part only. Calculate the current drawn from a 240 V supply for each connection.
16 A cell of emf 2 V has an internal resistance 0.1 Ω. Calculate the terminal p.d. when (a) there is no load connected and (b) a 2.9 Ω resistor is connected across the terminals. Explain why these two answers are different.
17 A battery has a terminal voltage of 1.8 V when supplying a current of 9 A. This voltage rises to 2.02 V when the load is removed. Calculate the internal resistance.
18 Four resistors of values 10 Ω, 20 Ω, 40 Ω and 40 Ω are connected in parallel across the terminals of a generator having an emf of 48 V and internal resistance 0.5 Ω. Sketch the circuit diagram and calculate (a) the current drawn from the generator, (b) the p.d. across each resistor and (c) the current flowing through each resistor.
19 Calculate the p.d. across the 3 Ω resistor shown in Figure 2.29 given that V_{AB} is 11 V.
20 Calculate the p.d. V_{AB} in Figure 2.30.
21 For the network shown in Figure 2.31, calculate (a) the total circuit resistance, (b) the supply current, (c) the p.d. across the 12 Ω resistor, (d) the total power dissipated in the whole circuit and (e) the power dissipated by the 12 Ω resistor.
22 A circuit consists of a 15 Ω and a 30 Ω resistor connected in parallel across a battery of internal resistance 2 Ω. If 60 W is dissipated by the 15 Ω resistor, calculate (a) the current in the 30 Ω resistor, (b) the terminal p.d. and emf of the battery, (c) the total energy dissipated in the external circuit in 1 minute and (d) the quantity of electricity through the battery in 1 minute.
23 Use Kirchhoff's laws to determine the three branch currents and the p.d. across the 5 Ω resistor in the network of Figure 2.32.
24 Determine the value and direction of current in each branch of the network of Figure 2.33, and the power dissipated by the 4 Ω load resistor.
25 Two batteries A and B are connected in parallel (positive to positive) with each other and this combination is connected in parallel with a battery C; this is in series with a 25 Ω resistor, the negative terminal of C being connected to the positive terminals of A and B. Battery A has an emf of 108 V and internal resistance 3 Ω, and the corresponding values for B are 120 V and 2 Ω. Battery C has an emf of 30 V and negligible internal resistance. Sketch the circuit and calculate (a) the value and direction of current in each battery and (b) the terminal p.d. of A.
26 For the circuit of Figure 2.34 determine (a) the current supplied by each battery, (b) the current through the 15 Ω resistor and (c) the p.d. across the 10 Ω resistor.
27 For the network of Figure 2.35, calculate the value and direction of all the branch currents and the p.d. across the 80 Ω load resistor.
28 The slidewire potentiometer instrument shown in Figure 2.36 when used to measure the emf of cell E_x yielded the following results:

a. galvo current was zero when connected to the standard cell and the movable contact was 552 mm from A;
b. galvo current was zero when connected to E_s and the movable contact was 647 mm from A.

Calculate the value of E_x, given E_s = 1.0183 V.
It was found initially that E_x was connected the opposite way round and a balance could not be obtained. Explain this result.

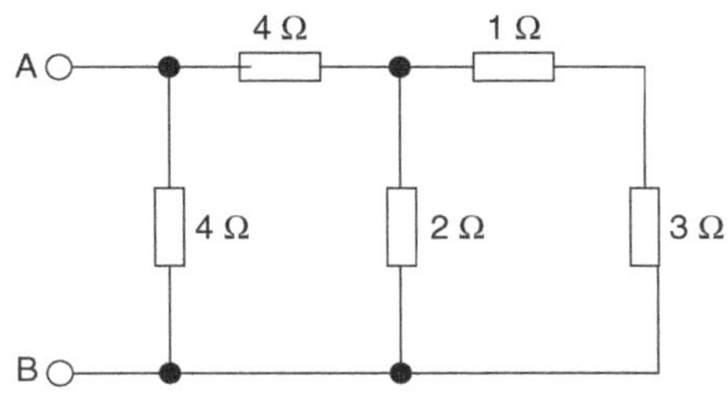

Figure 2.29 The circuit diagram for Assignment Question 19

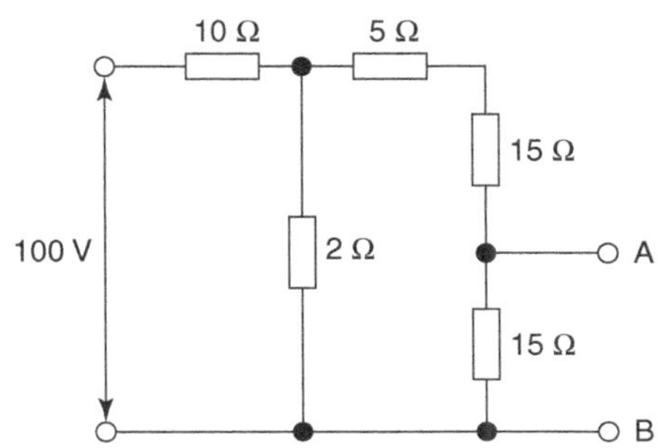

Figure 2.30 The circuit diagram for Assignment Question 20

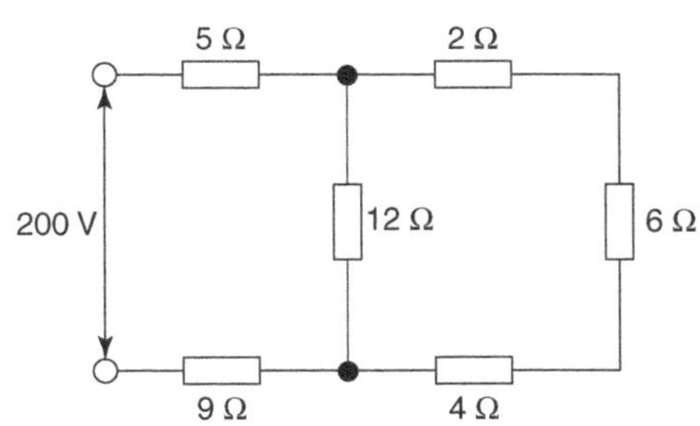

Figure 2.31 The circuit diagram for Assignment Question 21

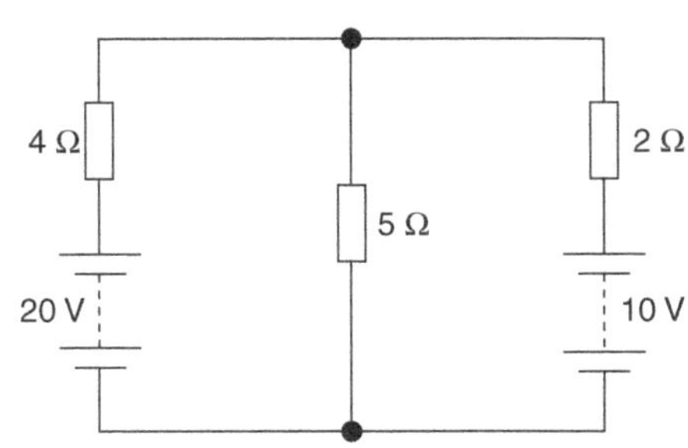

Figure 2.32 The circuit diagram for Assignment Question 23

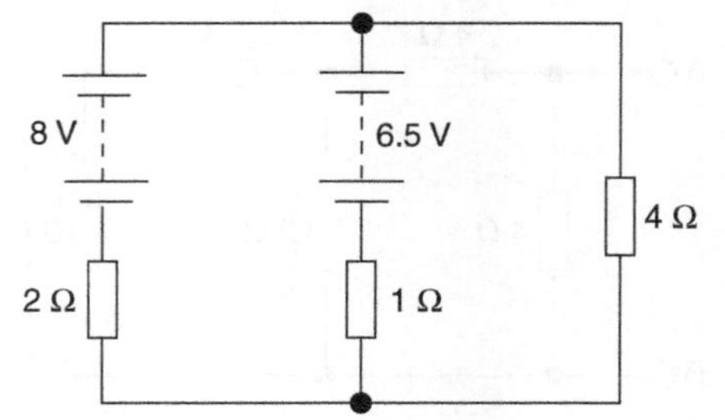

Figure 2.33 The circuit diagram for Assignment Question 24

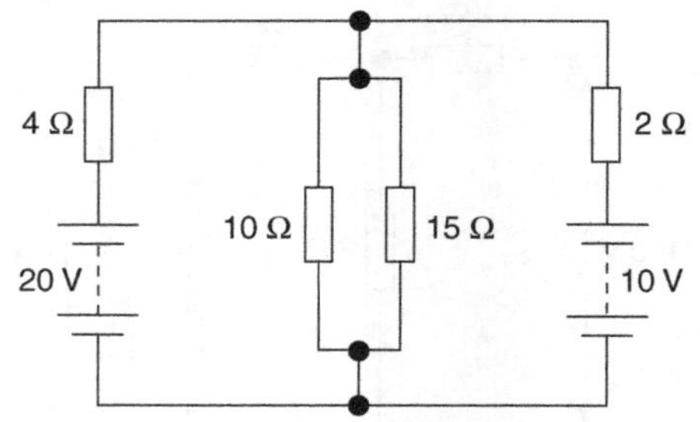

Figure 2.34 The circuit diagram for Assignment Question 26

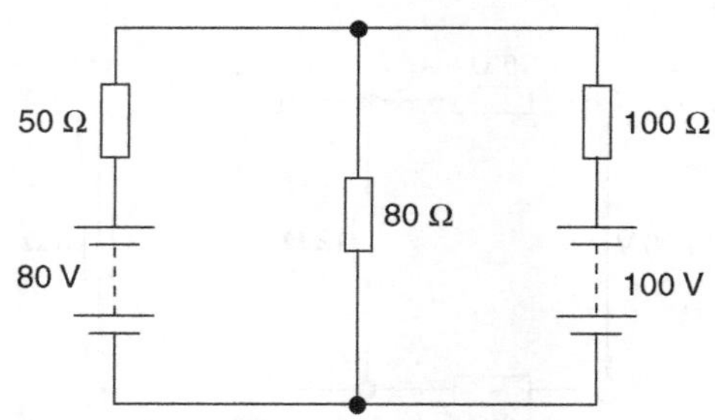

Figure 2.35 The circuit diagram for Assignment Question 27

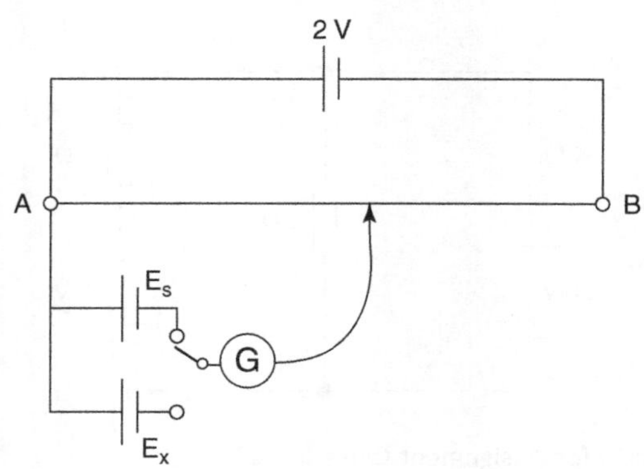

Figure 2.36 The circuit diagram for Assignment Question 28

SUGGESTED PRACTICAL ASSIGNMENTS

Note: Component values and specific items of equipment when quoted here are only suggestions. Those used in practice will of course depend upon availability within your institution.

Assignment 1

To investigate Ohm's law and Kirchhoff's laws as applied to series and parallel circuits.

Apparatus

Three resistors of different values

1 × variable d.c. power supply unit (psu)
1 × ammeter
1 × voltmeter

Method

1 Connect the three resistors in series across the terminals of the psu with the ammeter connected in the same circuit. Adjust the current (as measured with the ammeter) to a suitable value. Measure the applied voltage and the p.d. across each resistor. Note these values and compare the p.d.s to the theoretical (calculated) values.
2 Reconnect your circuit so that the resistors are now connected in parallel across the psu. Adjust the psu to a suitable voltage and measure, in turn, the current drawn from the psu and the three resistor currents. Note these values and compare to the theoretical values.
3 Write an assignment report and in your conclusions justify whether the assignment confirms Ohm's law and Kirchhoff's laws, allowing for experimental error and resistor tolerances.

Assignment 2

To investigate the application of Kirchhoff's laws to a network containing more than one source of emf.

Apparatus

2 × variable d.c. psu
3 × different value resistors
1 × ammeter
1 × voltmeter

Method

1 Connect the circuit as shown in Figure 2.37. Set psu 1 to 2 V and psu 2 to 4 V. Measure, in turn, the current in each limb of the circuit, and the p.d. across each resistor. For each of the

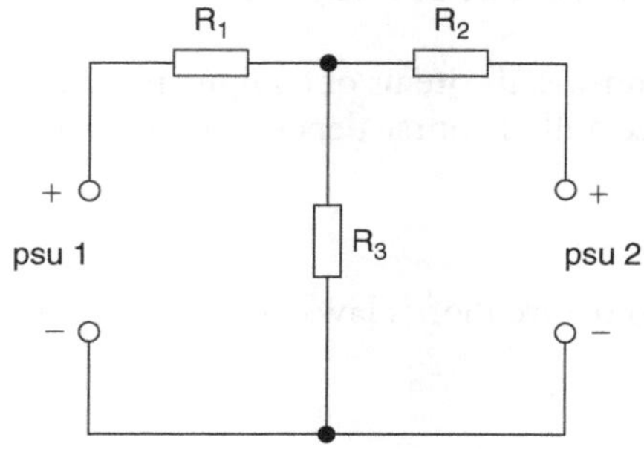

Figure 2.37 The circuit diagram for Practical Assignment 2

three possible loops in the circuit compare the sum of the p.d.s measured with the sum of the emfs. Carry out a similar exercise regarding the three currents.

2 Reverse the polarity of psu 2 and repeat the above.
3 Write the assignment report and in your conclusions justify whether or not Kirchhoff's laws have been verified for the network.

Assignment 3

To investigate potential and current dividers.

Apparatus

2 × decade resistance boxes
1 × ammeter
1 × voltmeter
1 × d.c. psu

Method

1 Connect the resistance boxes in series across the psu. Adjust one of them (R_1) to 3 kΩ and the other (R_2) to 7 kΩ. Set the psu to 10 V and measure the p.d. across each resistor. Compare the measured values with those predicted by the voltage divider theory.
2 Reset both R_1 and R_2 to two or more different values and repeat the above procedure.
3 Reconnect the two resistance boxes in parallel across the psu and adjust the current drawn from the psu to 10 mA. Measure the current flowing through each resistance and compare to those values predicted by the current division theory.
4 Repeat the procedure of 3 above for two more settings of R_1 and R_2, but let one of these settings be such that $R_1 = R_2$.

Assignment 4

Use a slidewire potentiometer to measure the emf of a number of primary cells (nominal emf no more than 1.5 V).

Chapter 3

Electric Fields and Capacitors

LEARNING OUTCOMES

This chapter deals with the laws and properties of electric fields and their application to electric components known as capacitors.

On completion of this chapter you should be able to:

1 Understand the properties of electric fields and insulating materials.
2 Carry out simple calculations involving these properties.
3 Carry out simple calculations concerning capacitors, and capacitors connected in series, parallel and series/parallel combinations.
4 Describe the construction and electrical properties of the different types of capacitor.
5 Understand the concept of energy storage in an electric field, and perform simple related calculations.

3.1 COULOMB'S LAW

A force exists between charged bodies. A force of attraction exists between opposite charges and a force of repulsion between like polarity charges. Coulomb's law states that the force, expressed in newtons, is directly proportional to the product of the charges and is inversely proportional to the square of the distance between their centres. So for the two bodies shown in Figure 3.1, this would be expressed as

$$F \propto \frac{Q_1 Q_2}{d^2}$$

In order to obtain a value for the force, a constant of proportionality must be introduced. In this case it is the permittivity of free space, ε_0. This concept of permittivity is dealt with later in this chapter, and need not concern you for the time being. The expression for the force in newtons becomes

Figure 3.1 Attraction force between two bodies

DOI: 10.1201/9781003308294-3

$$F = \frac{Q_1 Q_2}{\varepsilon_0 d^2} \tag{3.1}$$

This type of relationship is said to follow an inverse square law because $F \propto 1/d^2$. The consequence of this is that if the distance of separation is doubled, then the force will be reduced by a factor of 4 times. If the distance is increased by a factor of 4 times, the force will be reduced by a factor of 16 times, etc. The practical consequence is that although the force can never be reduced to zero, it diminishes very rapidly as the distance of separation is increased. This will continue until a point is reached where the force is negligible relative to other forces acting within the system.

Isaac Newton (1643–1727) was an English physicist and mathematician, studying the differential and integral calculus and developing a theory of colours based on the prism, turning white light into a visible spectrum. He also described gravity and Newton's three laws, describing the movement and equilibrium of forces. Therefore he is called the founding father of the classical mechanics.

3.2 ELECTRIC FIELDS

You are probably more familiar with the concepts and effects of magnetic and gravitational fields. For example, you have probably conducted simple experiments using bar magnets and iron filings to discover the shape of magnetic fields, and you are aware that forces exist between magnetised bodies. You also experience the effects of gravitational forces constantly, even though you probably do not consciously think about them.

Both of these fields are simply a means of transmitting the forces involved, from one body to another. However, the fields themselves cannot be detected by the human senses, since you cannot see, touch, hear or smell them. This tends to make it more difficult to understand their nature. An electric field behaves in the same way as these other two examples, except that it is the method by which forces are transmitted between charged bodies. In all three cases we can represent the appropriate field by means of arrowed lines. These lines are usually referred to as the lines of force.

To illustrate these points, consider Figure 3.2 which shows two oppositely charged spheres with a small positively charged particle placed on the surface of Q_1. Since like charges repel and unlike charges attract each other, the small charged particle will experience a force of repulsion from Q_1 and one of attraction from Q_2.

The force of repulsion from Q_1 will be very much stronger than the force of attraction from Q_2 because of the relative distance involved. The other feature of the forces is that they will act so as to be at right angles to the charged surfaces. Hence, there will be a resultant

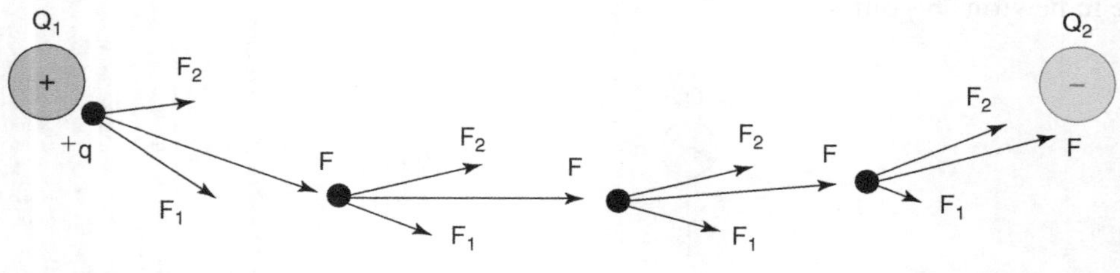

Figure 3.2 Two oppositely charged spheres

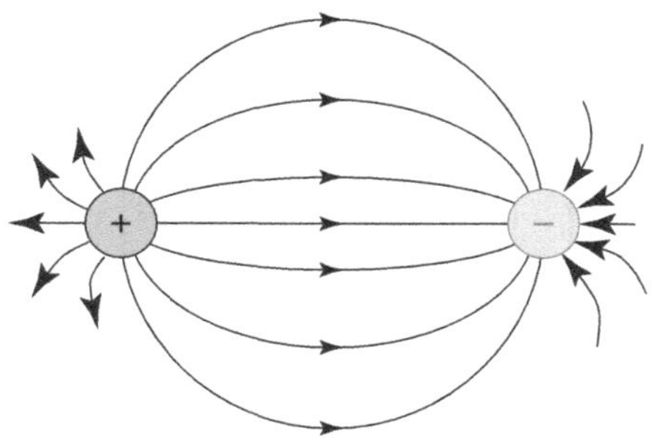

Figure 3.3 Flux lines between two oppositely charged spheres

force acting on the particle. Assuming that it is free to move, it will start to move in the direction of this resultant force. For the sake of clarity, the distance moved in this direction is greatly exaggerated in the diagram. However, when the particle moves to the new position, force F_1 will have decreased and F_2 will have increased. In addition, the direction of action of each force will have changed. Thus the direction of action of the resultant will have changed, but its magnitude will have remained constant. The particle will now respond to the new resultant force F. This is a continuous process and the particle will trace out a curved path until it reaches the surface of Q_2. If this 'experiment' was carried out for a number of starting points at the surface of Q_1, the paths taken by the particle would be as shown in Figure 3.3.

The following points should be noted:

1 The lines shown represent the possible paths taken by the positively charged particle in response to the force acting on it. Thus they are called the lines of electric force. They may also be referred to as the lines of electric flux, ψ.
2 The total electric flux makes up the whole electric field existing between and around the two charged bodies.
3 The lines themselves are imaginary and the field is three-dimensional. The whole of the space surrounding the charged bodies is occupied by the electric flux, so there are no 'gaps' in which a charged particle would not be affected.
4 The lines of force (flux) radiate outwards from the surface of a positive charge and terminate at the surface of a negative charge.
5 The lines always leave (or terminate) at right angles to a charged surface.
6 Although the lines drawn on a diagram do not actually exist as such, they are a very convenient way to represent the existence of the electric field. They therefore aid the understanding of its properties and effects.
7 Since force is a vector quantity any line representing it must be arrowed. The convention used here is that the arrows point from the positive to the negative charge.

It is evident from Figure 3.3 that the spacing between the lines of flux varies depending upon which part of the field you consider. This means that the field shown is non-uniform. A uniform electric field may be obtained between two parallel charged plates as shown in Figure 3.4.

Note that the electric field will exist in *all* of the space surrounding the two plates, but the uniform section exists only in the space between them. Some non-uniformity is shown by the curved lines at the edges (fringing effect). At this stage we are concerned only with the uniform field between the plates. If a positively charged particle was placed between

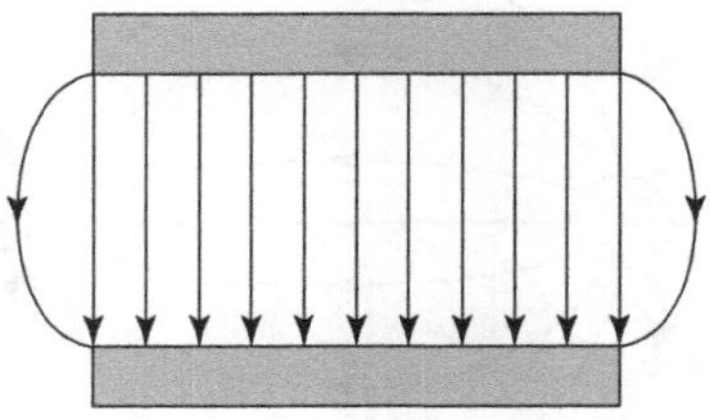

Figure 3.4 A uniform electric field between two parallel charged plates

the plates, it would experience a force that would cause it to move from the positive to the negative plate. The value of force acting on the particle depends upon what is known as the electric field strength, which will be explained in the next section.

3.3 ELECTRIC FIELD STRENGTH (E), ELECTRIC FLUX (Ψ) AND FLUX DENSITY (*D*)

The electric field strength is defined as the force (expressed in newton) per unit charge (expressed in coulomb) exerted on a test charge placed inside the electric field. (An outdated name for this property is 'electric force'.) Hence,

$$\text{field strength} = \frac{\text{force}}{\text{charge}}$$

$$\mathbf{E} = \frac{F}{q} \tag{3.2}$$

$$F = \mathbf{E}q \tag{3.3}$$

where q is the charge on the *particle*, and not the plates.

In the SI system one 'line' of flux is assumed to radiate from the surface of a positive charge of one coulomb and terminate at the surface of a negative charge of one coulomb. Hence the electric flux has the same numerical value as the charge that produces it. Therefore the coulomb is used as the unit of electric flux. In addition, the Greek letter ψ (pronounced as psi) is usually replaced by the symbol for charge, namely Q.

The electric flux density D is defined as the amount of flux per square metre of the electric field. This area is measured at right angles to the lines of force. This gives the following equation for the electric flux density D, expressed as coulomb/metre^2,

$$D = \frac{\Psi}{A}$$

$$D = \frac{Q}{A} \tag{3.4}$$

WORKED EXAMPLE 3.1

Q Two parallel plates of dimensions 30 mm by 20 mm are oppositely charged to a value of 50 mC. Calculate the density of the electric field existing between them.

$$Q = 50 \times 10^{-3} \text{ C}; A = 30 \times 10^{-3} \times 20 \times 10^{-3} \text{ m}^2$$

$$D = \frac{Q}{A} = \frac{50 \times 10^{-3}}{600 \times 10^{-6}} D = 83.3 \text{ C/m}^2$$

WORKED EXAMPLE 3.2

Q Two parallel metal plates, each having a cross-sectional area of 400 mm², are charged from a constant current source of 50 µA for a time of 3 seconds. Calculate (a) the charge on the plates and (b) the density of the electric field between them.

$$A = 400 \times 10^{-6} \text{ m}^2; I = 50 \times 10^{-6} \text{ A}; t = 3 \text{ s}$$

(a) $Q = It = 50 \times 10^{-6} \times 3 = 150\ \mu\text{C}$

(b) $D = \frac{Q}{A} = \frac{150 \times 10^{-6}}{400 \times 10^{-6}} = 0.375 \text{ C/m}^2$

3.4 THE CHARGING PROCESS AND POTENTIAL GRADIENT

We have already met the concept of a potential gradient when considering a uniform conductor (wire) carrying a current. This concept formed the basis of the slidewire potentiometer discussed in Chapter 2. However, we are now dealing with static charges that have been induced on to plates (the branch of science known as electrostatics). Current flow is only applicable during the charging process. The material between the plates is some form of insulator (a dielectric) which could be vacuum, air, rubber, glass, mica, PVC, etc. So under ideal conditions there will be no current flow from one plate to the other via the dielectric. Nonetheless, there will be a potential gradient throughout the dielectric.

Consider a pair of parallel plates (initially uncharged) that can be connected to a battery via a switch, as shown in Figure 3.5. Note that the number of electrons and protons shown for each plate are in no way representative of the actual numbers involved. They are shown to aid the explanation of the charging process that will take place when the switch is closed. On closing the switch, some electrons from plate A will be attracted to the positive terminal of the battery. In this case, since plate A has lost electrons it will acquire a positive charge. This results in an electric field radiating out from plate A. The effect of this field is to induce a negative charge on the top surface of plate B, by attracting electrons in the plate towards this surface. Consequently, the lower surface of plate B must have a positive charge. This in turn will attract electrons from the negative terminal of the battery. Thus for every electron that is removed from plate A one is transferred to plate B. The two plates will therefore become equally but oppositely charged.

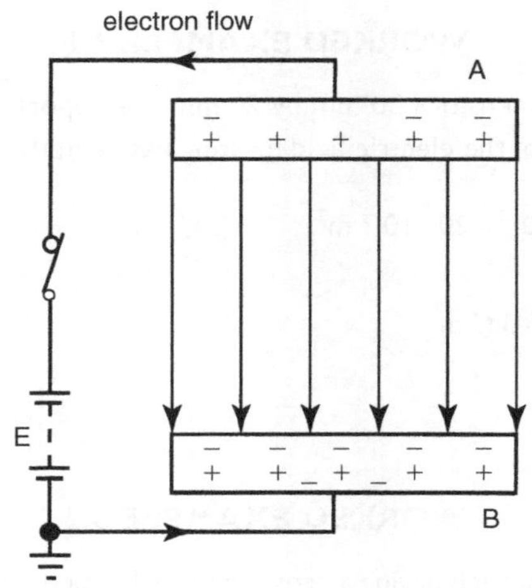

Figure 3.5 A pair of parallel plates connected to a battery

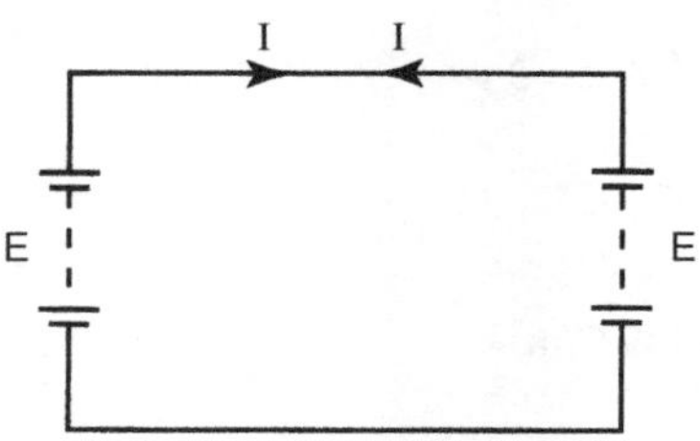

Figure 3.6 Two batteries of equal emf connected in parallel

This charging process will not carry on indefinitely (in fact it will last for only a very short space of time). This is because as the charge on the plates increases so too does the voltage developed between them. Thus the charging process continues only until the p.d. between the plates, V is equal to the emf, E of the battery. The charging current at this time will become zero, because plates A and B are positive and negative respectively. Thus, this circuit is equivalent to two batteries of equal emf connected in parallel as shown in Figure 3.6. In this case each battery would be trying to drive an equal value of current around the circuit, but in opposite directions. Hence, the two batteries 'balance out' each other, and no current will flow.

With suitable instrumentation, it would be possible to measure the p.d. between plate B and any point in the dielectric. If this was done, then a graph of the voltage versus distance from B would look like that in Figure 3.7. The slope of this graph is uniform and has units of potential gradient, expressed in volts/metre. So potential gradient equals

$$\text{potential gradient} = \frac{V}{d} \tag{3.5}$$

Now the energy, expressed in joule,

$$\text{energy} = VIt, \text{ and } I = \frac{Q}{t}$$

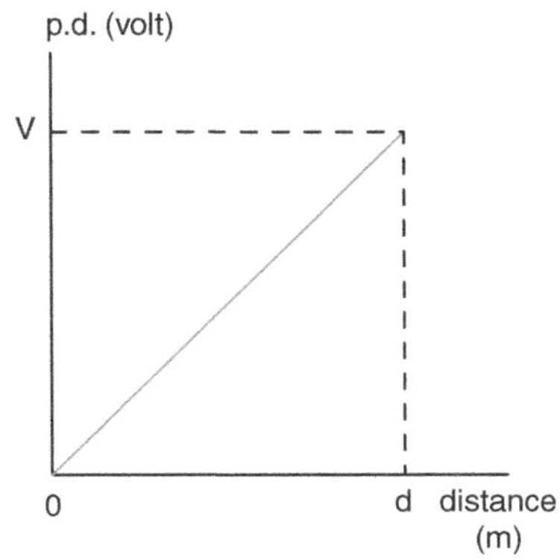

Figure 3.7 The voltage as function of the distance

$$\text{energy} = \frac{VQt}{t} = VQ, \text{ and transposing this}$$

$$V = \frac{\text{energy}}{Q}$$

$$\text{i.e. } 1 \text{ volt} = 1\frac{\text{J}}{\text{C}} \cdots\cdots\cdots\cdots\cdots [1]$$

but the joule is the unit used for work done, and work is force × distance, i.e. newton metre,

$$\text{so } 1 \text{ J} = 1 \text{ Nm} \ldots\ldots\ldots\ldots\ldots [2]$$

Substituting [2] into [1]:

$$1 \text{ volt} = 1 \text{ Nm/C or } V \equiv \frac{\text{Nm}}{\text{C}} \cdots\cdots\cdots\cdots [3]$$

Dividing both sides of [3] by distance of separation d:

$$\frac{V}{d} \equiv \frac{\text{Nm}}{\text{Cm}} = \frac{\text{N}}{\text{C}}$$

Referring back to Equation (3.2), we know that electric field strength **E** is measured in N/C. So potential gradient and electric field strength must be one and the same thing. Now, electric field strength is defined in terms of the ratio of the force exerted on a charge to the value of the charge. This is actually an extremely difficult thing to measure. However, it is a very simple matter to measure the p.d. and distance between the charged plates. Hence, for practical purposes, electric field strength is from now on quoted in the units volt/metre

$$\mathbf{E} = \frac{V}{d} \tag{3.6}$$

Notice that the symbol **E** (in bold) has been used for electric field strength. This is in order to avoid confusion with the symbol E (in italic) used for emf.

WORKED EXAMPLE 3.3

Q Two parallel plates separated by a dielectric of thickness 3 mm acquire a charge of 35 mC when connected to a 150 V source. If the effective cross-sectional area of the field between the plates is 144 mm², calculate (a) the electric field strength and (b) the flux density.

$$d = 3\times10^{-3}\text{ m};\ Q = 35\times10^{-3}\text{ C};\ V = 150\text{ V};\ A = 144\times10^{-6}\text{ m}^2$$

(a) $$\mathbf{E} = \frac{V}{d} = \frac{150}{3\times10^{-3}} = 50\text{ kV/m}$$

(b) $$D = \frac{Q}{A} = \frac{35\times10^{-3}}{144\times10^{-6}} = 243.1\text{ C/m}^2$$

3.5 CAPACITANCE (C)

We have seen that in order for one plate to be at a different potential to the other one, there is a need for a charge. This requirement is known as the capacity of the system. For a given system the ratio of the charge required to achieve a given p.d. is a constant for that system. This is called the *capacitance* (C) of the system

$$C = \frac{Q}{V} \tag{3.7}$$

$$Q = VC \tag{3.8}$$

From Equation (3.7) it may be seen that the unit for capacitance is the farad (F). This is defined as the capacitance of a system that requires a charge of one coulomb in order to raise its potential by one volt. The farad is a very large unit, so in practice it is more usual to express capacitance values in microfarads (uF), nanofarads (nF) or picofarads (pF).

Capacitors are usually very small, so there isn't much room to print all the crucial information on the packaging. Therefore, no more than three digits are used, with the third digit indicating how many zeros should be added to the first two digits. The corresponding number then expresses the capacitance value in pF. If only two digits are used, this directly indicates the capacitance value in pF. For example, 101 represents 100 pF and 22 corresponds to 22 pF. If it is followed by a letter, this letter then indicates the tolerance (like with the resistors): J for 5%, K for 10% and M for 20%, for example.

The tolerance is the deviation from the nominal value. This is normally expressed as a percentage. Thus a capacitor of nominal value 2 μF and a tolerance of ±10%, should have an actual value of between 1.8 μF and 2.2 μF.

Michael Faraday (1791–1867) was an English physicist and chemist, experimenting on whether electromagnetism can influence the polarisation of light by different types of glass. This magneto-optical effect is called the Faraday effect, showing for the first time the connection between light, magnetism and electricity, now known as electromagnetic radiation. Another contribution is the Faraday Cage. It is a construction of electrically conductive material that keeps static electric fields outside the cage. Static magnetic fields can penetrate the cage.

WORKED EXAMPLE 3.4

Q Two parallel plates, separated by an air space of 4 mm, receive a charge of 0.2 mC when connected to a 125 V source. Calculate (a) the electric field strength between the plates, (b) the cross-sectional area of the field between the plates if the flux density is 15 C/m² and (c) the capacitance of the plates.

$d = 4\times10^{-3}$ m; $Q = 2\times10^{-4}$ C; $V = 125$ V; $D = 15$ C/m²

(a) $$\mathbf{E} = \frac{V}{d} = \frac{125}{4\times10^{-3}} = 31.25 \text{ kV/m}$$

(b) $$D = \frac{Q}{A}$$

$$A = \frac{Q}{D} = \frac{2\times10^{-4}}{15} = 13.3\times10^{-6} \text{ m}^2 \text{ or } 13.3 \text{ mm}^2$$

(c) $$Q = CV$$

$$C = \frac{Q}{V} = \frac{2\times10^{-4}}{125} = 1.6\ \mu\text{F}$$

A capacitor is an electrical component that is designed to have a specified value of capacitance. In its simplest form it consists of two parallel plates separated by a dielectric; i.e. exactly the system we have been dealing with so far.

In order to be able to design a capacitor we need to know what dimensions are required for the plates, the thickness of the dielectric (the distance of separation d) and the other properties of the dielectric material chosen. Let us consider first the properties associated with the dielectric.

3.6 PERMITTIVITY

When an electric field exists in a vacuum then the ratio of the electric flux density to the electric field strength is a constant, known as the permittivity ε_0 of free space, expressed in F/m.

$$\varepsilon_0 = 8.854\times10^{-12}$$

Since a vacuum is a well-defined condition, the permittivity of free space is chosen as the reference or datum value from which the permittivity of all other dielectrics is measured. This is a similar principle to using Earth potential as the datum for measuring voltages.

The capacitance of two plates will be increased if, instead of a vacuum between the plates, some other dielectric is used. This difference in capacitance for different dielectrics is accounted for by the relative permittivity ε_r of each dielectric. Thus relative permittivity is defined as the ratio of the capacitance with that dielectric to the capacitance with a vacuum dielectric

$$\varepsilon_r = \frac{C_2}{C_1} \tag{3.9}$$

where C_1 is with a vacuum and C_2 is with the other dielectric.

Note: Dry air has the same effect as a vacuum so the relative permittivity ε_r for air dielectrics equals 1.

For a given system the ratio of the electric flux density to the electric field strength is a constant, known as the absolute permittivity ε of the dielectric being used, expressed in F/m.

$$\varepsilon = \frac{D}{\mathrm{E}} \tag{3.10}$$

but we have just seen that a dielectric (other than air) is more effective than a vacuum by a factor of ε_r times, so the absolute permittivity is given by:

$$\varepsilon = \varepsilon_0 \varepsilon_r \tag{3.11}$$

3.7 CALCULATING CAPACITOR VALUES

From Equation (3.10):

$$\varepsilon = \frac{D}{\mathbf{E}}$$

but $D = Q/A$ and $\mathbf{E} = V/d$

$$\varepsilon = \frac{Qd}{VA}$$

$$Q/V = C$$

$$\varepsilon = C\frac{d}{A}$$

and transposing this for C we have

$$C = \frac{\varepsilon A}{d} = \frac{\varepsilon_0 \varepsilon_r A}{d} \tag{3.12}$$

WORKED EXAMPLE 3.5

Q A capacitor is made from two parallel plates of dimensions 3 cm by 2 cm, separated by a sheet of mica 0.5 mm thick and of relative permittivity 5.8. Calculate (a) the capacitance and (b) the electric field strength if the capacitor is charged to a p.d. of 200 V.

$A = 3 \times 10^{-2}\ \text{m} \times 2 \times 10^{-2}\ \text{m};\ d = 5 \times 10^{-4}\ \text{m};\ \varepsilon_r = 5.8;\ V = 200\ \text{V}$

(a) $C = \frac{\varepsilon_0 \varepsilon_r A}{d} = \frac{8.854 \times 10^{-12} \times 5.8 \times 6 \times 10^{-4}}{5 \times 10^{-4}} = 61.62 \text{ pF}$

(b) $\mathbf{E} = \frac{V}{d} = \frac{200}{5 \times 10^{-4}} = 400 \text{ kV/m}$

WORKED EXAMPLE 3.6

Q A capacitor of value 0.224 nF is to be made from two plates each 75 mm by 75 mm, using a waxed paper dielectric of relative permittivity 2.5. Determine the thickness of paper required.

$C = 0.224 \times 10^{-9}$ F; $A = 75 \times 10^{-3}$ m $\times 75 \times 10^{-3}$ m; $\varepsilon_r = 2.5$

$$C = \frac{\varepsilon_0 \varepsilon_r A}{d}$$

$$d = \frac{\varepsilon_0 \varepsilon_r A}{C} = \frac{8.854 \times 10^{-12} \times 2.5 \times 5.625 \times 10^{-3}}{0.224 \times 10^{-9}} = 5.558 \times 10^{-4} \text{ m} = 0.5558 \text{ mm}$$

WORKED EXAMPLE 3.7

Q A capacitor of value 47 nF is made from two plates having an effective cross-sectional area of 4 cm² and separated by a ceramic dielectric 0.1 mm thick. Calculate the relative permittivity.

$C = 4.7 \times 10^{-8}$ F; $A = 4 \times 10^{-4}$ m^2; $d = 1 \times 10^{-4}$ m

$$C = \frac{\varepsilon_0 \varepsilon_r A}{d}$$

$$\varepsilon_r = \frac{Cd}{\varepsilon_0 A} = \frac{4.7 \times 10^{-8} \times 10^{-4}}{8.854 \times 10^{-12} \times 4 \times 10^{-4}} = 1327$$

WORKED EXAMPLE 3.8

Q A p.d. of 180 V creates an electric field in a dielectric of relative permittivity 3.5, thickness 3 mm and of effective cross-sectional area of 4.2 cm². Calculate the flux and flux density thus produced.

$V = 180$ V; $d = 3 \times 10^{-3}$ m; $\varepsilon_r = 3.5$; $A = 4.2 \times 10^{-4}$ m^2

There are two possible methods of solving this problem; either determine the capacitance and use $Q = VC$ or determine the electric field strength and use $D = \varepsilon_o \varepsilon_r \mathbf{E}$. Both solutions will be shown.

$$C = \frac{\varepsilon_0 \varepsilon_r A}{d} = \frac{8.854 \times 10^{-12} \times 3.5 \times 4.2 \times 10^{-4}}{3 \times 10^{-3}} = 4.338 \text{ pF}$$

since $Q = VC$ then

$$Q = 180 \times 4.338 \times 10^{-12} = 0.7809 \text{ nC}$$

$$D = \frac{Q}{A} = \frac{7.809 \times 10^{-10}}{4.2 \times 10^{-4}} = 1.859 \ \mu\text{C/m}^2$$

Alternatively:

$$\mathbf{E} = \frac{V}{d} = \frac{180}{3 \times 10^{-3}}$$

$\mathbf{E} = 60$ kV/m and using $D = \varepsilon_0 \varepsilon_r \mathbf{E}$

$$D = 8.854 \times 10^{-12} \times 3.5 \times 60 \times 10^3 = 1.859 \ \mu\text{C/m}^2$$

$$Q = DA = 1.859 \times 10^{-6} \times 4.2 \times 10^{-4} = 0.7809 \text{ nC}$$

3.8 CAPACITORS IN PARALLEL

Consider two capacitors that are identical in every way (same plate dimensions, same dielectric material and same distance of separation between plates) as shown in Figure 3.8. Let them now be moved vertically until the top and bottom edges respectively of their plates make contact. We will now effectively have a single capacitor of twice the cross-sectional area of one of the original capacitors, but all other properties will remain unchanged.

Since $C = \varepsilon A/d$, the 'new' capacitor formed will have twice the capacitance of one of the original capacitors. The same effect could have been achieved if we had simply connected the appropriate plates together by means of a simple electrical connection. In other words connect them in parallel with each other. Both of the original capacitors have the same capacitance, and this figure is doubled when they are connected in parallel. Thus we can draw the conclusion that with this connection the total capacitance of the combination is given simply by adding the capacitance values. However, this might be considered as a special case. Let us verify this conclusion by considering the general case of three different value capacitors connected in parallel to a d.c. supply of V volts as in Figure 3.9.

Each capacitor will take a charge from the supply according to its capacitance:

$$Q_1 = VC_1; Q_2 = VC_2; Q_3 = VC_3$$

but the total charge drawn from the supply must be:

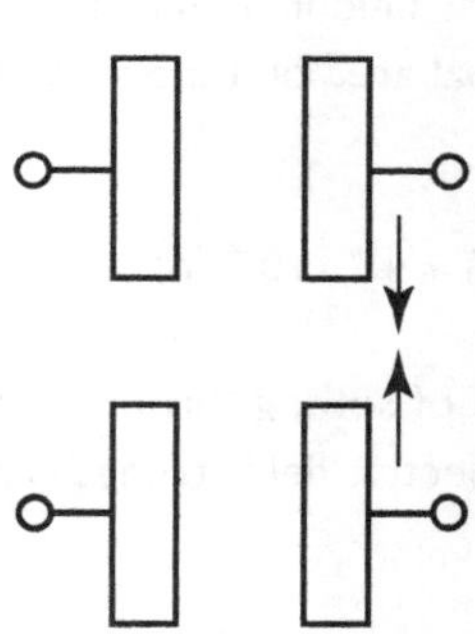

Figure 3.8 Capacitors in parallel

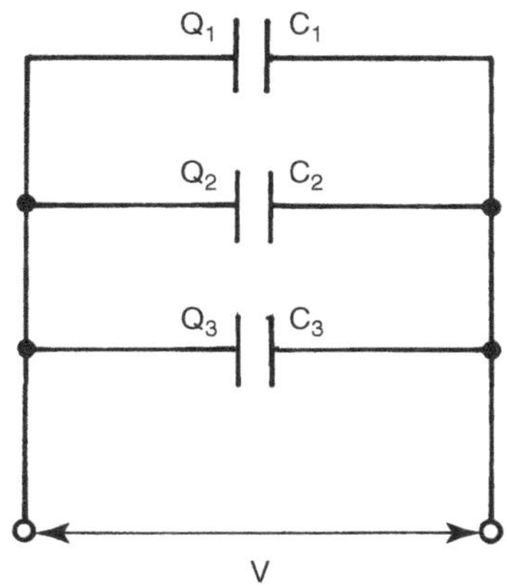

Figure 3.9 Three capacitors connected in parallel

$Q = Q_1 + Q_2 + Q_3$
total charge, $Q = VC$

where C is the total circuit capacitance.
Thus, $VC = VC_1 + VC_2 + VC_3$, and dividing through by V

$$C = C_1 + C_2 + C_3 \tag{3.13}$$

Note: This result is exactly the *opposite in form* to that for *resistors* in parallel.

WORKED EXAMPLE 3.9

Q Three capacitors of value 4.7 μF, 3.9 μF and 2.2 μF are connected in parallel. Calculate the resulting capacitance of this combination.

$C = C_1 + C_2 + C_3$ $C = 4.7 + 3.9 + 2.2$
$C = 10.8$ μF

Most practical capacitors consist of more than one pair of parallel plates, and in these cases they are referred to as multiplate capacitors. The sets of plates are often interleaved as shown in Figure 3.10. The example illustrated has a total of five plates. It may be seen that this effectively forms four identical capacitors, in which the three inner plates are common to the two 'inner' capacitors. Since all the positive plates are joined together, and so too are the negative plates, this arrangement is equivalent to four identical capacitors connected in parallel, as shown in Figure 3.11. The total capacitance of four identical capacitors connected in parallel is simply four times the capacitance of one of them. Thus, this value will be the effective capacitance of the complete capacitor.

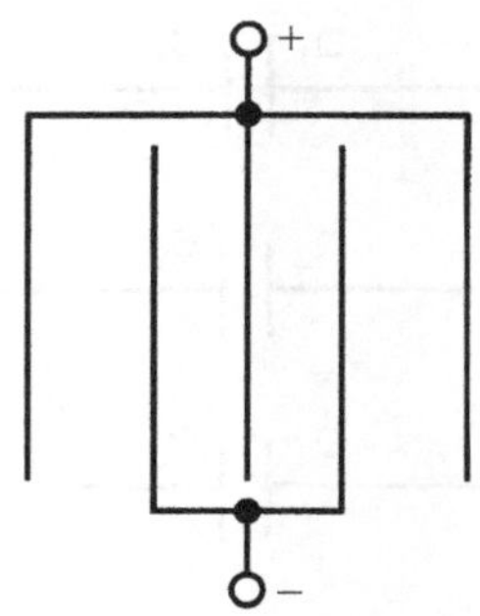

Figure 3.10 A multiplate capacitor

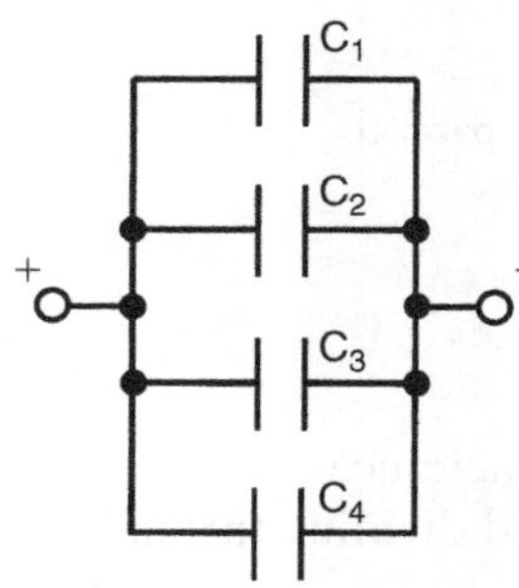

Figure 3.11 Equivalence of multiplate capacitor

The capacitance between one adjacent pair of plates will be

$$C_1 = \frac{\varepsilon_0 \varepsilon_r A}{d}$$

so, the total for the complete arrangement = $C_1 \times 4$, but we can express 4 as (5 – 1) so the total capacitance is

$$C_1 = \frac{\varepsilon_0 \varepsilon_r A(5-1)}{d}$$

In general therefore, if a capacitor has N plates, the capacitance is given by the expression:

$$C_1 = \frac{\varepsilon_0 \varepsilon_r A(N-1)}{d} \quad (3.14)$$

Since the above equation applies generally, then it must also apply to capacitors having just one pair of plates as previously considered. This is correct, since if $N = 2$ then $(N - 1) = 1$, and the above equation becomes identical to Equation (3.12) previously used.

WORKED EXAMPLE 3.10

Q A capacitor is made from 20 interleaved plates each 80 mm by 80 mm separated by mica sheets 1.5 mm thick. If the relative permittivity for mica is 6.4, calculate the capacitance.

$N = 20$; $A = 80\times10^{-3}\ \text{m}\times80\times10^{-3}\ \text{m}$; $d = 1.5\times10^{-3}\ \text{m}$; $\varepsilon_r = 6.4$

$$C_1 = \frac{\varepsilon_0\varepsilon_r A(N-1)}{d} = \frac{8.854\times10^{-12}\times6.4\times6400\times10^{-6}\times19}{1.5\times10^{-3}} = 4.6\ \text{nF}$$

WORKED EXAMPLE 3.11

Q A 300 pF capacitor has nine parallel plates, each 40 mm by 30 mm, separated by mica of relative permittivity 5. Determine the thickness of the mica.

$N = 9$; $C = 3\times10^{-10}\ \text{F}$; $A = 40\times10^{-3}\ \text{m}\times30\times10^{-3}\ \text{m}$; $\varepsilon_r = 5$

$$C = \frac{\varepsilon_0\varepsilon_r A(N-1)}{d}$$

$$d = \frac{\varepsilon_0\varepsilon_r A(N-1)}{C} = \frac{8.854\times10^{-12}\times5\times1200\times10^{-6}}{3\times10^{-10}} = 1.42\ \text{mm}$$

WORKED EXAMPLE 3.12

Q A parallel plate capacitor consists of 11 circular plates, each of radius 25 mm, with an air gap of 0.5 mm between each pair of plates. Calculate the value of the capacitor.

$N = 11$; $r = 25\times10^{-3}\ \text{m}$; $d = 5\times10^{-4}\ \text{m}$; $\varepsilon_r = 1(\text{air})$

$$A = \pi r^2\ \text{m}^2 = \pi\times\left(25\times10^{-3}\right)^2 = 1.9635\times10^{-3}\ \text{m}^2$$

$$C = \frac{\varepsilon_0\varepsilon_r A(N-1)}{d} = \frac{8.854\times10^{-12}\times1.9635\times10^{-3}\times10}{5\times10^{-4}} = 3.48\times10^{-10}\ \text{F or } 348\ \text{pF}$$

3.9 CAPACITORS IN SERIES

Three parallel plate capacitors are shown connected in series in Figure 3.12. Each capacitor will receive a charge. However, you may wonder how capacitor C_2 can receive any charging current since it is sandwiched between the other two, and of course the charging current cannot flow through the dielectrics of these. The answer lies in the explanation of the charging process described in Section 3.4 earlier. To assist the explanation, the plates of capacitors C_1 to C_3 have been labelled with letters.

Plate A will lose electrons to the positive terminal of the supply, and so acquires a positive charge. This creates an electric field in the dielectric of C_1 which will cause plate B to attract electrons from plate C of C_2. The resulting electric field in C_2 in turn causes plate D to attract electrons from plate E. Finally, plate F attracts electrons from the negative terminal of the supply. Thus all three capacitors become charged to the same value.

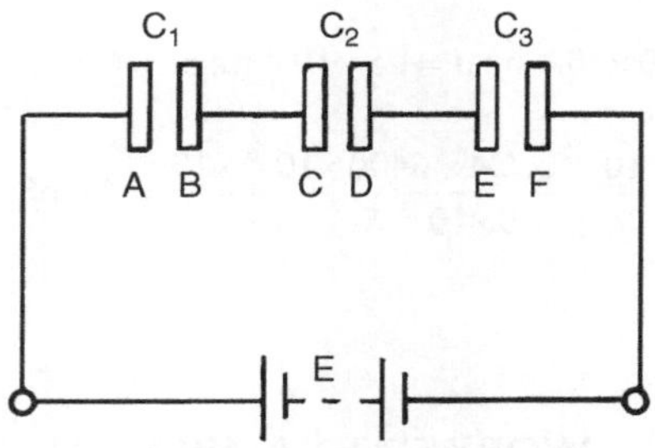

Figure 3.12 Capacitors in series

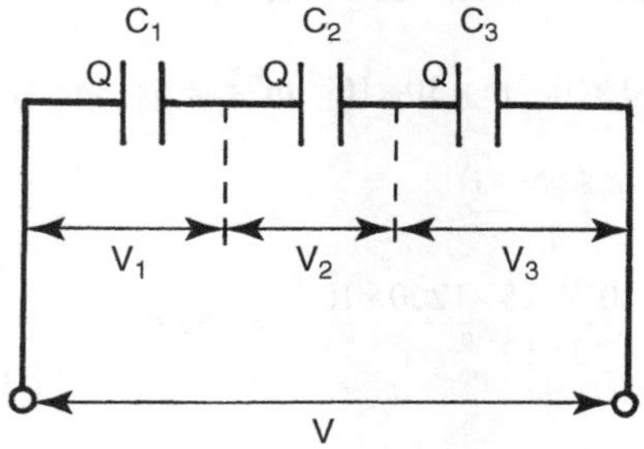

Figure 3.13 Three capacitors connected in series

Having established that all three capacitors will receive the same amount of charge, let us now determine the total capacitance of the arrangement. Since the capacitors are of different values, each will acquire a different p.d. between its plates. This is illustrated in Figure 3.13.

$$V_1 = \frac{Q}{C_1};\ V_2 = \frac{Q}{C_2};\ V_3 = \frac{Q}{C_3}$$

$V = V_1 + V_2 + V_3$ (Kirchhoff's voltage law)

$V = \frac{Q}{C}$ with C is the total circuit capacitance.

$$\frac{Q}{C} = \frac{Q}{C_1} + \frac{Q}{C_2} + \frac{Q}{C_3}$$

$$\frac{1}{C} = \frac{1}{C_1} + \frac{1}{C_2} + \frac{1}{C_3} \qquad (3.15)$$

Note: The above equation does not give the total capacitance directly. To obtain the value for *C* the reciprocal of the answer obtained from Equation (3.15) must be found. However, if *only two* capacitors are connected in series the total capacitance may be obtained directly by using the 'product/sum' form

$$C = \frac{C_1 C_2}{C_1 + C_2} \qquad (3.16)$$

WORKED EXAMPLE 3.13

Q A 6 μF and a 4 μF capacitor are connected in series across a 150 V supply. Calculate (a) the total capacitance, (b) the charge on each capacitor and (c) the p.d. developed across each.

Figure 3.14 shows the appropriate circuit diagram.

$C_1 = 6\ \mu F;\ C_2 = 4\ \mu F;\ V = 150\ V$

(a) $$C = \frac{C_1 C_2}{C_1 + C_2} = \frac{6 \times 4}{6 + 4} = \frac{24}{10}\ \mu F = 2.4\ \mu F$$

(b) $$Q = VC = 150 \times 2.4 \times 10^{-6} = 360\ \mu C$$

(same charge on both)

Since capacitors in series all receive the same value of charge, then this must be the total charge drawn from the supply,

$$Q = VC$$

This is equivalent to a series resistor circuit where the current drawn from the supply is common to all the resistors.

(c) $$V_1 = \frac{Q}{C_1} = \frac{360 \times 10^{-6}}{6 \times 10^{-6}} = 60\ V$$

$$V_2 = \frac{Q}{C_2} = \frac{360 \times 10^{-6}}{4 \times 10^{-6}} = 90\ V$$

Note that $V_1 + V_2 = 150\ V = V$

WORKED EXAMPLE 3.14

Q Capacitors of 3 μF, 6 μF and 12 μF are connected in series across a 400 V supply. Determine the p.d. across each capacitor.

Figure 3.15 shows the relevant circuit diagram.

$C_1 = 3\ \mu F;\ C_2 = 6\ \mu F;\ C_3 = 12\ \mu F;\ V = 400\ V$

$$\frac{1}{C} = \frac{1}{C_1} + \frac{1}{C_2} + \frac{1}{C_3} = \frac{1}{3} + \frac{1}{6} + \frac{1}{12} = \frac{4 + 2 + 1}{12} = \frac{7}{12}$$

$$C = \frac{12}{7} = 1.714\ \mu F$$

$$Q = VC = 400\ V \times 1.714 \times 10^{-6}\ F$$

$$Q = 685.7\ \mu C$$

$$V_1 = \frac{Q}{C_1} = \frac{685.7 \times 10^{-6}}{3 \times 10^{-6}} = 228.6\ V$$

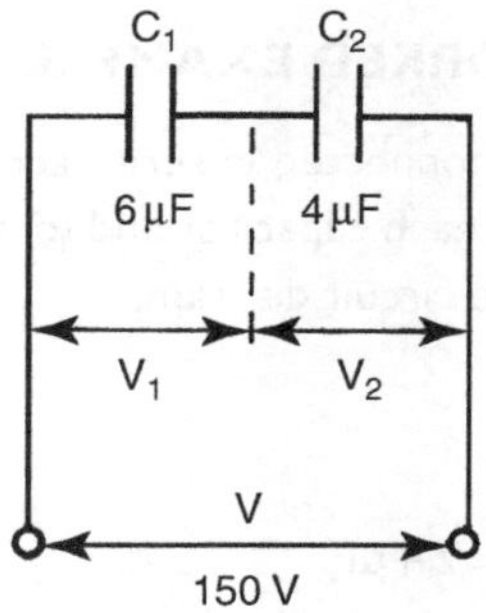

Figure 3.14 The circuit diagram for Worked Example 3.13

$$V_2 = \frac{685.7 \times 10^{-6}}{6 \times 10^{-6}} = 114.3\text{ V}$$

$$V_3 = \frac{685.7 \times 10^{-6}}{12 \times 10^{-6}} = 57.1\text{ V}$$

3.10 SERIES/PARALLEL COMBINATIONS

The techniques required for the solution of this type of circuit are again best demonstrated by means of a worked example.

WORKED EXAMPLE 3.15

Q For the circuit shown in Figure 3.16, determine (a) the charge drawn from the supply, (b) the charge on the 8 μF capacitor, (c) the p.d. across the 4 μF capacitor and (d) the p.d. across the 3 μF capacitor.

The first task is to label the diagram as shown in Figure 3.17.

(a) $C_{BCD} = \frac{3 \times 6}{3+6} = 2\ \mu\text{F}$ (see Figure 3.18)

$C_{BD} = 2 + 4 = 6\ \mu\text{F}$ (see Figure 3.19)

$C_{AD} = \frac{6 \times 2}{6+2} = 1.5\ \mu\text{F}$ (see Figure 3.20)

$C = C_{AD} + C_{EF} = 1.5 + 8 = 9.5\ \mu\text{F}$

$Q = VC = 200 \times 9.5 \times 10^{-6} = 1.9\text{ mC}$

(b) $Q_{EF} = VC_{EF} = 200 \times 8 \times 10^{-6} = 1.6\text{ mC}$

Total charge $Q = 1.9$ mC and $Q_{EF} = 1.6$ mC

(c) so $Q_{AD} = 1.9 - 1.6 = 0.3$ mC (see Figure 3.20)

and referring to Figure 3.19, this will be the charge on both the capacitors shown, i.e.

$Q_{AB} = Q_{BD} = 0.3$ mC.

Thus,

V_{BD} = p.d. across

4 μF capacitor (see Figure 3.18)

$$V_{BD} = \frac{Q_{BD}}{C_{BD}} = \frac{0.3 \times 10^{-3}}{6 \times 10^{-6}} = 50 \text{ V}$$

(d) $Q_{BCD} = V_{BD} C_{BCD}$ (see Figures 3.18 and 3.17)

$= 50 \times 2 \times 10^{-6} = 100\ \mu\text{C}$

and this will be the charge on both the 3 μF and 6 μF capacitors, i.e.

$Q_{BC} = Q_{CD} = 100\ \mu\text{C}$

$$V_{BC} = \frac{Q_{BC}}{C_{BC}} = \frac{1 \times 10^{-4}}{3 \times 10^{-6}}$$

$V_{BC} = 33.33$ V

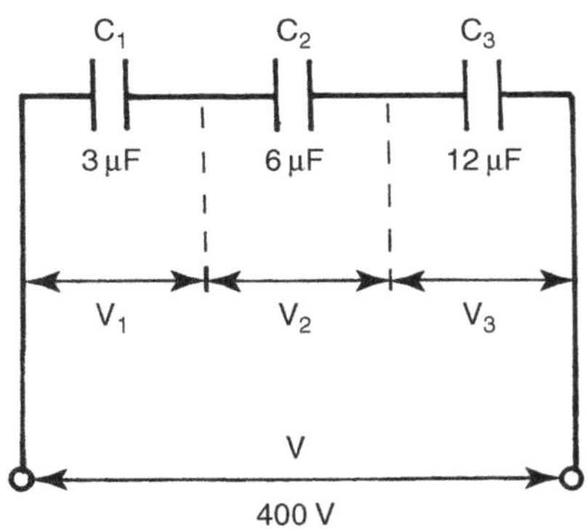

Figure 3.15 The circuit diagram for Worked Example 3.14

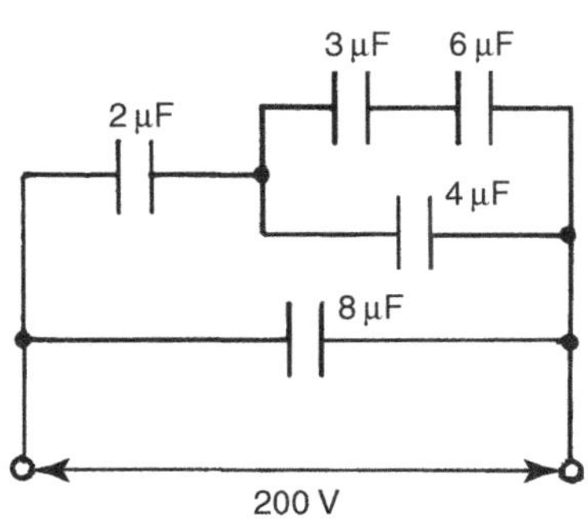

Figure 3.16 The circuit diagram for Worked Example 3.15

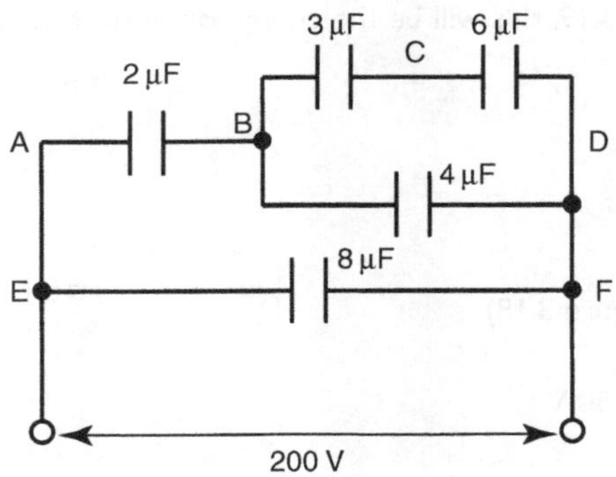

Figure 3.17 The circuit diagram for Worked Example 3.15, labelled

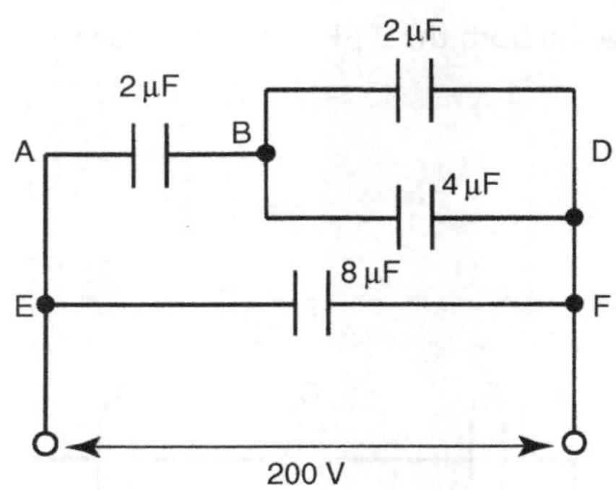

Figure 3.18 The circuit diagram for Worked Example 3.15, reduced

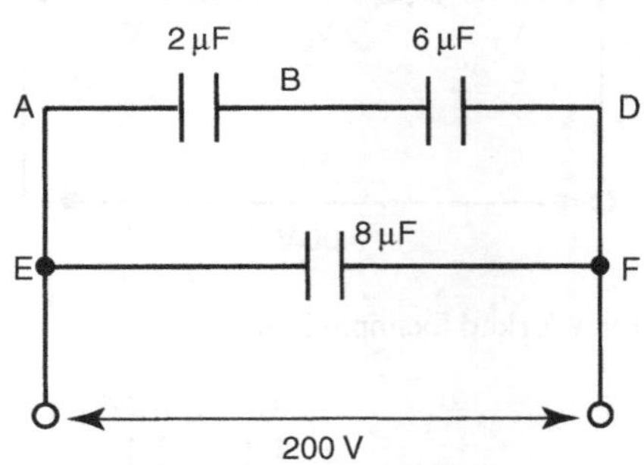

Figure 3.19 The circuit diagram for Worked Example 3.15, further reduced

3.11 ENERGY STORED

When a capacitor is connected to a voltage source of V volts we have seen that it will charge up until the p.d. between the plates is also V volts. If the capacitor is now disconnected from the supply, the charge and p.d. between its plates will be retained.

Consider such a charged capacitor, as shown in Figure 3.21, which now has a resistor connected across its terminals. In this case the capacitor will behave as if it were a source of

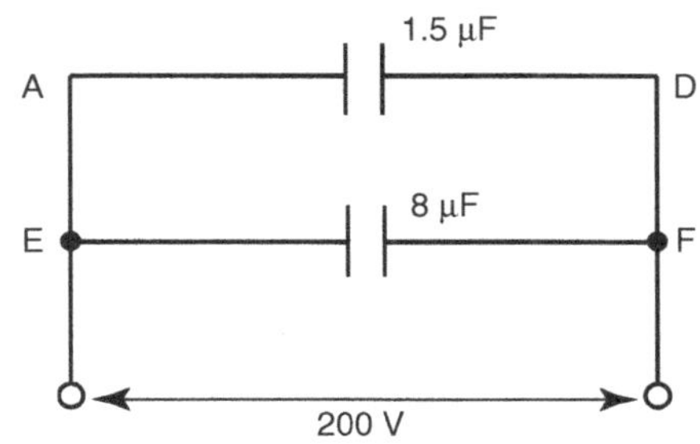

Figure 3.20 The circuit diagram for Worked Example 3.15, final

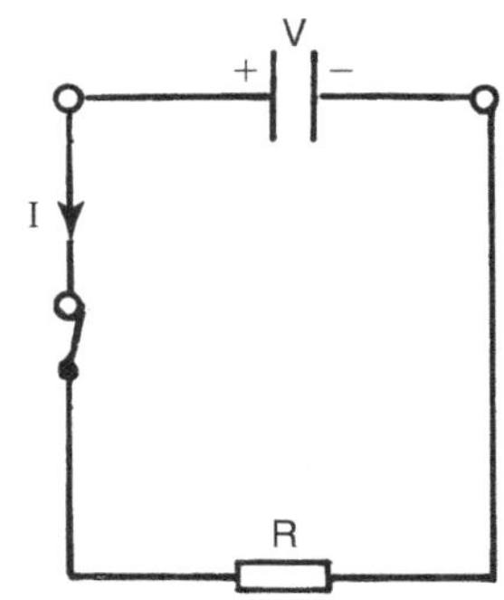

Figure 3.21 A charged capacitor connected to a resistor

emf. It will therefore drive current through the resistor. In this way the stored charge will be dissipated as the excess electrons on its negative plate are returned to the positive plate. This discharge process will continue until the capacitor becomes completely discharged (both plates electrically neutral). Note that the discharge current marked on the diagram indicates *conventional* current flow.

However, if a discharge current flows then work must be done (energy is being dissipated). The only possible source of this energy in these circumstances must be the capacitor itself. Thus the charged capacitor must store energy.

If a graph is plotted of capacitor p.d. to the charge it receives, the area under the graph represents the energy stored. Assuming a constant charging current, the graph will be as shown in Figure 3.22.

$$\text{Area under the graph} = \frac{1}{2}QV$$

but $Q = CV$ coulomb

$$\text{Stored energy, } W = \frac{1}{2}CV^2 \text{ joule} \qquad (3.17)$$

A capacitor hence acts as a rechargeable battery, by storing electrical energy and giving it back afterwards. By using a voltage source and by connecting both poles to a capacitor, this capacitor is charged. We can notice that the process is slowing down by the lower difference

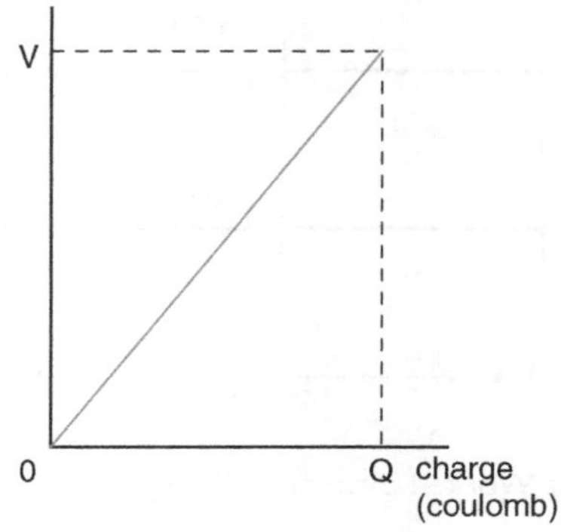

Figure 3.22 Capacitor p.d. as function of the charge

between the voltage source and the capacitor voltage, when more and more energy is stored in the capacitor. It does take some time until the capacitor and the source have the same voltage and at that moment the charge redistribution process stops. If the voltage source is replaced by a LED (light-emitting diode, a lighting device explained further in Chapter 6) in series with a resistor (to prevent too high current), the LED lights up and shows that the capacitor earlier stored charges. We can conclude that the less charge on the capacitor, the less voltage available and hence the less light intensity. This process can be repeated and repeated, behaving as a rechargeable battery, although not optimised for charging and recharging.

This way of charging and discharging is sometimes used with flashing lights, like a camera flash for instance. It can be seen as a resistor in series with a capacitor, and the lamp in parallel with the capacitor. The lamp here is a gas discharge lamp like the natrium lamps in street lighting in use at night time. An important property of this lamp is that it only lights up when a certain voltage is present and the lamp remains lit until the voltage drops to a minimum voltage. When lit the lamp can be modelled as a resistor. When not lit, it behaves like an open chain. When the lamp behaves as an open circuit, the supply voltage will charge the capacitor through the resistor up to the supply voltage. As soon as the voltage across the capacitor reaches the maximal voltage, the lamp lights up and the capacitor discharges to the minimum voltage. Then the lamp becomes an open circuit again and the cycle of charging and discharging restarts.

WORKED EXAMPLE 3.16

Q A 3 μF capacitor is charged from a 250 V d.c. supply. Calculate the charge and energy stored. The charged capacitor is now removed from the supply and connected across an uncharged 6 μF capacitor. Calculate the p.d. between the plates and the energy now stored by the combination.

$C_1 = 3\ \mu F;\ V_1 = 250\ V;\ C_2 = 6\ \mu F$

$$Q = V_1C_1 = 250 \times 3 \times 10^{-6} = 0.75\ mC$$

$$W = \frac{1}{2}C_1V_1^2 = \frac{1}{2} \times 3 \times 10^{-6} \times (250)^2 = 93.75\ mJ$$

When the two capacitors are connected in parallel the 3μF will share its charge with the 6μF capacitor. Thus the total charge in the system will remain unchanged, but the total capacitance will now be different:

$$\text{Total Capacitance, } C = C_1 + C_2 = 3 + 6$$
$$C = 9\ \mu\text{F}$$
$$V = \frac{Q}{C} = \frac{7.5 \times 10^{-4}}{9 \times 10^{-6}} = 83.33\ \text{V}$$
$$\text{Total energy stored, } W = \frac{1}{2}CV^2 = \frac{1}{2} \times 9 \times 10^{-6} \times (83.33)^2 = 31.25\ \text{mJ}$$

Note: The above example illustrates the law of conservation of charge, since the charge placed on the first capacitor is simply redistributed between the two capacitors when connected in parallel. The total charge therefore remains the same. However, the p.d. now existing between the plates has fallen, and so too has the total energy stored. But there is also a law of conservation of energy, so what has happened to the 'lost' energy? Well, in order for the 3μF capacitor to share its charge with the 6μF capacitor a charging current had to flow from one to the other. Thus this 'lost' energy was used in the charging process.

WORKED EXAMPLE 3.17

Q Consider the circuit of Figure 3.23, where initially all three capacitors are fully discharged, with the switch in position '1'.

(a) If the switch is now moved to position '2', calculate the charge and energy stored by C_1.

(b) Once C_1 is fully charged, the switch is returned to position '1'. Calculate the p.d. now existing across C_1 and the amount of energy used in charging C_2 and C_3 from C_1.

(a)
$$Q_1 = VC_1 = 200 \times 10 \times 10^{-6} = 2\ \text{mC}$$
$$W_1 = \frac{1}{2}C_1V^2 = 0.5 \times 10 \times 10^{-6} \times (200)^2 = 0.2\ \text{J}$$

(b) C_2 and C_3 in series are equivalent to

$$C_4 = \frac{C_2C_3}{C_2 + C_3} = \frac{6.8 \times 4.7}{6.8 + 4.7} = 2.78\ \mu\text{F}$$

and total capacitance of the whole circuit,

$$C = C_1 + C_4 = 10 + 2.78 = 12.78\ \mu\text{F}$$

Now, the charge received by the circuit remains constant, although the total capacitance has increased.

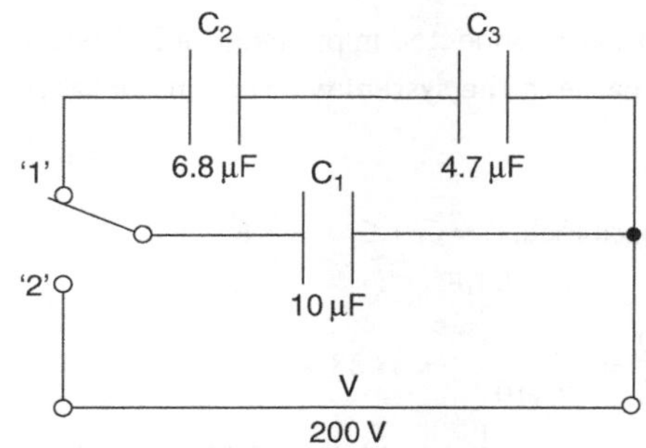

Figure 3.23 The circuit diagram for Worked Example 3.17

$$V = \frac{Q}{C} = \frac{2\times10^{-3}}{12.78\times10^{-6}} = 156.5\text{ V}$$

Total energy remaining in the circuit,

$$W = \frac{1}{2}CV^2 = 0.5\times12.78\times10^{-6}\times(156.5)^2 = 0.156\text{ J}$$

The energy used up must be the difference between the energy first stored by C_1 and the final energy stored in the system, hence

$$\text{Energy used} = 0.2\text{ J} - 0.156\text{ J} = 0.044\text{ J} = 44\text{ mJ}$$

3.12 DIELECTRIC STRENGTH AND WORKING VOLTAGE

There is a maximum potential gradient that any insulating material can withstand before dielectric breakdown occurs. There are of course some applications where dielectric breakdown is deliberately produced, e.g. a sparking plug in a car engine, which produces an arc between its electrodes when subjected to a p.d. of several kilovolts. This then ignites the air/fuel mixture. However, it is obviously not a condition that is desirable in a capacitor, since it results in its destruction.

Capacitors normally have marked on them a maximum working voltage. When in use you must ensure that the voltage applied between its terminals does not exceed this value, otherwise dielectric breakdown will occur.

Dielectric breakdown is the effect produced in an insulating material when the voltage applied across it is more than it can withstand. The result is that the material is forced to conduct. However, when this happens, the sudden surge of current through it will cause it to burn, melt, vaporise or be permanently damaged in some other way. If you search on www.youtube.com 'exploding capacitors' or 'exploding capacitors slow motion', you can see many different examples, where first the capacitor is bulging and this is followed by an explosion. Please don't try this at home by yourself, as it can be very dangerous.

Another way of referring to this maximum working voltage is to quote the dielectric strength. This is the maximum voltage gradient that the dielectric can withstand, quoted in kV/m or in V/mm.

WORKED EXAMPLE 3.18

Q A capacitor is designed to be operated from a 400 V supply, and uses a dielectric which (allowing for a factor of safety), has a dielectric strength of 0.5 MV/m. Calculate the minimum thickness of dielectric required.

$V = 400$ V; $\mathbf{E} = 0.5 \times 10^6$ V/m

$$\mathbf{E} = \frac{V}{d}$$

$$d = \frac{V}{\mathbf{E}} = \frac{400}{0.5 \times 10^6} = 0.8 \text{ mm}$$

WORKED EXAMPLE 3.19

Q A 270 pF capacitor is to be made from two metallic foil sheets, each of length 20 cm and width 3 cm, separated by a sheet of Teflon having a relative permittivity of 2.1. Determine (a) the thickness of Teflon sheet required, and (b) the maximum possible working voltage for the capacitor if the Teflon has a dielectric strength of 350 kV/m.

$C = 270 \times 10^{-12}$ F; $A = 20 \times 10^{-2}$ m $\times 3 \times 10^{-2}$ m; $\varepsilon_r = 2.1$; $\mathrm{E} = 350 \times 10^3$ V/m

(a)

$$C = \frac{\varepsilon_0 \varepsilon_r A}{d}$$

$$d = \frac{\varepsilon_0 \varepsilon_r A}{C} = \frac{8.854 \times 10^{-12} \times 2.1 \times 60 \times 10^{-4}}{270 \times 10^{-12}} = 0.413 \text{ mm}$$

(b) Dielectric strength is the same thing as electric field strength, expressed in volt/metre, so

$$\mathbf{E} = \frac{V}{d}$$

$$V = \mathbf{E}d = 350 \times 10^3 \times 0.413 \times 10^{-3} = 144.6 \text{ V}$$

Note: This figure is the voltage at which the dielectric will start to break down, so, for practical purposes, the maximum working voltage would be specified at a lower value. For example, if a factor of safety of 20% was required, then the maximum working voltage in this case would be specified as 115 V.

3.13 CAPACITOR TYPES

The main difference between capacitor types is in the dielectric used. There are a number of factors that will influence the choice of capacitor type for a given application. Amongst these are the capacitance value, the working voltage, the tolerance, the stability, the leakage resistance, the size and the price.

Since $C = \varepsilon A/d$, any or all of these factors can be varied to suit particular requirements. Thus, if a large value of capacitance is required, a large cross-sectional area and/or a small distance of separation will be necessary, together with a dielectric of high relative permittivity ε_r. However, if the area is to be large, then this can result in a device that is unacceptably large. Additionally, the dielectric cannot be made too thin lest its dielectric strength is exceeded. The various capacitor types overcome these problems in a number of ways.

3.13.1 Paper

This is the simplest form of capacitor. It utilises two strips of aluminium foil separated by sheets of waxed paper. The whole assembly is rolled up into the form of a cylinder (like a Swiss roll). Metal end caps make the electrical connections to the foils, and the whole assembly is then encapsulated in a case. By rolling up the foil and paper a comparatively large cross-sectional area can be produced with reasonably compact dimensions. This type is illustrated in Figure 3.24.

3.13.2 Air

Air dielectric capacitors are the most common form of variable capacitor, as a realisation of a multiplate capacitor. The construction is shown in Figure 3.25. One set of plates is fixed, and the other set can be rotated to provide either more or less overlap between the two. This causes variation of the effective cross-sectional area and hence variation of capacitance. This is the type of device connected to the station-tuning control of a radio.

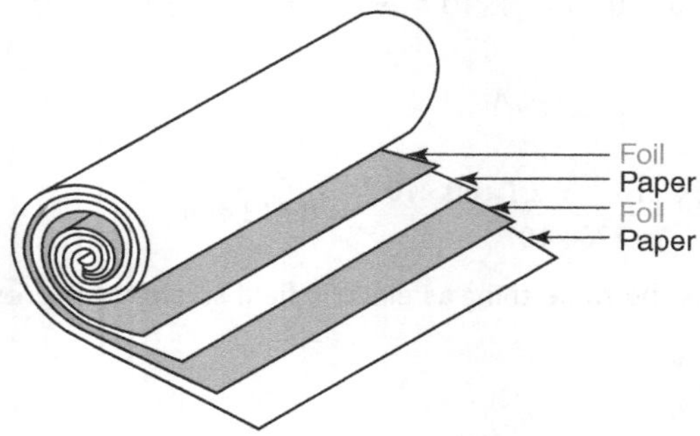

Figure 3.24 Paper capacitor

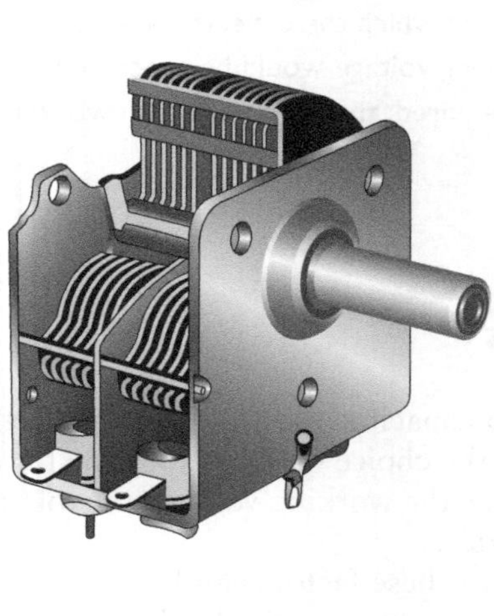

Figure 3.25 Air capacitor

3.13.3 'Plastic'

With these capacitors the dielectric can be of polyester, polystyrene, polycarbonate or polypropylene. Each material has slightly different electrical characteristics which can be used to advantage, depending upon the proposed application. The construction takes much the same form as that for paper capacitors. Examples of these types are shown in Figure 3.26, and their different characteristics are listed in Table 3.1.

3.13.4 Silvered Mica

These are the most accurate and reliable of the capacitor types, having a low tolerance figure. These features are usually reflected in their cost. They consist of a disc or hollow cylinder of ceramic material which is coated with a silver compound. Electrical connections are affixed to the silver coatings and the whole assembly is placed into a casing or (more usually) the assembly is encased in a waxy substance.

3.13.5 Mixed Dielectric

This dielectric consists of paper impregnated with polyester which separates two aluminium foil sheets as in the paper capacitor. This type makes a good general-purpose capacitor, and an example is shown in Figure 3.27.

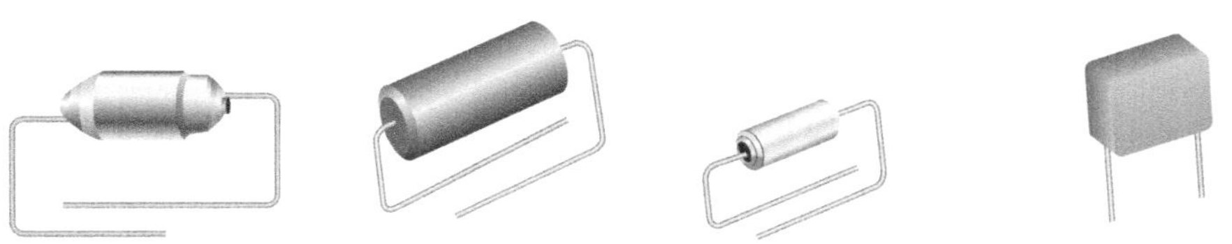

Figure 3.26 Plastic capacitor

Table 3.1 Capacitor characteristics

Type	*Capacitance*	*Tolerance (%)*	*Other characteristics*
Paper	1 nF – 40 μF	±2	Cheap. Poor stability
Air	5 pF – 1 nF	±1	Variable. Good stability
Polycarbonate	100 pF – 10 μF	±10	Low loss. High temperature
Polyester	1 nF – 2 μF	±20	Cheap. Low frequency
Polypropylene	100 pF – 10 nF	±5	Low loss. High frequency
Polystyrene	10 pF – 10 nF	±2	Low loss. High frequency
Mixed	1 nF – 1 μF	±20	General purpose
Silvered mica	10 pF – 10 nF	±1	High stability. Low loss
Electrolytic (aluminium)	1 – 100 000 μF	−20 to +80	High loss. High leakage. d.c. circuits only
Electrolytic (tantalum)	0.1 – 150 μF	±20	As for aluminium above
Ceramic	2pF – 100 nF	±10	Low temperature coefficient. High frequency

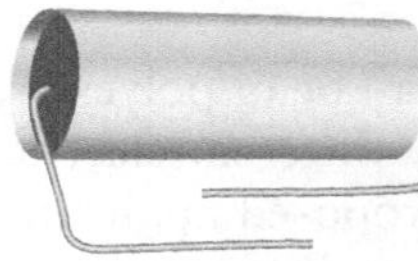

Figure 3.27 Mixed dielectric capacitor

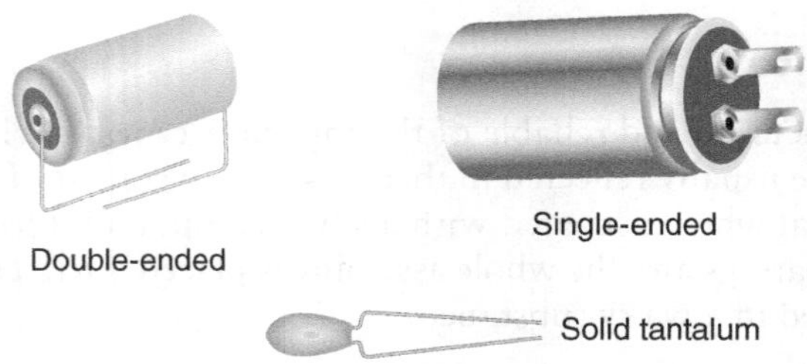

Figure 3.28 Oxide capacitor

3.13.6 Electrolytic

This is the form of construction used for the largest value capacitors. However, they also have the disadvantages of reduced working voltage, high leakage current and the requirement to be polarised. Their terminals are marked + and –, and these polarities must be observed when the device is connected into a circuit. Capacitance values up to 100 000 μF are possible.

A polarised capacitor is one in which the dielectric is formed by passing a d.c. current through it. The polarity of the d.c. supply used for this purpose must be subsequently observed in any circuit in which the capacitor is then used. Thus, they should be used only in d.c. circuits.

The dielectric consists of either an aluminium oxide or tantalum oxide film that is just a few micrometres thick. It is this fact that allows such high capacitance values, but at the same time reduces the possible maximum working voltage. Tantalum capacitors are usually very much smaller than the aluminium types. They therefore cannot obtain the very high values of capacitance possible with the aluminium type. The latter consist of two sheets of aluminium separated by paper impregnated with an electrolyte. These are then rolled up like a simple paper capacitor. This assembly is then placed in a hermetically sealed aluminium canister. The oxide layer is formed by passing a charging current through the device, and it is the polarity of this charging process that determines the resulting terminal polarity that must be subsequently observed. If the opposite polarity is applied to the capacitor the oxide layer is destroyed. Examples of electrolytic capacitors are shown in Figure 3.28.

SUMMARY OF EQUATIONS

Force between charges: $F = \frac{Q_1 Q_2}{\varepsilon_o d^2}$

Electric field strength: $\mathbf{E} = \frac{F}{q} = \frac{V}{d}$

Electric flux density: $D = \frac{Q}{A} = \varepsilon_o \varepsilon_r E$

Permittivity: $\varepsilon = \varepsilon_o \varepsilon_r = \dfrac{D}{E}$

Capacitance: $C = \dfrac{Q}{V} = \dfrac{\varepsilon_o \varepsilon_r A(N-1)}{d}$

Capacitors in parallel: $C = C_1 + C_2 + C_3 + \cdots$

Capacitors in series: $\dfrac{1}{C} = \dfrac{1}{C_1} + \dfrac{1}{C_2} + \dfrac{1}{C_3} + \cdots$

For *only* two in series: $C = \dfrac{C_1 C_2}{C_1 + C_2}\left(\dfrac{\text{product}}{\text{sum}}\right)$

Energy stored: $W = 0.5\,QV = 0.5\,CV^2$

ASSIGNMENT QUESTIONS

1. Two parallel plates 25 cm by 35 cm receive a charge of 0.2 μC from a 250 V supply. Calculate (a) the electric flux and (b) the electric flux density.
2. The flux density between two plates separated by a dielectric of relative permittivity 8 is 1.2 μC/m^2. Determine the potential gradient between them.
3. Calculate the electrical field strength between a pair of plates spaced 10 mm apart when a p.d. of 0.5 kV exists between them.
4. Two plates have a charge of 30 μC. If the effective area of the plates is 5 cm^2, calculate the flux density.
5. A capacitor has a dielectric 0.4 mm thick and operates at 50 V. Determine the electric field strength.
6. A 100 μF capacitor has a p.d. of 400 V across it. Calculate the charge that it has received.
7. A 47 μF capacitor stores a charge of 7.8 mC when connected to a d.c. supply. Calculate the supply voltage.
8. Determine the p.d. between the plates of a 470 nF capacitor if it stores a charge of 0.141 mC.
9. Calculate the capacitance of a pair of plates having a p.d. of 600 V when charged to 0.3 μC.
10. The capacitance of a pair of plates is 40 pF when the dielectric between them is air. If a sheet of glass is placed between the plates (so that it completely fills the space between them), calculate the capacitance of the new arrangement if the relative permittivity of the glass is 6.
11. A dielectric 2.5 mm thick has a p.d. of 440 V developed across it. If the resulting flux density is 4.7 μC/m^2 determine the relative permittivity of the dielectric.
12. State the factors that affect the capacitance of a parallel plate capacitor, and explain how the variation of each of these factors affects the capacitance. Calculate the value of a two-plate capacitor with a mica dielectric of relative permittivity 5 and thickness 0.2 mm. The effective area of the plates is 250 cm^2.

13 A capacitor consists of two plates, each of effective area 500 cm^2, spaced 1 mm apart in air. If the capacitor is connected to a 400 V supply, determine (a) the capacitance, (b) the charge stored and (c) the potential gradient.

14 A paper dielectric capacitor has two plates, each of effective cross-sectional area of 0.2 m^2. If the capacitance is 50 nF calculate the thickness of the paper, given that its relative permittivity is 2.5.

15 A two-plate capacitor has a value of 47 nF. If the plate area was doubled and the thickness of the dielectric was halved, what then would be the capacitance?

16 A parallel plate capacitor has 20 plates, each 50 mm by 35 mm, separated by a dielectric 0.4 mm thick. If the capacitance is 1000 pF determine the relative permittivity of the dielectric.

17 Calculate the number of plates used in a 0.5 nF capacitor if each plate is 40 mm square, separated by dielectric of relative permittivity 6 and thickness 0.102 mm.

18 A capacitor is to be designed to have a capacitance of 4.7 pF and to operate with a p.d. of 120 V across its terminals. The dielectric is to be Teflon ($\varepsilon_r = 2.1$) which, after allowing for a safety factor, has a dielectric strength of 25 kV/m. Calculate (a) the thickness of Teflon required and (b) the area of a plate.

19 Capacitors of 4 μF and 10 μF are connected (a) in parallel and (b) in series. Calculate the equivalent capacitance in each case.

20 Determine the equivalent capacitance when the following capacitors are connected (a) in series and (b) in parallel

i 3 μF, 4 μF and 10 μF
ii 0.02 μF, 0.05 μF and 0.22 μF
iii 20 pF and 470 pF
iv 0.01 μF and 220 pF

21 Determine the value of capacitor which when connected in series with a 2 nF capacitor produces a total capacitance of 1.6 nF.

22 Three 15 μF capacitors are connected in series across a 600 V supply. Calculate (a) the total capacitance, (b) the p.d. across each and (c) the charge on each.

23 Three capacitors, of 6 μF, 8 μF and 10 μF respectively are connected in parallel across a 60 V supply. Calculate (a) the total capacitance, (b) the charge stored in the 8 μF capacitor and (c) the total charge taken from the supply.

24 For the circuit of Figure 3.29, calculate (a) the p.d. across each capacitor and (b) the charge stored in the 3 nF.

25 Calculate the values of C_2 and C_3 shown in Figure 3.30.

26 Calculate the p.d. across, and charge stored in, each of the capacitors shown in Figure 3.31.

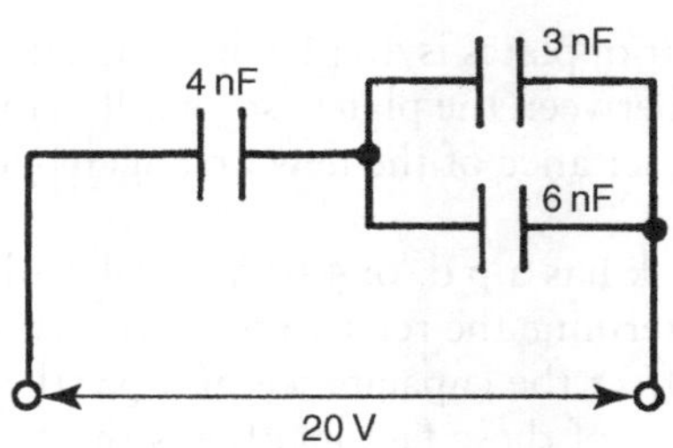

Figure 3.29 The circuit diagram for Assignment Question 24

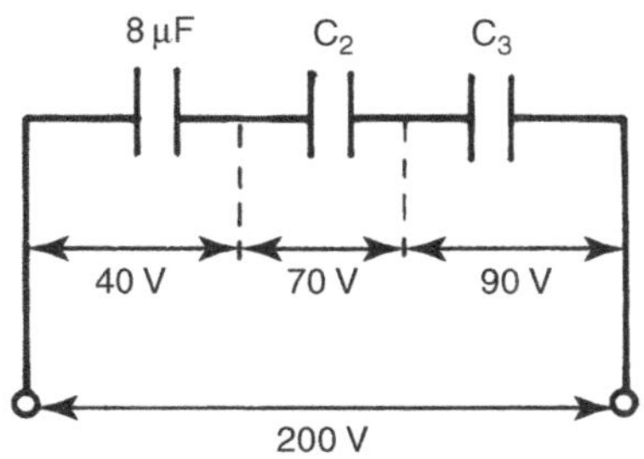

Figure 3.30 The circuit diagram for Assignment Question 25

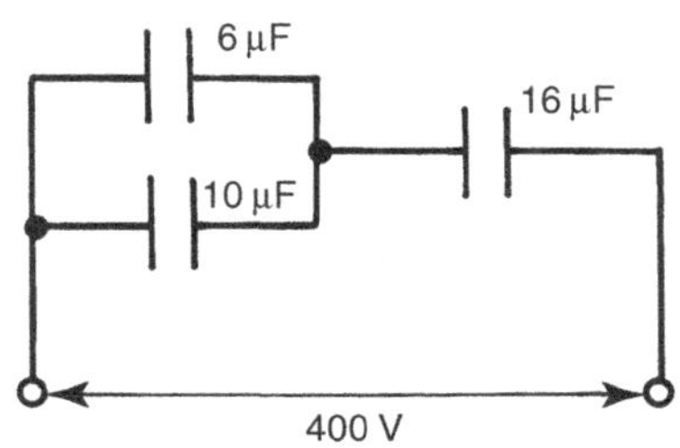

Figure 3.31 The circuit diagram for Assignment Question 26

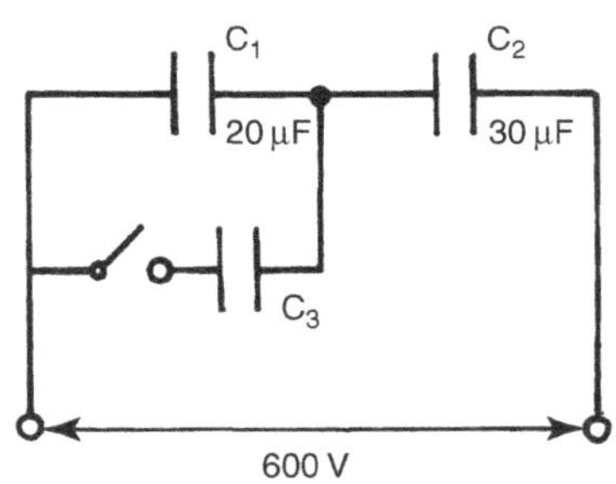

Figure 3.32 The circuit diagram for Assignment Question 27

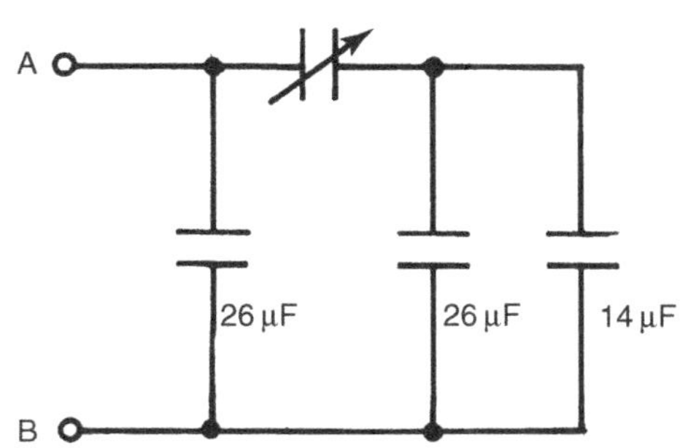

Figure 3.33 The circuit diagram for Assignment Question 28

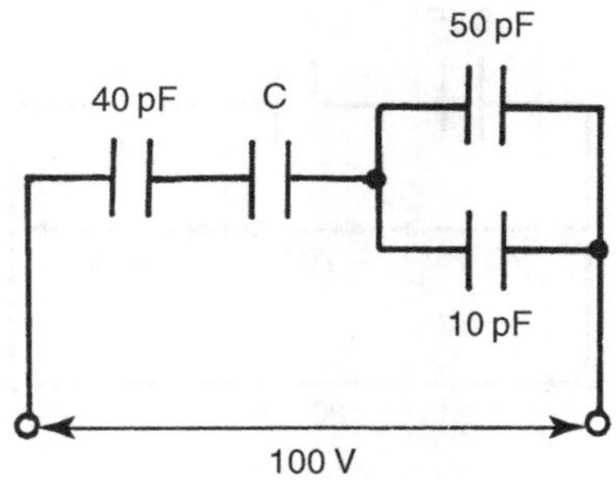

Figure 3.34 The circuit diagram for Assignment Question 30

27 A capacitor circuit is shown in Figure 3.32. With the switch in the open position, calculate the p.d.s across capacitors C_1 and C_2. When the switch is closed, the p.d. across C_2 becomes 400 V. Calculate the value of C_3.
28 In the circuit of Figure 3.33, the variable capacitor is set to 60 μF. Determine the p.d. across this capacitor if the supply voltage between terminals AB is 500 V.
29 A 50 pF capacitor is made up of two plates separated by a dielectric 2 mm thick and of relative permittivity 1.4. Calculate the effective plate area.
30 For the circuit shown in Figure 3.34 the total capacitance is 16 pF. Calculate (a) the value of the unmarked capacitor, (b) the charge on the 10 pF capacitor and (c) the p.d. across the 40 pF capacitor.
31 A 20 μF capacitor is charged to a p.d. of 250 V. Calculate the energy stored.
32 The energy stored by a 400 pF capacitor is 8 μJ. Calculate the p.d. between its plates.
33 Determine the capacitance of a capacitor that stores 4 mJ of energy when charged to a p.d. of 40 V.
34 When a capacitor is connected across a 200 V supply it takes a charge of 8 μC. Calculate (a) its capacitance, (b) the energy stored and (c) the electric field strength if the plates are 0.5 mm apart.
35 A 4 μF capacitor is charged to a p.d. of 400 V and then connected across an uncharged 2 μF capacitor. Calculate (a) the original charge and energy stored in the 4 μF and (b) the p.d. across, and energy stored in, the parallel combination.
36 Two capacitors, of 4 μF and 6 μF, are connected in series across a 250 V supply. (a) Calculate the charge and p.d. across each. (b) The capacitors are now disconnected from the supply and reconnected in parallel with each other, with terminals of similar polarity being joined together. Calculate the p.d. and charge for each.
37 A ceramic capacitor is to be made so that it has a capacitance of 100 pF and is to be operated from a 750 V supply. Allowing for a safety factor, the dielectric has a strength of 500 kV/m. Determine (a) the thickness of the ceramic, (b) the plate area if the relative permittivity of the ceramic is 3.2, (c) the charge and energy stored when the capacitor is connected to its rated supply voltage and (d) the flux density under these conditions.
38 A large electrolytic capacitor of value 100 μF has an effective plate area of 0.942 m^2. If the aluminium oxide film dielectric has a relative permittivity of 6, calculate its thickness.

SUGGESTED PRACTICAL ASSIGNMENT

Assignment 1

To determine the total capacitance of capacitors, when connected in series, and in parallel.

Apparatus

Various capacitors, of known values
1 × capacitance meter, or capacitance bridge

Method

1 Using either the meter or bridge, measure the actual value of each capacitor.
2 Connect different combinations of capacitors in parallel, and measure the total capacitance of each combination.
3 Repeat the above procedure, for various series combinations.
4 Calculate the total capacitance for each combination, and compare these values to those previously measured.
5 Account for any difference between the actual and nominal values, for the individual capacitors.

Chapter 4

Magnetic Fields and Circuits

LEARNING OUTCOMES

This chapter introduces the concepts and laws associated with magnetic fields and their application to magnetic circuits and materials.

On completion of this chapter you should be able to:

1 Describe the forces of attraction and repulsion between magnetised bodies.
2 Understand the various magnetic properties and quantities, and use them to solve simple series magnetic circuit problems.
3 Appreciate the effect of magnetic hysteresis, and the properties of different types of magnetic material.

4.1 MAGNETIC MATERIALS

All materials may be broadly classified as being in one of two groups. They may be magnetic or non-magnetic, depending upon the degree to which they exhibit magnetic effects. The vast majority of materials fall into the latter group, which may be further classified into diamagnetic and paramagnetic materials. The magnetic properties of these materials are very slight, and extremely difficult even to detect. Thus, for practical purposes, we can say that they are totally non-magnetic. The magnetic materials (based on iron, cobalt and ferrites) are the ferromagnetic materials, all of which exhibit very strong magnetic effects. It is with these materials that we will be principally concerned.

4.2 MAGNETIC FIELDS

Magnetic fields are produced by permanent magnets and by electric current flowing through a conductor. Like the electric field, a magnetic field may be considered as being the medium by which forces are transmitted and, in this case, the forces between magnetised materials. A magnetic field is also represented by lines of force or magnetic flux. These are attributed with certain characteristics, listed below:

1 They always form complete closed loops. Unlike lines of electric flux, which radiate from and terminate at the charged surfaces, lines of magnetic flux also exist all the way through the magnet.

DOI: 10.1201/9781003308294-4

2 They behave as if they are elastic. That is, when distorted they try to return to their natural shape and spacing.
3 In the space surrounding a magnet, the lines of force radiate from the north (N) pole to the south (S) pole.
4 They never intersect (cross).
5 Like poles repel and unlike poles attract each other.

Characteristics (1) and (3) are illustrated in Figure 4.1 which shows the magnetic field pattern produced by a bar magnet. Characteristics (2) and (4) are used to explain characteristic (5), as illustrated in Figures 4.2 and 4.3. In the case of the arrangement of Figure 4.2, since the lines behave as if they are elastic, then those lines linking the two magnets try to shorten themselves. This tends to bring the two magnets together.

The force of repulsion shown in Figure 4.3 is a result of the unnatural compression of the lines between the two magnets. Once more, acting as if they are elastic, these lines will expand to their normal shape. This will tend to push the magnets apart.

A permanent magnet consists of ferromagnetic material in which all magnetic dipoles permanently face the same direction. Permanent magnets have the advantage that no electrical supply is required to produce the magnetic field. However, they also have several disadvantages. The strength of the field cannot be varied. Over a period of time they tend to lose some of their magnetism (especially if subjected to physical shock or vibration). For many practical applications these disadvantages are unacceptable. Therefore a more convenient method of producing a magnetic field is required.

Permanent magnets were originally made of steel, but today there are all kinds of alloys such as Alnico that are extremely suitable for this purpose, such as alloys of iron, nickel and cobalt to which some aluminium, manganese and copper is added, or also ceramic materials such as barium oxide and iron(III) oxide. Very strong permanent magnets are nowadays made from sintered combinations with rare earths, such as samarium-cobalt (SmCo5) or neodymium-iron-boron (Nd2Fe14B).

In addition to the heating effect, an electric current also produces a magnetic field. The strength of this field is directly proportional to the value of the current. Thus a magnetic field produced in this way may be turned on and off, reversed and varied in strength very simply. A magnetic field is a vector quantity, as indicated by the arrows in the previous diagrams. The field pattern produced by a current flowing through a straight conductor is illustrated in Figure 4.4(a) and (b). Note that *conventional* current flow is considered. The convention adopted to represent conventional current flowing away from the observer is a cross, and current towards the observer is marked by a dot. The direction of the arrows on the flux lines can easily be determined by considering the X as the head of a cross-head

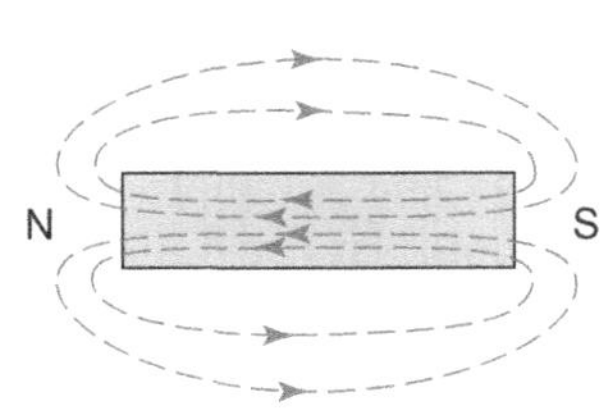

Figure 4.1 The magnetic field pattern produced by a bar magnet

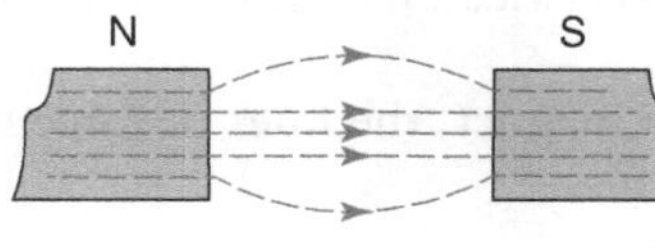

Figure 4.2 Lines attracting between two magnets

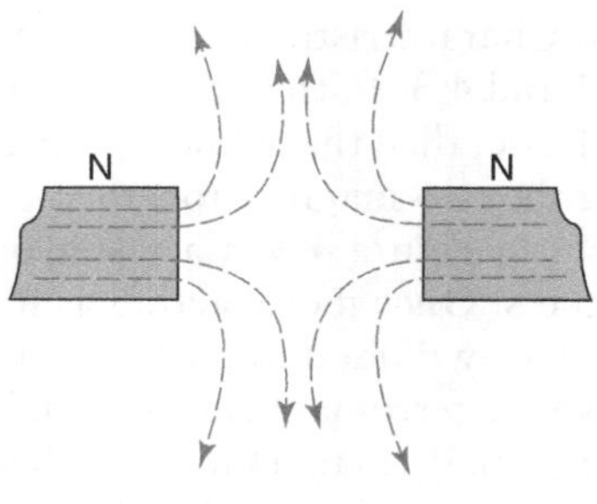

Figure 4.3 Lines repulsing between two magnets

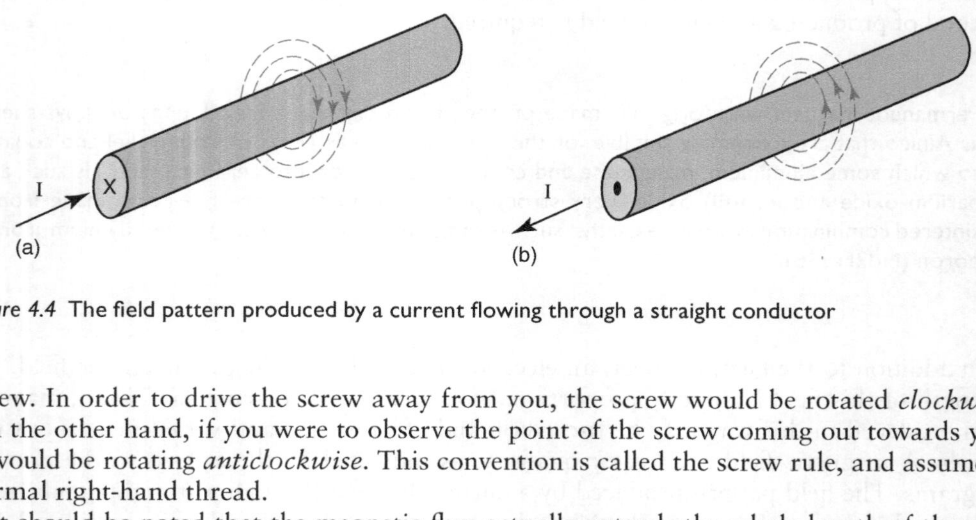

Figure 4.4 The field pattern produced by a current flowing through a straight conductor

screw. In order to drive the screw away from you, the screw would be rotated *clockwise*. On the other hand, if you were to observe the point of the screw coming out towards you, it would be rotating *anticlockwise*. This convention is called the screw rule, and assumes a normal right-hand thread.

It should be noted that the magnetic flux actually extends the whole length of the conductor, in the same way that the insulation on a cable covers the whole length. In addition, the flux pattern extends outwards in concentric circles to infinity. However, as with electric and gravitational fields, the force associated with the field follows an inverse square law. It therefore diminishes very rapidly with distance.

The flux pattern produced by a straight conductor can be adapted to provide a field pattern like a bar magnet. This is achieved by winding the conductor in the form of a coil. This arrangement is known as a solenoid. The principle is illustrated in Figure 4.5(a) and (b), which show a cross-section of a solenoid. Figure 4.5(a) shows the flux patterns produced by two adjacent turns of the coil. However, since lines of flux will not intersect, the flux distorts to form complete loops around the whole coil as shown in Figure 4.5(b).

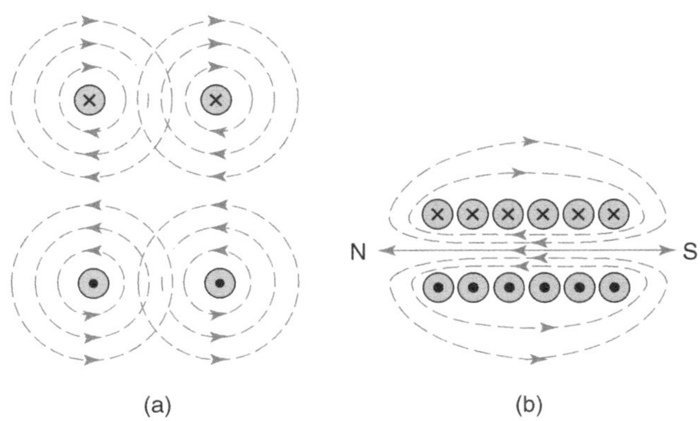

Figure 4.5 Cross-section of a solenoid

This magnetic field is sometimes used to detect cars waiting in front of red traffic lights or at the barriers at a car park. This detection can be recognised by the traces of sawn open asphalt, where a copper wire with five or six windings is placed in it. Due to the amount of metal in a car, the magnetic field will be changed and the current will change accordingly. This change is detected and causes the green light or the opening of barrier for the waiting car. If your car does not contain enough metal or if you pass by on your bicycle, you will not be detected at all. To solve this, a backpack full of metal is not really useful. However, you can change the magnetic field with some super magnets (made of an alloy of neodymium). These are used in, among other things, the brakes of a computer hard disk. But beware, they can also unintentionally (de)magnetise other objects, such as bank cards, pacemakers, etc. Another way to influence the behaviour of that coil is to make your own coil with the same size and with approximately the same number of windings and short circuit it on top of the coil embedded in asphalt. The short circuit will cause the magnetic field to behave differently and that is exactly what is being detected.

4.3 THE MAGNETIC CIRCUIT

A magnetic circuit is all of the space occupied by the magnetic flux. Figure 4.6 shows an iron-cored solenoid, supplied with direct current, and the resulting flux pattern. This is what is known as a composite magnetic circuit, since the flux exists both in the iron core and in the surrounding air space. In addition, it can be seen that the spacing of the lines within the iron core is uniform, whereas it varies in the air space. Thus there is a uniform magnetic field in the core and a non-uniform field in the rest of the magnetic circuit.

In order to make the design and analysis of a magnetic circuit easier, it is more convenient if a uniform field can be produced. This may be achieved by the use of a completely enclosed magnetic circuit. One form of such a circuit is an iron toroid that has a current-carrying coil wound round it. A toroid is a ‘doughnut’ shape having either a circular or a rectangular cross-section. Such an arrangement is shown in Figure 4.7, and from this it can be seen that only the toroid itself forms the magnetic circuit. Provided that it has a uniform cross-section then the field contained within it will be uniform.

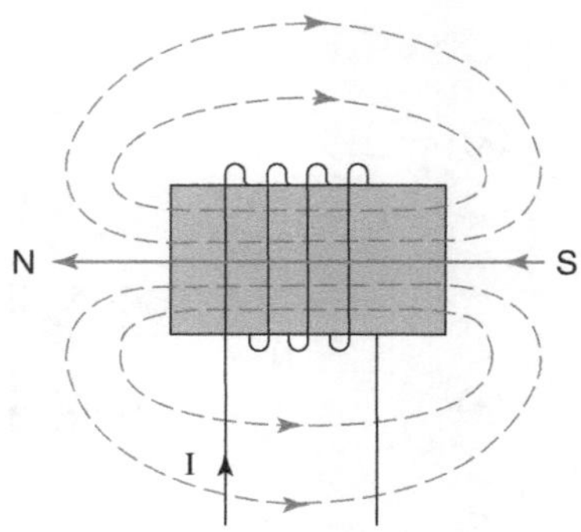

Figure 4.6 An iron-cored solenoid

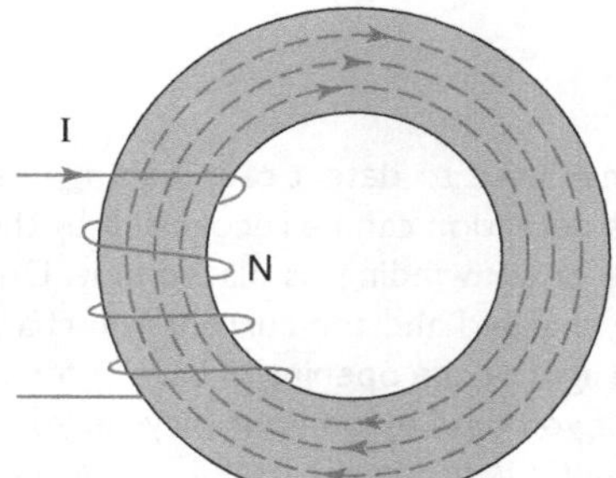

Figure 4.7 A toroid

4.4 MAGNETIC FLUX AND FLUX DENSITY

The magnetic flux is what causes the observable magnetic effects such as attraction, repulsion, etc. The unit of magnetic flux is the weber (Wb). The number of webers of flux per square metre of cross-section of the field is defined as the magnetic flux density (B), which is measured in tesla (T). This sometimes causes some confusion at first, since the logical unit would appear to be weber/metre^2. Indeed, this is the way in which it is calculated: the value of flux must be divided by the appropriate area. On reflection, it should not be particularly confusing, since the logical unit for electrical current would be coulomb/second; but it seems quite natural to use the term ampere. The quantity symbols for magnetic flux and flux density are Φ and B respectively. Hence, flux density is given by the equation and is expressed in tesla:

$$B = \frac{\Phi}{A} \tag{4.1}$$

Note: References have been made to iron as a core material and as the material used for toroids, etc. This does not necessarily mean that pure iron is used. It could be mild steel, cast iron, silicon iron, ferrite, etc. The term 'iron circuit', when used in this context, is merely a simple way in which to refer to that part of the circuit that consists of a magnetic material. It is used when some parts of the circuit may be formed from non-magnetic materials.

Wilhelm Eduard Weber (1804–1891) was a German physicist and philosopher, best known for his work in the field of magnetism. He worked closely with Gauss and conducted important research in the fields of magnetism, induction, electrical units and other phenomena. He also built the first magnetometer to measure and map the Earth's magnetic field.

Nikola Tesla (1856–1943) was an inventor, electrical engineer and physicist. He is known as the inventor of the alternating current generator, the alternating current electric motor and other important components of the current electrical grid. In rudimentary form, these devices were mostly developed by others before Tesla's time. Tesla's credit was that he developed the alternating current principle much further and greatly improved or designed almost all the necessary devices for an alternating current-based reliable power grid.

WORKED EXAMPLE 4.1

Q The pole face of a magnet is 3 cm by 2 cm and it produces a flux of 30 μWb. Calculate the flux density at the pole face.

$$A = 3 \times 2 \times 10^{-4}\ \text{m}^2;\ \Phi = 30 \times 10^{-6}\ \text{Wb}$$

$$B = \frac{\Phi}{A} = \frac{30 \times 10^{-6}}{6 \times 10^{-4}} = 50\ \text{mT}$$

WORKED EXAMPLE 4.2

Q A magnetic field of density 0.6 T has an effective cross-sectional area of $45 \times 10^{-6}\ \text{m}^2$. Determine the flux.

$$B = 0.6\ \text{T};\ A = 45 \times 10^{-6}\ \text{m}^2$$

$$B = \frac{\Phi}{A},\ \text{then}\ \Phi = BA$$

$$\Phi = 0.6 \times 45 \times 10^{-6} = 27\ \mu\text{Wb}$$

4.5 MAGNETOMOTIVE FORCE (MMF)

In an electric circuit, any current that flows is due to the existence of an emf. Similarly, in a magnetic circuit, the magnetic flux is due to the existence of a magnetomotive force (shortened to mmf). The concept of an mmf for permanent magnets is a difficult one. Fortunately it is simple when we consider the flux being produced by current flowing through a coil. This is the case for most practical magnetic circuits.

In Section 4.2, we saw that each turn of the coil made a contribution to the total flux produced, so the flux must be directly proportional to the number of turns on the coil. The flux is also directly proportional to the value of current passed through the coil. Putting these two facts together we can say that the mmf is the product of the current and the number of turns. The quantity symbol for mmf is F (the same as for mechanical force). The number of turns is just a number and therefore dimensionless. The SI unit for mmf is therefore simply ampere. However, this tends to cause considerable confusion to students new to the subject. For this reason, throughout *this book*, the unit will be quoted as ampere turns (At).

$$\text{mmf, } F = NI \tag{4.2}$$

WORKED EXAMPLE 4.3

Q A 1500-turn coil is uniformly wound around an iron toroid of uniform cross-sectional area of 5 cm². Calculate the mmf and flux density produced, if the resulting flux is 0.2 mWb when the coil current is 0.75 A.

$N = 1500$; $A = 5 \times 10^{-4}$ m²; $\Phi = 0.2 \times 10^{-3}$ Wb; $I = 0.75$ A

$F = NI = 1500 \times 0.75 = 1125$ At

$$B = \frac{\Phi}{A} = \frac{0.2 \times 10^{-3}}{5 \times 10^{-4}} = 0.4 \text{ T}$$

WORKED EXAMPLE 4.4

Q Calculate the excitation current required in a 600 turn coil in order to produce an mmf of 1500 At.

$N = 600$; $F = 1500$ At

$F = NI$, then $I = \dfrac{F}{N}$

$$I = \frac{1500}{600} = 2.5 \text{ A}$$

4.6 MAGNETIC FIELD STRENGTH

The magnetic field strength is the magnetic equivalent to electric field strength in electrostatics. It was found that electric field strength is the same as potential gradient, and is measured in volt/metre. Now, the volt is the unit of emf, and we have just seen that mmf and emf are comparable quantities, i.e. mmf can be considered as the magnetic circuit equivalent of electric potential. Hence magnetic field strength is defined as the mmf per metre length of the magnetic circuit. The quantity symbol for magnetic field strength is H, the unit of measurement being ampere turn/metre.

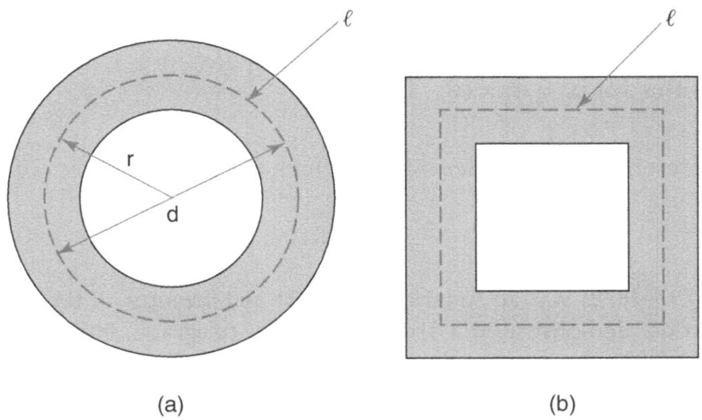

Figure 4.8 A circular and a rectangular toroid

$$H = \frac{F}{\ell} = \frac{NI}{\ell} \quad (4.3)$$

where ℓ is the *mean* or average length of the magnetic circuit. Thus, if the circuit consists of a circular toroid, the mean length is the mean circumference. This point is illustrated in Figure 4.8(a) and (b).

WORKED EXAMPLE 4.5

Q A current of 400 mA is passed through a 550 turn coil, wound on a toroid of mean diameter 8 cm. Calculate the magnetic field strength.

$I = 0.4$ A; $N = 550$; $d = 8 \times 10^{-2}$ m

$\ell = \pi d = \pi \times 8 \times 10^{-2}$ m $= 0.251$ m

$$H = \frac{NI}{\ell} = \frac{550 \times 0.4}{0.251} = 875.35 \text{ At/m}$$

4.7 PERMEABILITY

We have seen in electrostatics that the permittivity of the dielectric is a measure of the 'willingness' of the dielectric to allow an electric field to exist in it. In magnetic circuits the corresponding quantity is the permeability of the material.

If the magnetic field exists in a vacuum, then the ratio of the flux density to the magnetic field strength is a constant, called the permeability of free space and expressed in henry/metre:

$$\mu_0 = \frac{B}{H} \quad (4.4)$$

compare this to $\varepsilon_0 = \frac{D}{\boldsymbol{E}}$, expressed in farad/metre

The value for $\mu_0 = 4\pi \times 10^{-7}$ H/m

μ_0 is used as the reference or datum level from which the permeabilities of all other materials are measured.

Joseph Henry (1797–1878) was an American physicist and inventor of the relay. He described the law of induction, demonstrated a prototype of the telegraph, introduced the first linear electric motor based on the concept devised by Michael Faraday and invented the relay, which switches on and off a circuit in a safe and reliable way by using an electromagnet.

Consider an air-cored solenoid with a fixed value of current flowing it. The mmf will produce a certain flux density in this air core. If an iron core was now inserted, it would be found that the flux density would be very much increased. To account for these different results for different core materials, a quantity known as the relative permeability μ_r is used. This is defined as the ratio of the flux density produced in the iron, to that produced in the air, for the same applied mmf.

$$\text{i.e. } \mu_r = \frac{B_2}{B_1} \tag{4.5}$$

where B_2 is the flux density produced in the iron and B_1 is the flux density produced in the air.

Compare this to the equation $\varepsilon_r = \frac{C_2}{C_1}$ used in electrostatics. As with ε_r, μ_r has no units, since it is simply a ratio.

Note: For air or any other *non-magnetic* material, $\mu_r = 1$. In other words, all non-magnetic materials have the same magnetic properties as a vacuum.

The absolute permeability of a material is the ratio of the flux density to magnetic field strength, for a given mmf expressed in henry/metre as follows.

$$\text{Thus, } \mu = \frac{B}{H} \tag{4.6}$$

but since μ_0 is the reference value, then $\mu = \mu_0\mu_r$, which can be compared to the equation $\varepsilon = \varepsilon_0\varepsilon_r$. Therefore,

$$\mu_0\mu_r = \frac{B}{H} \text{ so, } B = \mu_0\mu_r H \tag{4.7}$$

This equation compares directly with $D = \varepsilon_0\varepsilon_r E$ coulomb/m^2.

WORKED EXAMPLE 4.6

Q A solenoid with a core of cross-sectional area of 15 cm^2 and relative permeability 65 produces a flux of 200 μWb. If the core material is changed to one of relative permeability 800, what will be the new flux and flux density?

$$A = 15 \times 10^{-4}\ \text{m}^2;\ \mu_{r1} = 65;\ \Phi_1 = 2 \times 10^{-4}\ \text{Wb};\ \mu_{r2} = 800$$

$$B_1 = \frac{\Phi_1}{A} = \frac{2 \times 10^{-4}}{15 \times 10^{-4}} = 0.133\ \text{T}$$

Now, the original core is 65 times more effective than air. The second core is 800 times more effective than air. Therefore, we can say that the second core will produce a greater flux density. The ratio of the two flux densities will be 800/65 = 12.31:1. Thus the second core will result in a flux density 12.31 times greater than produced by the first core.

$$B_2 = 12.31\,B_1 = 12.31 \times 0.133 = 1.641\ \text{T}$$
$$\Phi_2 = B_2\,A = 1.641 \times 15 \times 10^{-4} = 2.462\ \text{mWb}$$

WORKED EXAMPLE 4.7

Q A toroid of mean radius 40 mm, effective cross-sectional area of 3 cm^2 and relative permeability 150, is wound with a 900 turn coil that carries a current of 1.5 A. Calculate (a) the mmf, (b) the magnetic field strength and (c) the flux and flux density.

$$r = 0.04\ \text{m};\ A = 3 \times 10^{-4}\ \text{m}^2;\ \mu_r = 150;\ N = 900;\ I = 1.5\ \text{A}$$

(a) $F = NI = 900 \times 1.5 = 1350\ \text{At}$

(b) $H = \dfrac{F}{\ell} = \dfrac{1350}{2\pi \times 0.041} = 5371.5\ \text{At/m}$

(c)
$$B = \mu_0 \mu_r H = 4\pi \times 10^{-7} \times 150 \times 5371.5 = 1.0125\ \text{T}$$
$$\Phi = BA = 1.0125 \times 3 \times 10^{-4}$$
$$\Phi = 303.75\ \mu\text{Wb}$$

WORKED EXAMPLE 4.8

Q A steel toroid of the dimensions shown in Figure 4.9 is wound with a 500 turn coil of wire. What value of current needs to be passed through this coil in order to produce a flux of 250 μWb in the toroid, if under these conditions the relative permeability of the toroid is 300?

$$r = 3 \times 10^{-2}\ \text{m};\ A = 4.5 \times 10^{-4}\ \text{m}^2;\ N = 500;\ \Phi = 250 \times 10^{-6}\ \text{Wb};\ \mu_r = 300$$

Effective length of the toroid, $l = 2\pi r = 2\pi \times 3 \times 10^{-2}\ \text{m} = 0.188\ \text{m}$

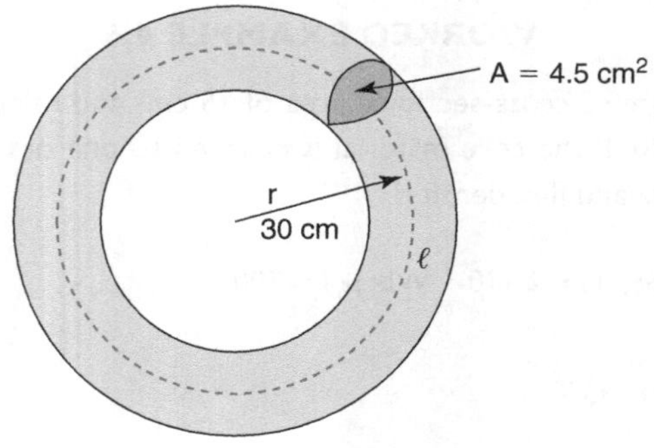

Figure 4.9 The circuit diagram for Worked Example 4.8

$$B = \frac{\Phi}{A} = \frac{250 \times 10^{-6}}{4.5 \times 10^{-4}} = 0.556 \text{ T}$$

Now, $B = \mu_0 \mu_r H$

$$H = \frac{B}{\mu_0 \mu_r} = \frac{0.556}{4\pi \times 10^{-7} \times 300} = 1474$$

$$F = H\ell = 1474 \frac{\text{At}}{\text{m}} \times 0.188 = 277 \text{ At}$$

$$I = \frac{F}{N} = \frac{277}{500}$$

$$I = 0.55 \text{ A}$$

WORKED EXAMPLE 4.9

Q A coil is made by winding a single layer of 0.5 mm diameter wire onto a cylindrical wooden dowel, which is 5 cm long and has a cross-sectional area of 7 cm². When a current of 0.2 A is passed through the coil, calculate (a) the mmf produced, (b) the flux density and (c) the flux produced.

(a) $I = 0.2\text{A};\ \ell = 5 \times 10^{-2} \text{ m};\ A = 7 \times 10^{-4} \text{ m}^2;\ d = 0.5 \times 10^{-3} \text{ m};\ \mu_r = 1(\text{wood})$

Since $F = NI$, we first need to calculate the number of turns of wire on the coil. Consider Figure 4.10 which represents the coil wound onto the dowel. From Figure 4.10 it may be seen that the number of turns may be obtained by dividing the length of the dowel by the diameter (thickness) of the wire.

$$N = \frac{\ell}{d} = \frac{50 \times 10^{-2}}{0.5 \times 10^{-3}}$$

$$N = 100$$

$$F = 100 \times 0.2 = 20 \text{ At}$$

(b) $B = \dfrac{\Phi}{A}$ or $B = \mu_0 \mu_r H$

but since we do not yet know the value for the flux, but can calculate the value for H, then the second equation needs to be used.

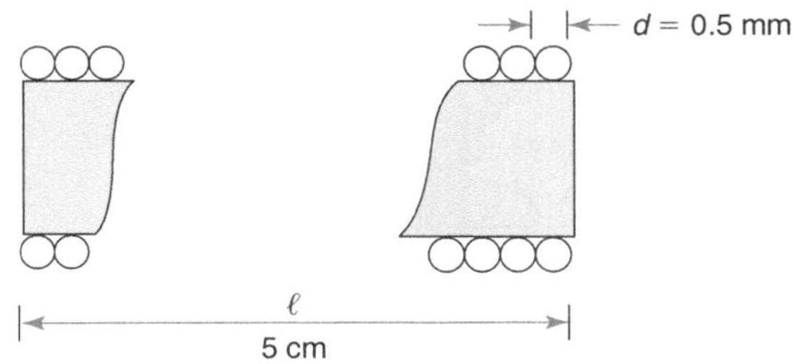

Figure 4.10 The circuit diagram for Worked Example 4.9

$$H = \frac{F}{\ell} = \frac{20}{5 \times 10^{-2}} \quad H = 400 \text{ At/m}$$

$$B = 4\pi \times 10^{-7} \times 1 \times 400 = 5.026 \times 10^{-4} \text{ T} = 503\ \mu\text{T}$$

(c) $\Phi = BA = 503 \times 10^{-6} \times 7 \times 10^{-4} = 0.352\ \mu\text{Wb}$

4.8 MAGNETISATION (*B*/*H*) CURVE

A magnetisation curve is a graph of the flux density produced in a magnetic circuit as the magnetic field strength is varied. Since $H = NI/\ell$, then for a given magnetic circuit, the field strength may be varied by varying the current through the coil. If the magnetic circuit consists entirely of air, or any other non-magnetic material, the resulting graph will be a straight line passing through the origin. The reason for this is that since $\mu_r = 1$ for all non-magnetic materials, the ratio B/H remains constant.

Unfortunately, the relative permeability of magnetic materials does not remain constant for all values of applied field strength, which results in a curved graph. This non-linearity is due to an effect known as magnetic saturation. The complete explanation of this effect is beyond the scope of this book, but a much simplified version of this is afforded by Ewing's molecular theory. This states that each molecule in a magnetic material may be considered as a minute magnet in its own right. When the material is unmagnetised, these molecular magnets are orientated in a completely random fashion. Thus, the material has no overall magnetic polarisation. This is similar to a conductor in which the free electrons are drifting in a random manner. Thus, when no emf is applied, no current flows. This random orientation of the molecular magnets is illustrated in Figure 4.11 where the arrows represent the north poles. However, as the coil magnetisation current is slowly increased, so the molecular magnets start to rotate towards a particular orientation. This results in a certain degree of polarisation of the material, as shown in Figure 4.12. As the coil current continues to be increased, so the molecular magnets continue to become more aligned. Eventually, the coil current will be sufficient to produce complete alignment. This means that the flux will have reached its maximum possible value. Further increase of the current will produce no further increase of flux. The material is then said to have reached magnetic saturation, as illustrated in Figure 4.13.

Typical magnetisation curves for air and a magnetic material are shown in Figure 4.14. Note that the flux density produced for a given value of H is very much greater in the

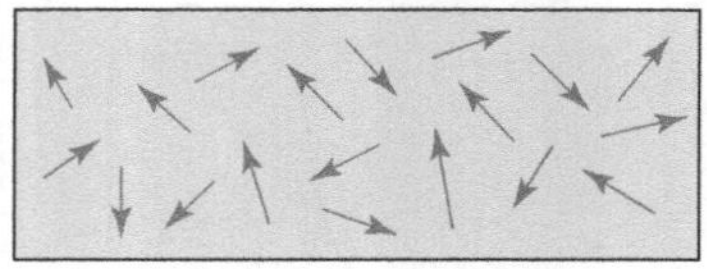

Figure 4.11 Random orientation of the molecular magnets

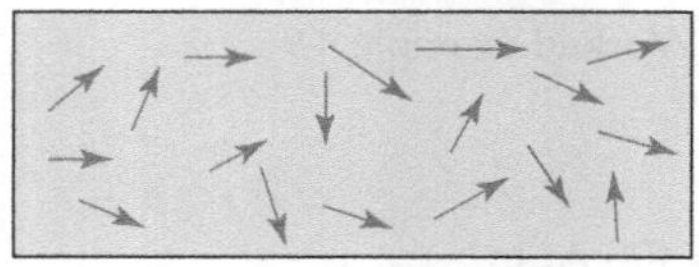

Figure 4.12 Random orientation of the molecular magnets, partly magnetised

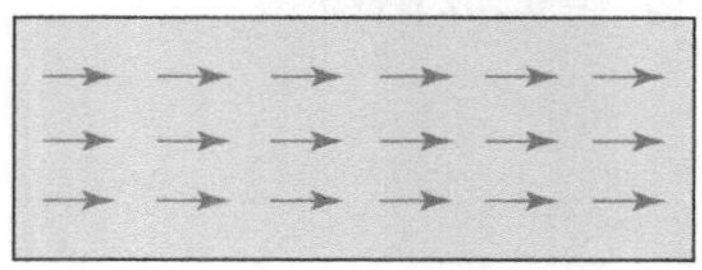

Figure 4.13 Random orientation of the molecular magnets, saturated

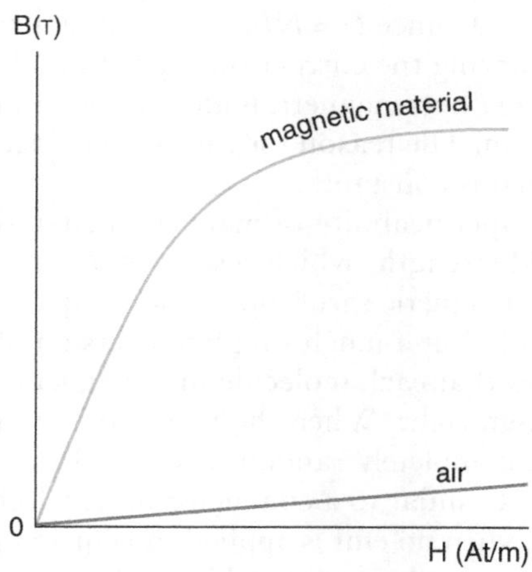

Figure 4.14 Magnetisation curves for air and magnetic material

magnetic material. The slope of the graph is $B/H = \mu_0\mu_r$, and this slope varies. Since μ_0 is a constant, then the value of μ_r for the magnetic material must vary as the slope of the graph varies.

The variation of μ_r with variation of H may be obtained from the B/H curve, and the resulting μ_r/H graph is shown in Figure 4.15. The magnetisation curves for a range of magnetic materials are given in Figure 4.16.

For a practical magnetic circuit, a single value for μ_r cannot be specified unless it is quoted for a specified value of B or H. Thus B/H data must be available. These may be presented

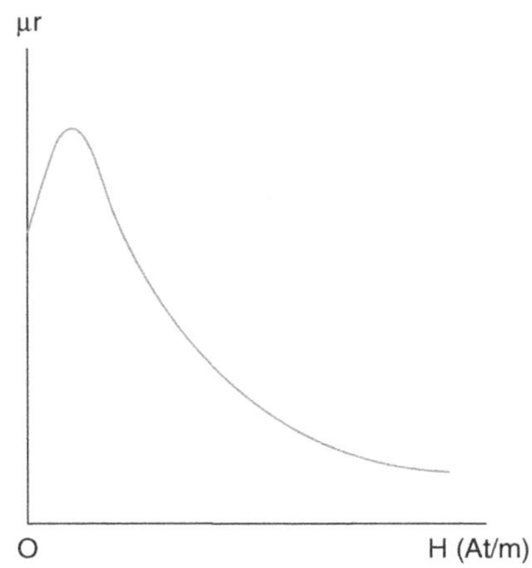

Figure 4.15 μr as function of H

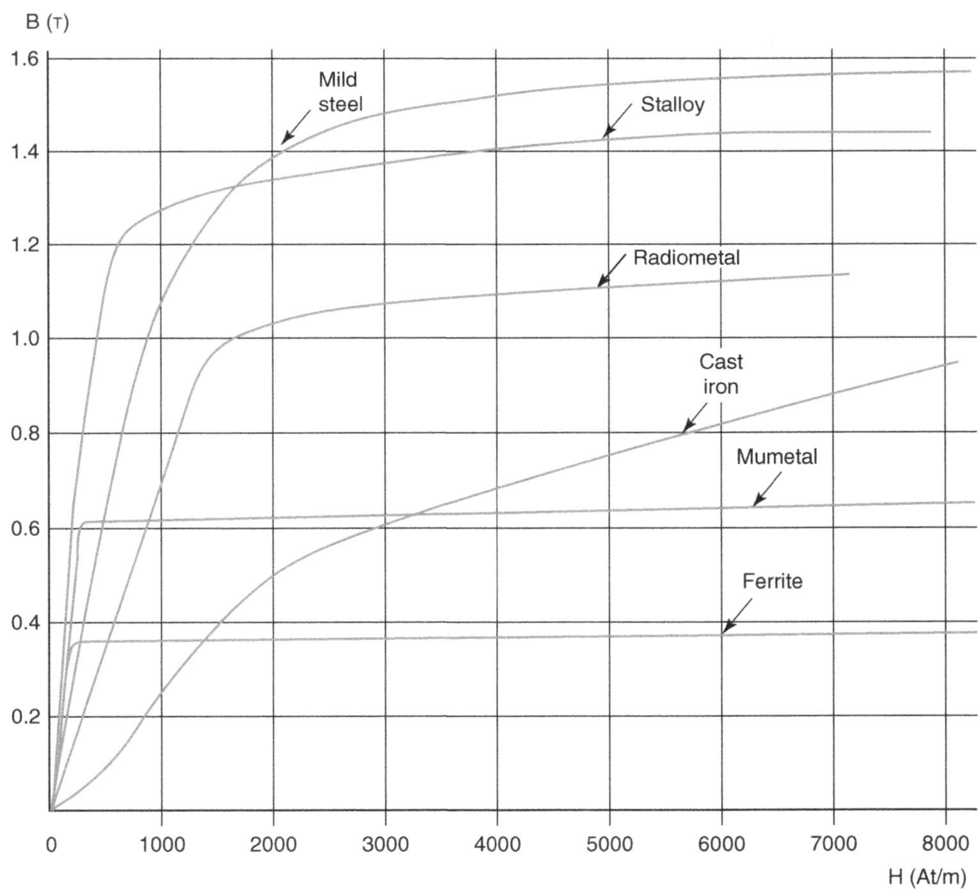

Figure 4.16 Magnetisation curves for a range of magnetic materials

either in the form of a graph as in Figure 4.14, or in the form of tabulated data, from which the relevant section of the *B/H* curve may be plotted.

WORKED EXAMPLE 4.10

Q An iron toroid having a mean radius of 0.1 m and cross-sectional area of π cm^2 is wound with a 1000-turn coil. The coil current results in a flux of 0.1775 mWb in the toroid. Using the following data, determine (a) the coil current and (b) the relative permeability of the toroid under these conditions.

H (At/m)	80	85	90	95	100
B(T)	0.50	0.55	0.58	0.59	0.6

The first step in the solution of the problem is to plot the section of *B/H* graph from the given data.

Note: This *must* be plotted as accurately as possible on *graph* paper. The values used in this example have been obtained from such a graph.

$$B = \frac{\Phi}{A} = \frac{0.1775 \times 10^{-3}}{\pi \times 10^{-4}} = 0.565 \text{ T}$$

and from the plotted graph, when $B = 0.565$ T, $H = 88$ At/m.

(a) Now, the length of the toroid, $\ell = 2\pi r = 0.27\ \pi$m

$$H = \frac{NI}{\ell}$$

$$I = \frac{H\ell}{N} = \frac{88 \times 0.2\ \pi \text{ m}}{1000} = 55.3 \text{ mA}$$

$$B = \mu_0 \mu_r H$$

(b) $$\mu_r = \frac{B}{\mu_0 H} = \frac{0.565}{4\pi \times 10^{-7} \times 88} \quad \mu_r = 5109$$

WORKED EXAMPLE 4.11

Q A cast iron toroid of mean length 15 cm is wound with a 2500-turn coil, through which a magnetising current of 0.3 A is passed. Calculate the resulting flux density and relative permeability of the toroid under these conditions.

Since *B/H* data are necessary for the solution, but none have been quoted, the *B/H* curve for cast iron shown in Figure 4.16 will be used.

$\ell = 0.15$ m; $N = 2500$; $I = 0.3$ A

$$H = \frac{NI}{\ell} = \frac{2500 \times 0.3}{0.15} = 5000 \text{ At/m}$$

and from the graph for cast iron in Figure 4.16, the corresponding flux density is

$$B = 0.75\ \text{T}$$
$$B = \mu_0\mu_r H$$
$$\mu_r = \frac{B}{\mu_0 H} = \frac{0.75}{4\pi \times 10^{-7} \times 5000} = 119.4$$

4.9 COMPOSITE SERIES MAGNETIC CIRCUITS

Most practical magnetic circuits consist of more than one material in series. This may be deliberate, as in the case of an electric motor or generator, where there have to be air gaps between the stationary and rotating parts. Sometimes an air gap may not be required, but the method of construction results in small but unavoidable gaps. In other circumstances it may be a requirement that two or more different magnetic materials form a single magnetic circuit. Let us consider the case where an air gap is deliberately introduced into a magnetic circuit, for example, making a sawcut through a toroid, at right angles to the flux path.

WORKED EXAMPLE 4.12

Q A mild steel toroid of mean length 18.75 cm and cross-sectional area of 0.8cm^2 is wound with a 750 turn coil. (a) Calculate the coil current required to produce a flux of 112 μWb in the toroid. (b) If a 0.5 mm sawcut is now made across the toroid, calculate the coil current required to maintain the flux at its original value.

$l = 0.1875$ m; $A = 8 \times 10^{-5}$ m^2; $N = 750$;

$\Phi = 112 \times 10^{-6}$ Wb; $\ell_{gap} = 0.5 \times 10^{-3}$ m

(a) $$B = \frac{\Phi}{A} = \frac{112 \times 10^{-6}}{8 \times 10^{-5}} = 1.4\ \text{T}$$

From the graph for mild steel in Figure 4.16, the corresponding value for H is 2000 At/m

$$F_{Fe} = Hl = 2000 \times 0.1875 = 375\ \text{At}$$

Fe is the chemical symbol for iron. In Worked Example 4.12 the mmf required to produce the flux in the 'iron' part of the circuit has been referred to as F_{Fe}. This will distinguish it from the mmf required for the air gap which is shown as F_{gap}

$$I = \frac{F_{Fe}}{N} = \frac{375}{750}$$
$$I = 0.5\ \text{A}$$

(b) When the air gap is introduced into the steel the effective length of the steel circuit changes by only 0.27%. This is a negligible amount, so the values obtained in part (a) above for

H and F_{Fe} remain unchanged. However, the introduction of the air gap will produce a considerable reduction of the circuit flux. Thus we need to calculate the extra mmf, and hence current, required to restore the flux to its original value. Since the relative permeability for air is a constant (=1), a B/H graph is not required. The cross-sectional area of the gap is the same as that for the steel, and the same flux exists in it. Thus, the flux density in the gap must also be the same as that calculated in part (a) above. Hence the value of H required to maintain this flux density in the gap can be calculated from:

$$B = \mu_0 H_{gap}$$

$$H_{gap} = \frac{B}{\mu_0} = \frac{1.4}{4\pi \times 10^{-7}}$$

$$H_{gap} = 1.11 \times 10^6 \text{ At/m}$$

also, since $F_{gap} = H_{gap}\ell_{gap}$

$$F_{gap} = 1.11 \times 10^6 \times 0.5 \times 10^{-3} = 557 \text{ At}$$

Total circuit mmf, $F = F_{Fe} + F_{gap} = 375\ At + 557 \text{ At} = 932 \text{ At}$

$$I = \frac{F}{N} = \frac{932}{750}$$

$$I = 1.243 \text{ A}$$

WORKED EXAMPLE 4.13

Q A magnetic circuit consists of two stalloy sections A and B as shown in Figure 4.17. The mean length and cross-sectional area for A are 25 cm and 11.5 cm^2, whilst the corresponding values for B are 15 cm and 12 cm^2 respectively. A 1000-turn coil wound on section A produces a circuit flux of 1.5 mWb. Calculate the coil current required.

$\ell_A = 0.25$ m; $A_A = 11.5 \times 10^{-4}$ m^2; $\ell_B = 0.15$ m

$A_B = 12 \times 10^{-4}$ m^2; $\Phi = 1.5 \times 10^{-3}$ Wb; $N = 1000$

$$B_A = \frac{\Phi}{A_A} \quad \text{and} \quad B_B = \frac{\Phi}{A_B}$$

$$= \frac{1.5 \times 10^{-3}}{11.5 \times 10^{-4}} = 1.3 \text{ T} \qquad = \frac{1.5 \times 10^{-3}}{12 \times 10^{-4}} = 1.25 \text{ T}$$

From the B/H curve for stalloy in Figure 4.16, the corresponding H values are:

$$H_A = 1470 \text{ At/m} \quad \text{and} \quad H_B = 845 \text{ At/m}$$

$$F_A = H_A l_A = 1470 \times 0.25 = 367.5 \text{ At} \quad \text{and} \quad F_B = H_B l_B = 845 \times 0.15 = 126.75 \text{ At}$$

total circuit mmf, $F = F_A + F_B$

$$F = 494.25 \text{ At}$$

$$I = \frac{F}{N} = \frac{494.25}{1000} = 0.494 \text{ A}$$

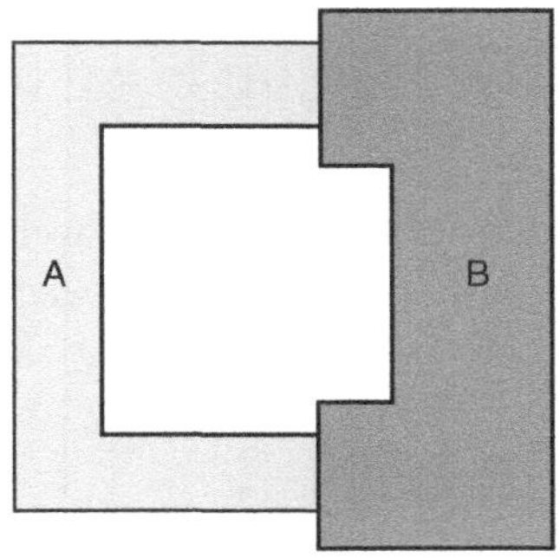

Figure 4.17 The circuit diagram for Worked Example 4.13

From the last two examples it should now be apparent that in a series magnetic circuit the only quantity that is common to both (all) sections is the magnetic flux Φ. This common flux is produced by the current flowing through the coil, i.e. the total circuit mmf F. Also, if the lengths, cross-sectional area and/or the materials are different for the sections, then their flux densities and H values must be different. For these reasons it is not legitimate to add together the individual H values. It *is* correct, however, to add together the individual mmfs to obtain the total circuit mmf F. This technique is equivalent to adding together the p.d.s across resistors connected in series in an electrical circuit. The sum of these p.d.s then gives the value of emf required to maintain a certain current through the circuit.

For example, if a current of 4 A is to be maintained through two resistors of 10 Ω and 20 Ω connected in series, then the p.d.s would be 40 V and 80 V respectively. Thus, the emf required would be 120 V.

4.10 RELUCTANCE (*S*)

Comparisons have already been made between the electric circuit and the magnetic circuit. We have compared mmf to emf; current to flux; and potential gradient to magnetic field strength. A further comparison may be made, as follows.

The resistance of an electric circuit limits the current that can flow for a given applied emf. Similarly, in a magnetic circuit, the flux produced by a given mmf is limited by the reluctance of the circuit. Thus, the reluctance of a magnetic circuit is the opposition it offers to the existence of a magnetic flux within it.

Current is a movement of electrons around an electric circuit. A magnetic flux merely *exists* in a magnetic circuit; it does not involve a flow of particles. However, both current and flux are the direct result of some form of applied force.

$$\text{In an electric circuit, current} = \frac{\text{emf}}{\text{resistance}}$$

$$\text{so in a magnetic circuit, flux} = \frac{\text{mmf}}{\text{reluctance}}$$

Thus, $\Phi = \dfrac{F}{S} = \dfrac{NI}{S}$

Reluctance $S = \dfrac{F}{\Phi} = \dfrac{NI}{\Phi}$ and expressed in amp turn/weber (4.8)

but $NI = Hl$

$$S = \frac{Hl}{\Phi}$$

$$\Phi = BA$$

$$\frac{H}{B} = \frac{1}{\mu} = \frac{1}{\mu_0\mu_r}$$

then $S = \dfrac{\ell}{\mu_0\mu_r A}$ (4.9)

Let us continue the comparison between series electrical circuits and series magnetic circuits. We know that the total resistance in the electrical circuit is obtained simply by adding together the resistor values. The same technique may be used in magnetic circuits, such that the total reluctance of a series magnetic circuit, S is given by

$$S = S_1 + S_2 + S_3 + \cdots \quad (4.10)$$

Assume that the physical dimensions of the sections, and the relative permeabilities (for the given operating conditions) of each section are known. In this case, equations (4.9), (4.10) and (4.8) enable an alternative form of solution.

WORKED EXAMPLE 4.14

Q An iron ring of cross-sectional area of 8 cm² and mean diameter 24 cm contains an air gap of 3 mm wide. It is required to produce a flux of 1.2 mWb in the air gap. Calculate the mmf required, given that the relative permeability of the iron is 1200 under these operating conditions.

$A_{Fe} = A_{gap} = 8\times10^{-4}\ \text{m}^2;\ \Phi = 1.2\times10^{-3};\ \mu_r = 1200;\quad \ell_{gap} = 3\times10^{-3}\ \text{m};\ \ell_{Fe} = 0.24\times\pi\text{m}$

For the iron circuit: $S_{Fe} = \dfrac{\ell_{Fe}}{\mu_0\mu_r A} = \dfrac{0.24\times\pi\ \text{m}}{4\pi\times10^{-7}\times1200\times8\times10^{-4}} = 6.25\times10^5$ At/Wb

For the air gap: $S_{gap} = \dfrac{\ell_{gap}}{\mu_0\mu_r A} = \dfrac{3\times10^{-3}}{4\pi\times10^{-7}\times1\times8\times10^{-4}} = 2.984\times10^6$ At/Wb

Total circuit reluctance, $S = S_{Fe}+S_{gap} = 6.25\times10^5 + 2.984\times10^6 = 3.61\times10^6$ At/Wb

Since $\Phi = \dfrac{F}{S}$

$F = \Phi S$ (compare to $V = IR$)

$F = 4331$ At

The above example illustrates the quite dramatic increase of circuit reluctance produced by even a very small air gap. In this example, the air gap length is only 0.4% of the total circuit length. Yet its reluctance is almost five times greater than that of the iron section. For this reason, the design of a magnetic circuit should be such as to try to minimise any unavoidable air gaps.

WORKED EXAMPLE 4.15

Q A magnetic circuit consists of three sections, the data for which is given below. Calculate (a) the circuit reluctance and (b) the current required in a 500 turn coil, wound onto section 1, to produce a flux of 2mWb.

Section	*Length (cm)*	*Cross-sectional area (cm²)*	μ_r
1	85	10	600
2	65	15	950
3	0.1	12.5	1

(a) $$S_1 = \frac{\ell_1}{\mu_0\mu_r A_1} = \frac{0.85\text{ m}}{4\pi\times10^{-7}\times600\times10^{-3}} = 1.127\times10^6 \text{ At/Wb}$$

$$S_2 = \frac{\ell_2}{\mu_0\mu_r A_2} = \frac{0.65}{4\pi\times10^{-7}\times950\times15\times10^{-4}} = 3.63\times10^5 \text{ At/Wb}$$

$$S_3 = \frac{\ell_3}{\mu_0\mu_r A_3} = \frac{10^{-3}}{4\pi\times10^{-7}\times1\times12.5\times10^{-4}} = 6.37\times10^5 \text{ At/Wb}$$

Total reluctance, $S = S_1 + S_2 + S_3 = 1.27\times10^6 + 3.63\times10^5 + 6.37\times10^5 = 2.13\times10^6$ At/Wb

(b) $$F = \Phi S = 2\times10^{-3}\text{ Wb}\times2.13\times10^6\text{ At/Wb}$$

$$F = 4254\text{ At}$$

$$I = \frac{F}{N} = \frac{4254}{500} = 8.51\text{ A}$$

Magnetic flux, like most other things in nature, tends to take the easiest path available. For flux this means the lowest reluctance path. This is illustrated in Figure 4.18. The reluctance of the soft iron bar is very much less than the surrounding air. For this reason, the flux will opt to distort from its normal pattern, and make use of this lower reluctance path.

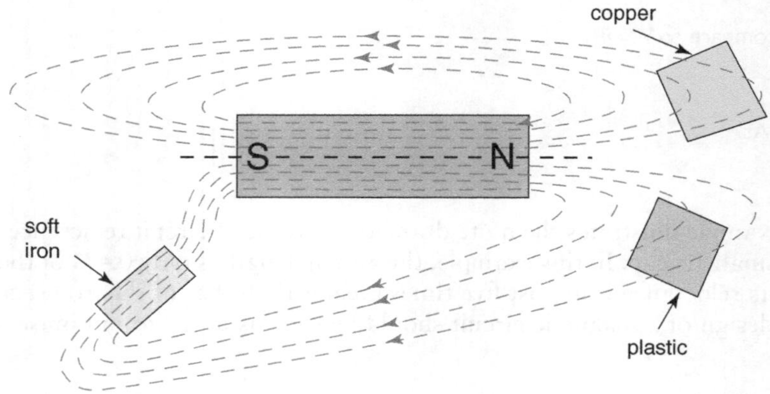

Figure 4.18 Magnetic flux taking the lowest reluctance path

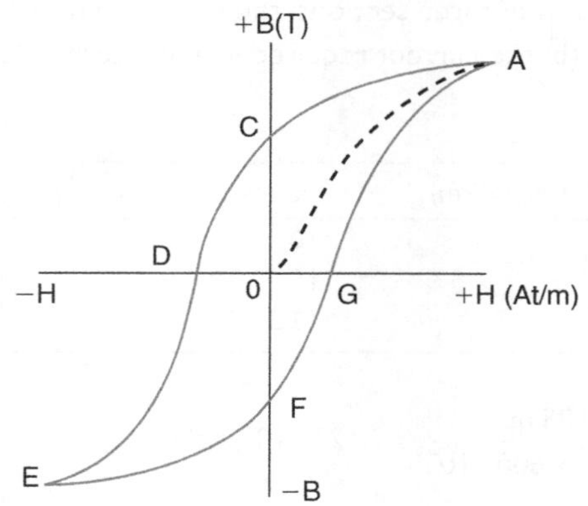

Figure 4.19 Magnetic hysteresis

4.11 MAGNETIC HYSTERESIS

Hysteresis comes from a Greek word meaning 'to lag behind'. It is found that when the magnetic field strength in a magnetic material is varied, the resulting flux density displays a lagging effect.

Consider such a specimen of magnetic material that initially is completely unmagnetised. If no current flows through the magnetising coil then both H and B will initially be zero. The value of H is now increased by increasing the coil current in discrete steps. The corresponding flux density is then noted at each step. If these values are plotted on a graph until magnetic saturation is achieved, the dotted curve (the initial magnetisation curve) shown in Figure 4.19 results.

Let the current now be reduced (in steps) to zero, and the corresponding values for B again noted and plotted. This would result in the section of graph from A to C. This shows that when the current is zero once more (so $H = 0$), the flux density has not reduced to zero. The flux density remaining is called the remanent flux density (OC). This property of a magnetic material, to retain some flux after the magnetising current is removed, is known as the remanence or retentivity of the material.

Let the current now be reversed, and increased in the opposite direction. This will have the effect of opposing the residual flux. Hence, the latter will be reduced until at some value of –H it reaches zero (point D on the graph). The amount of reverse magnetic field strength required to reduce the residual flux to zero is known as the coercive force. This property of a material is called its coercivity.

If we now continue to increase the current in this reverse direction, the material will once more reach saturation (at point E). In this case it will be of the opposite polarity to that achieved at point A on the graph.

Once again, the current may be reduced to zero, reversed and then increased in the original direction. This will take the graph from point E back to A, passing through points F and G on the way. Note that residual flux density shown as OC has the same value, but opposite polarity, to that shown as OF. Similarly, coercive force OD = OG.

In taking the specimen through the loop ACDEFGA we have taken it through one complete magnetisation cycle. The loop is referred to as the hysteresis loop. The degree to which a material is magnetised depends upon the extent to which the 'molecular magnets' have been aligned. Thus, in taking the specimen through a magnetisation cycle, energy must be expended. This energy is proportional to the area enclosed by the loop, and the rate (frequency) at which the cycle is repeated.

Magnetic materials may be subdivided into what are known as 'hard' and 'soft' magnetic materials. A hard magnetic material is one which possesses a large remanence and coercivity. It is therefore one which retains most of its magnetism, when the magnetising current is removed. It is also difficult to demagnetise. These are the materials used to form permanent magnets, and they will have a very 'fat' loop as illustrated in Figure 4.20(a).

A soft magnetic material, such as soft iron and mild steel, retains very little of the induced magnetism. It will therefore have a relatively 'thin' hysteresis loop, as shown in Figure 4.20(b). The soft magnetic materials are the ones used most often for engineering applications. Examples are the magnetic circuits for rotating electric machines (motors and generators), relays and the cores for inductors and transformers.

When a magnetic circuit is subjected to continuous cycling through the loop a considerable amount of energy is dissipated. This energy appears as heat in the material. Since this is normally an undesirable effect, the energy thus dissipated is called the hysteresis loss. Thus, the thinner the loop, the less wasted energy. This is why 'soft' magnetic materials are used for the applications listed above.

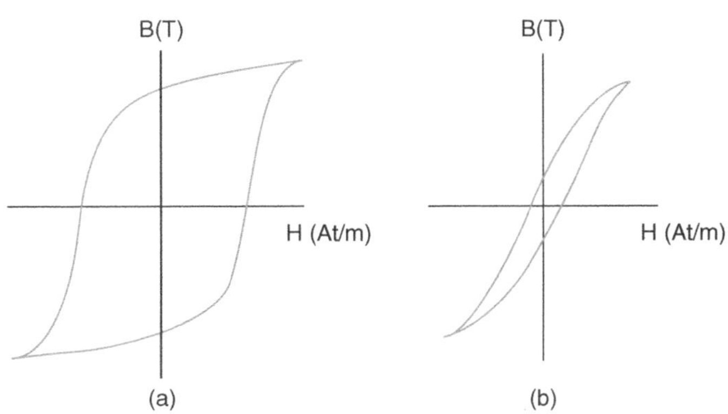

Figure 4.20 Magnetic hysteresis: permanent magnets versus soft magnets

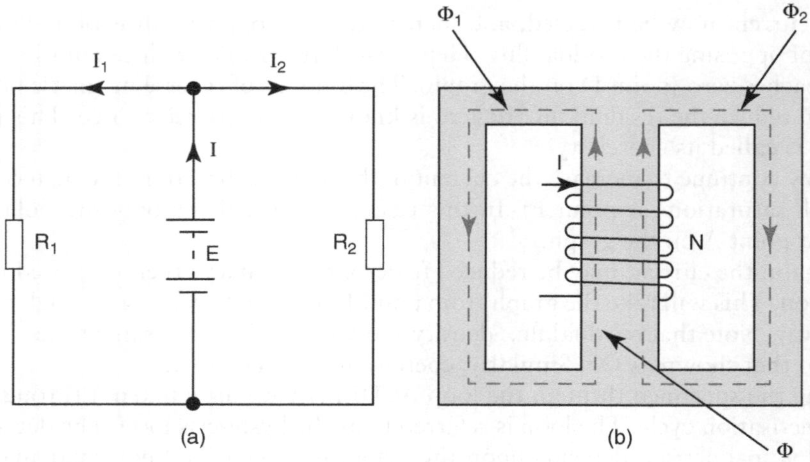

Figure 4.21 Parallel magnetic circuits

4.12 PARALLEL MAGNETIC CIRCUITS

We have seen that the magnetic circuit may be treated in much the same manner as its electrical circuit equivalent. The same is true for parallel circuits in the two systems. Two equivalent circuits are shown in Figure 4.21, and from this the following points emerge:

1. In the electrical circuit, the current supplied by the source of emf splits between the two outer branches according to the resistances offered. In the magnetic circuit, the flux produced by the mmf splits between the outer limbs according to the reluctances offered.
2. If the two resistors in the outer branches are identical, the current splits equally. Similarly, if the reluctances of the outer limbs are the same then the flux splits equally between them.

However, a note of caution. In the electric circuit it has been assumed that the source of emf is ideal (no internal resistance) and that the connecting wires have no resistance. The latter assumption cannot be applied to the magnetic circuit. All three limbs will have a value of reluctance that *must* be taken into account when calculating the total circuit reluctance.

SUMMARY OF EQUATIONS

Magnetic flux density: $B = \dfrac{\Phi}{A} = \mu_0\mu_r H$

Magnetomotive force (mmf): $F = NI = \Phi S$

Magnetic field strength: $H = \dfrac{F}{\ell} = \dfrac{NI}{\ell}$

Permeability: $\mu = \mu_0\mu_r \dfrac{B}{H}$

Reluctance: $S = \dfrac{\ell}{\mu_0\mu_{rA}} = \dfrac{NI}{\Phi} = \dfrac{F}{\Phi}$

Series magnetic circuit: $S = S_1 + S_2 + S_3 + \cdots$

Also listed below are some comparable equations.

$$E = IR;\ F = \mu S$$

$$J = \frac{I}{A};\ B = \frac{\Phi}{A};\ D = \frac{Q}{A}$$

$$R = R_1 + R_2 + \cdots;\ S = S_1 + S_2 + \cdots$$

$$\text{potential gradient} = \frac{E}{d};\ H = \frac{F}{\ell}$$

$$E = \frac{V}{d}$$

$$B = \mu_0 \mu_r H;\ D = \varepsilon_0 \varepsilon_r E$$

$$\mu_r = \frac{B_2}{B_1};\ \varepsilon_r = \frac{D_2}{D_1} \text{ or } \frac{C_2}{C_1}$$

$\mu_r = 1$ for all non-magnetic materials; $\varepsilon_r = 1$ air only

$$\mu_0 = 4\pi \times \frac{10^{-7}\ \text{H}}{m}$$

$$\varepsilon_r = 8.854 \times 10^{-12}\ \text{F/m}$$

Note: Although the concept of current density has not been covered previously, it may be seen from Table 4.1 that it is simply the value of current flowing through a conductor divided by the cross-sectional area of the conductor.

Table 4.1 Comparison of quantities

Electrical			*Magnetic*			*Electrostatic*		
Quantity	*Symbol*	*Unit*	*Quantity*	*Symbol*	*Unit*	*Quantity*	*Symbol*	*Unit*
emf	E	V	mmf	F	At	emf	E	V
current	I	A	flux	Φ	Wb	flux	Q	C
resistance	R	Ω	reluctance	S	At/Wb	resistance	R	Ω
resistivity	ρ	Ωm	permeability	μ	H/m	permittivity	ε	F/m
potential gradient	–	V/m	field strength	H	At/m	field strength	**E**	V/m
current density	J	A/m²	flux density	B	T	flux density	D	C/m²

ASSIGNMENT QUESTIONS

1 The pole faces of a magnet are 4 cm × 3 cm and produce a flux of 0.5 mWb. Calculate the flux density.

2 A flux density of 1.8 T exists in an air gap of effective cross-sectional area of 11 cm². Calculate the value of the flux.

3 If a flux of 5 mWb has a density of 1.25 T, determine the cross-sectional area of the field.

4 A magnetising coil of 850 turns carries a current of 25 mA. Determine the resulting mmf.

5 It is required to produce an mmf of 1200 At from a 1500-turn coil. What will be the required current?

6 A current of 2.5 A when flowing through a coil produces an mmf of 675 At. Calculate the number of turns on the coil.

7 A toroid has an mmf of 845 At applied to it. If the mean length of the toroid is 15 cm, determine the resulting magnetic field strength.

8 A magnetic field strength of 2500 At/m exists in a magnetic circuit of mean length 45 mm. Calculate the value of the applied mmf.

9 Calculate the current required in a 500 turn coil to produce an electric field strength of 4000 At/m in an iron circuit of mean length 25 cm.

10 A 400 turn coil is wound onto an iron toroid of mean length 18 cm and uniform cross-sectional area of 4.5 cm^2. If a coil current of 2.25 A results in a flux of 0.5 mWb, determine (a) the mmf, (b) the flux density and (c) the magnetic field strength.

11 An air-cored coil contains a flux density of 25 mT. When an iron core is inserted the flux density is increased to 1.6 T. Calculate the relative permeability of the iron under these conditions.

12 A magnetic circuit of mean diameter 12 cm has an applied mmf of 275 At. If the resulting flux density is 0.8 T, calculate the relative permeability of the circuit under these conditions.

13 A toroid of mean radius 35 mm, effective cross-sectional area of 4 cm^2 and relative permeability 200 is wound with a 1000-turn coil that carries a current of 1.2 A. Calculate (a) the mmf, (b) the magnetic field strength, (c) the flux density and (d) the flux in the toroid.

14 A magnetic circuit of square cross-section 1.5 cm × 1.5 cm and mean length 20 cm is wound with a 500 turn coil. Given the *B*/*H* data below, determine (a) the coil current required to produce a flux of 258.8 μWb and (b) the relative permeability of the circuit under these conditions.

B(T)	0.9	1.1	1.2	1.3
H(At/m)	250	450	600	825

15 For the circuit of Question 14 above, a 1.5 mm sawcut is made through it. Calculate the current now required to maintain the flux at its original value.

16 A cast steel toroid has the following *B*/*H* data. Complete the data table for the corresponding values of μ_r and hence plot the μ_r/H graph, and (a) from your graph determine the values of magnetic field strength at which the relative permeability of the steel is 520, and (b) the value of relative permeability when H = 1200 At/m.

B(T)	0.15	0.35	0.74	1.05	1.25	1.39
H(At/m)	250	500	1000	1500	2000	2500
μ_r						

17 A magnetic circuit made of radiometal is subjected to a magnetic field strength of 5000 At/m. Using the data given in Figure 4.16, determine the relative permeability under this condition.

18 A magnetic circuit consists of two sections as shown in Figure 4.22. Section 1 is made of mild steel and is wound with a 100 turn coil. Section 2 is made from cast iron.

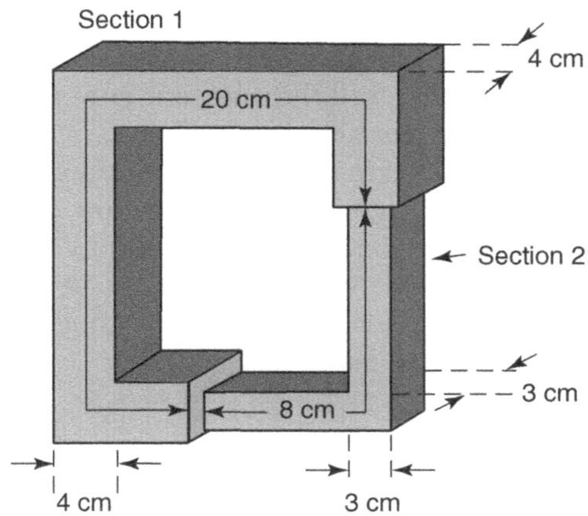

Figure 4.22 The circuit diagram for Assignment Question 18

Calculate the coil current required to produce a flux of 0.72 mWb in the circuit. Use the *B*/*H* data given in Figure 4.16.

19 A circular toroid of mean diameter 25 cm and cross-sectional area of 4 cm^2 has a 1.5 mm air gap in it. The toroid is wound with a 1200-turn coil and carries a flux of 0.48 mWb. If, under these conditions, the relative permeability of the toroid is 800, calculate the coil current required.

20 A closed magnetic circuit made from silicon steel consists of two sections, connected in series. One is of effective length 42 mm and cross-sectional area of 85 mm^2, and the other of length 17 mm and cross-sectional area of 65 mm^2. A 50 turn coil is wound on to the second section and carries a current of 0.4 A. Determine the flux density in the 17 mm length section if the relative permeability of the silicon iron under this condition is 3000.

21 A magnetic circuit of cross-sectional area of 0.45 cm^2 consists of one part 4 cm long and μ_r of 1200; and a second part 3 cm long and μ_r of 750. A 100 turn coil is wound onto the first part and a current of 1.5 A is passed through it. Calculate the flux produced in the circuit.

SUGGESTED PRACTICAL ASSIGNMENTS

Note: These assignments are qualitative in nature.

Assignment 1

To compare the effectiveness of different magnetic core materials.

Apparatus

1 × coil of wire of known number of turns
1 × d.c. psu
1 × ammeter
1 × set of laboratory weights
1 × set of different ferromagnetic cores, suitable for the coil used

Method

1 Connect the circuit as shown in Figure 4.23.
2 Adjust the coil current carefully until the magnetic core ***just*** holds the smallest weight in place. Note the value of current and weight.
3 Using larger weights, in turn, increase the coil current until each weight is just held by the core. Record all values of weight and corresponding current.
4 Repeat the above procedure for the other core materials.
5 Tabulate all results. Calculate and tabulate the force of attraction and mmf in each case.
6 Write an assignment report, commenting on your findings, and comparing the relative effectiveness of the different core materials.

Assignment 2

To plot a magnetisation curve, and initial section of a hysteresis loop, for a magnetic circuit.

Apparatus

1 × magnetic circuit of known length, and containing a coil(s) of known number of turns
1 × variable d.c. psu
1 × Hall effect probe
1 × ammeter
1 × voltmeter

Method

1 Ensure that the core is completely demagnetised before starting.
2 Zero the Hall probe, monitoring its output with the voltmeter.
3 Connect the circuit as in Figure 4.24.

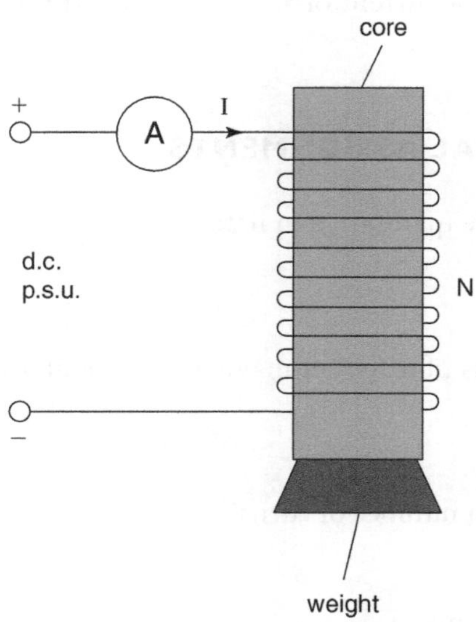

Figure 4.23 The circuit diagram for Practical Assignment 1

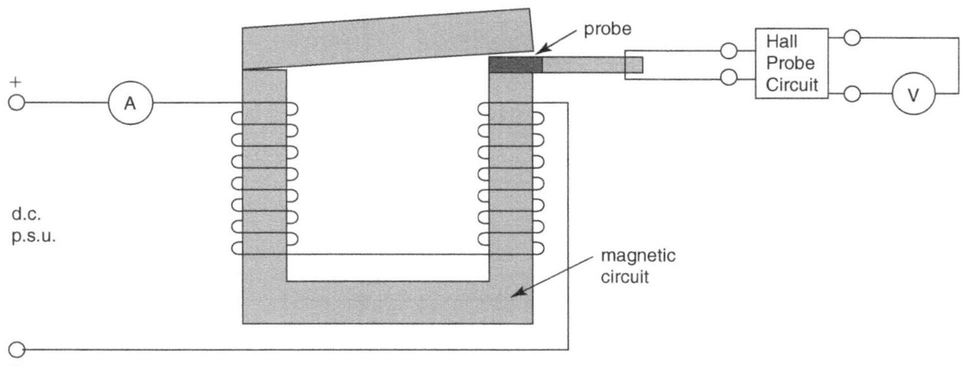

Figure 4.24 The circuit diagram for Practical Assignment 2

4 Increase the coil current in 0.1 A steps, up to 2 A. Record the voltmeter reading at each step.

Note: If you 'overshoot' a desired current setting, *do not* then reduce the current back to that setting. Record the value actually set, together with the corresponding voltmeter reading.

5 Reduce the current from 2 A to zero, in 0.1 A steps. Once more, if you overshoot a desired current setting, *do not* attempt to correct it.
6 Reverse the connections to the psu, and increase the reversed current in small steps until the voltmeter indicates zero.

Note: The Hall effect probe output (as measured by the voltmeter) *represents* the flux density in the core. The magnetic field strength, H, may be calculated from NI/ℓ.

7 Plot a graph of voltmeter readings (B) versus H.
8 Submit a full assignment report.

Chapter 5

Electromagnetism

LEARNING OUTCOMES

This chapter concerns the principles and laws governing electromagnetic induction and the concepts of self and mutual inductance.

On completion of this chapter you should be able to use these principles to:

1 Understand the basic operating principles of motors and generators.
2 Carry out simple calculations involving the generation of voltage, and the production of force and torque.
3 Appreciate the significance of eddy current loss.
4 Determine the value of inductors, and apply the concepts of self and mutual inductance to the operating principles of transformers.
5 Calculate the energy stored in a magnetic field.
6 Explain the principle of the moving coil metre and carry out simple calculations for the instrument.

5.1 FARADAY'S LAW OF ELECTROMAGNETIC INDUCTION

It is mainly due to the pioneering work of Michael Faraday, in the nineteenth century, that the modern technological world exists as we know it. Without the development of the generation of electrical power, such advances would have been impossible. Thus, although the concepts involved with electromagnetic induction are very simple, they have far-reaching influence. Faraday's law is best considered in two interrelated parts:

1 The value of emf induced in a circuit or conductor is directly proportional to the rate of change of magnetic flux linking with it.
2 The polarity of such an emf, induced by an increasing flux, is opposite to that induced by a decreasing flux.

The key to electromagnetic induction is contained in part 1 of the law quoted above. Here, the words 'rate of change' are used. If there is no change in flux, or the way in which this flux links with a conductor, then no emf will be induced. The proof of the law can be very simply demonstrated. Consider a coil of wire, a permanent bar magnet and a galvanometer as illustrated in Figures 5.1 and 5.2.

DOI: 10.1201/9781003308294-5

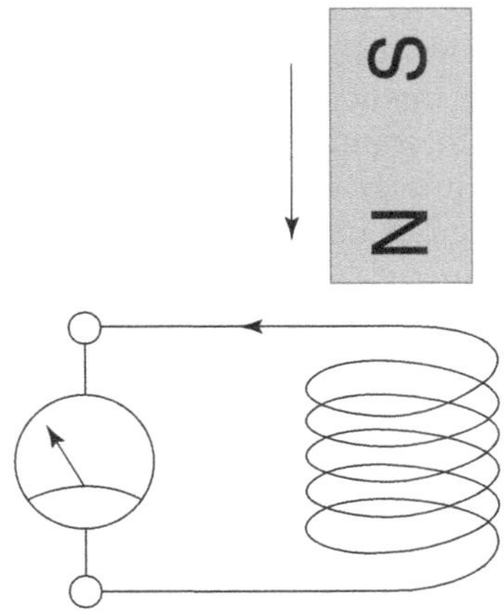

Figure 5.1 A coil of wire, a permanent bar magnet and a galvanometer with movement in one direction.

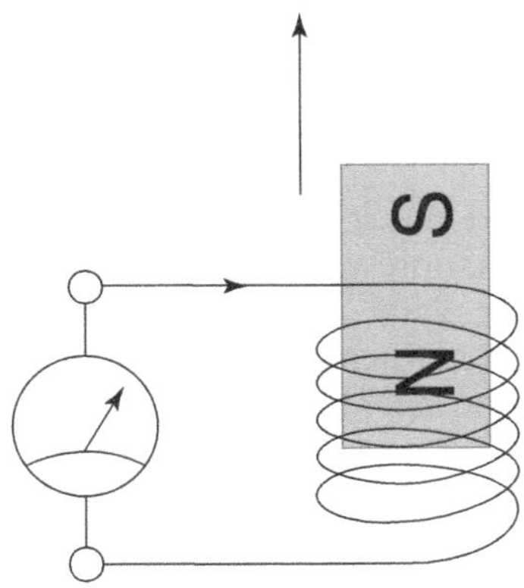

Figure 5.2 A coil of wire, a permanent bar magnet and a galvanometer with movement in another direction.

Consider the magnet being moved so that it enters the centre of the coil. When this is done it will be seen that the pointer of the galvo deflects in one direction. This deflection of the pointer is only momentary, since it only occurs whilst the magnet is moving. The galvo is of course a current-measuring device. However, any current flowing through it must be due to a voltage between its terminals. Since there is no other source of emf in the circuit, then it must be concluded that an emf has been induced or created in the coil itself. The resulting current indicated by the galvo depends on the value of this emf. It will also be observed that when the magnet is stationary (either inside or outside the coil) the galvo does not deflect. Hence, emf is induced into the coil only when the magnet is in motion.

When the magnet is withdrawn from the coil, the galvo will again be seen to deflect momentarily. This time, the deflection will be in the opposite direction. Provided that the magnet is removed at the same rate as it was inserted, then the magnitudes of the deflections will be the same. The polarities of the induced emfs will be opposite to each other, since the current flow is reversed. Thus far, we have confirmation that an emf is induced in the coil when a magnetic flux is moving relative to it. We also have confirmation of part 2 of the law.

In order to deduce the relationship between the value of induced emf and the rate of change of flux, the magnet needs to be moved at different speeds into and out of the coil.

When this is done, and the resulting magnitudes of the galvo deflection noted, it will be found that the faster the movement, the greater the induced emf.

This simple experiment can be further extended in three ways. If the magnet is replaced by a more powerful one, it will be found that for the same speed of movement, the corresponding emf will be greater. Similarly, if the coil is replaced with one having more turns, then for a given magnet and speed of movement, the value of the emf will again be found to be greater. Finally, if the magnet is held stationary within the coil, and the coil is then moved away, it will be found that an emf is once more induced in the coil. In this last case, it will also be found that the emf has the same polarity as obtained when the magnet was first inserted into the stationary coil. This last effect illustrates the point that it is the *relative movement* between the coil and the flux that induces the emf.

The experimental procedure described above is purely qualitative. However, if it was refined and performed under controlled conditions, it would yield the following results: the magnitude of the induced emf is directly proportional to the value of magnetic flux, the rate at which this flux links with the coil and the number of turns on the coil. Expressed as an equation we have:

$$e = \frac{-N\mathrm{d}\Phi}{\mathrm{d}t} \tag{5.1}$$

Notes:

1 The symbol for the induced emf is shown as a lower-case letter e and is expressed in volt. This is because it is only present for the short interval of time during which there is relative movement taking place, and so has only a momentary value.
2 The term $\mathrm{d}\Phi/\mathrm{d}t$ is simply a mathematical means of stating 'the rate of change of flux with time'. The combination $N\Phi/\mathrm{d}t$ is often referred to as the 'rate of change of flux linkages'.
3 Equation (5.1) forms the basis for the definition of the unit of magnetic flux, the weber, thus: the weber is that magnetic flux which, linking a circuit of one turn, induces in it an emf of 1 volt when the flux is reduced to zero at a uniform rate in 1 second.

In other words, 1 volt = 1 weber/second or 1 weber = 1 volt second.

4 The minus sign is a reminder that Lenz's law applies. This law states that the polarity of an induced emf is always such that it opposes the change which produced it. This is similar to the statement in mechanics, and that for every force there is an opposite reaction.

Heinrich Lenz (1804–1865) was a Baltic-German physicist and studied electromagnetism, by repeating and carefully expanding experiments by Michael Faraday. In addition to the law named after him, Lenz also independently discovered Joule's law.

5.2 FLEMING'S RIGHT-HAND RULE

This is a convenient means of determining the polarity of an induced emf in a conductor. Also, provided that the conductor forms part of a complete circuit, it will indicate the direction of the resulting current flow.

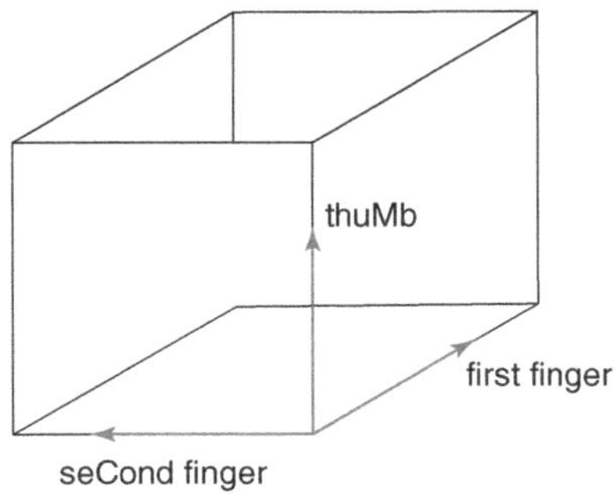

Figure 5.3 Fleming's right-hand rule: the cube

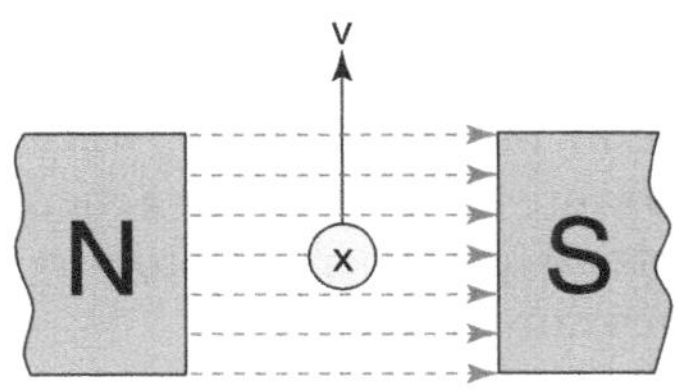

Figure 5.4 Fleming's right-hand rule: the cross-section of a conductor

The first finger, the second finger and the thumb of the *right* hand are held out mutually at right angles to each other (like the three edges of a cube as shown in Figure 5.3). The ***F***irst finger indicates the direction of the ***F***lux, the thu***M***b the direction of ***M***otion of the conductor relative to the flux, and the s***EC***ond finger indicates the polarity of the induced ***E***mf and the direction of ***C***urrent flow. This process is illustrated in Figure 5.4, which shows the cross-section of a conductor being moved vertically upwards at a constant velocity through the magnetic field.

Note: The thumb indicates the direction of motion of the *conductor relative to* the flux. Thus, the same result would be obtained from the arrangement of Figure 5.4 if the conductor was kept stationary and the magnetic field was moved down.

WORKED EXAMPLE 5.1

Q The flux linking a 100-turn coil changes from 5 mWb to 15 mWb in a time of 2 ms. Calculate the average emf induced in the coil; see Figure 5.5.

$$N = 100;\ d\Phi = (15 - 5) \times 10^{-3}\ \text{Wb};\ dt = 2 \times 10^{-3}\ \text{s}$$

$$e = \frac{-N d\Phi}{dt} = \frac{-100 \times (15 - 5) \times 10^{-3}}{2 \times 10^{-3}} = \frac{-100 \times 10 \times 10^{-3}}{2 \times 10^{-3}} = -500\ \text{V}$$

Note that if the flux was *reduced* from 15 mWb to 5 mWb, then the term shown in brackets above would be –10. The resulting emf would be +500 V. When quoting Equation (5.1),

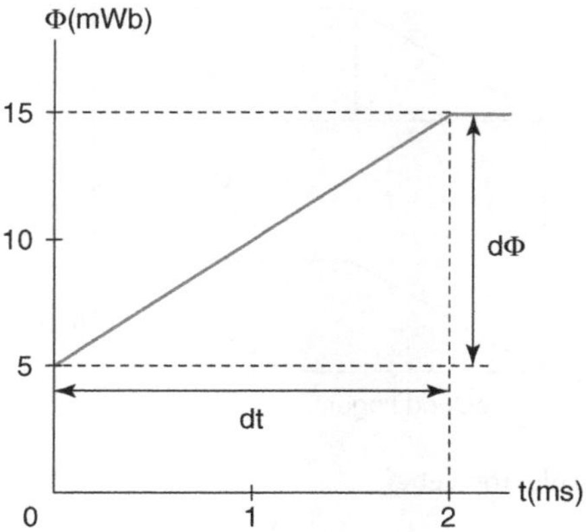

Figure 5.5 The flux as function of the time for Worked Example 5.1

the minus sign should always be included. However, since it is often the magnitude of the induced emf that is more important, it is normal practice to ignore the minus sign in the subsequent calculation. One of the major exceptions to this practice arises when considering the principles of operation of the transformer.

WORKED EXAMPLE 5.2

Q A 250-turn coil is linked by a magnetic flux that varies as follows: an increase from zero to 20 mWb in a time of 0.05 s; constant at this value for 0.02 s; followed by a decrease to 4 mWb in a time of 0.01 s. Assuming that these changes are uniform, draw a sketch graph (i.e. not to an accurate scale) of the variation of the flux and the corresponding emf induced in the coil, showing all principal values.

Firstly, the values of induced emf must be calculated for those periods when the flux changes.

$$d\Phi_1 = (20 - 0) \times 10^{-3} \text{ Wb; } dt_1 = 0.05 \text{ s}$$
$$d\Phi_3 = (4 - 20) \times 10^{-3} \text{ Wb; } dt_3 = 0.01 \text{ s}$$

$$e_1 = \frac{-Nd\Phi_1}{dt_1} = \frac{-250 \times 20 \times 10^{-3}}{0.05} = -100 \text{ V}$$

$$e_3 = \frac{-Nd\Phi_3}{dt_3} = \frac{-250 \times (-16) \times 10^{-3}}{0.01} = 400 \text{ V}$$

The resulting sketch graph is shown in Figure 5.6.

WORKED EXAMPLE 5.3

Q A coil when linked by a flux which changes at the rate of 0.1 Wb/s has induced in it an emf of 100 V. Determine the number of turns on the coil.

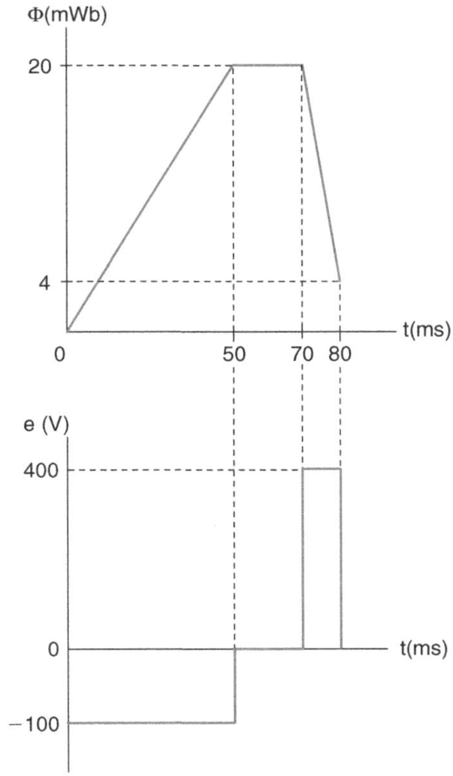

Figure 5.6 The flux and the emf as function of the time for Worked Example 5.2

$$e = 100\ \text{V};\ \frac{d\Phi}{dt} = 0.1\ \text{Wb/s}$$

$$e = \frac{-Nd\Phi}{dt}$$

$$N = \frac{e}{d\Phi / dt} = \frac{100}{0.1} = 1000\ \text{turns}$$

Note that the minus sign has been ignored in the calculation. A negative value for the number of turns makes no sense.

5.3 EMF INDUCED IN A SINGLE STRAIGHT CONDUCTOR

Consider a conductor moving at a constant velocity v metre per second at right angles to a magnetic field having the dimensions shown in Figure 5.7. The direction of the induced emf may be obtained using Fleming's right-hand rule, and is shown in the diagram. Equation (5.1) is applicable, and in this case, the value for N is 1.

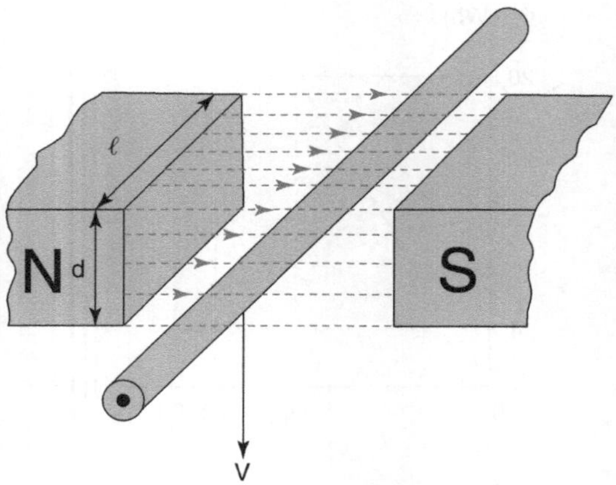

Figure 5.7 EMF Induced in a single straight conductor

Thus, $e = \frac{d\Phi}{dt}$, and since Φ is constant

$$e = \frac{\Phi}{t}$$

$$\Phi = BA$$

$$e = \frac{BA}{t}$$

also, the cross-sectional area of the field,

$$A = \ell \times d \text{ metre}^2$$

so

$$e = \frac{B\ell d}{t} \text{ and since } \frac{d}{t} = \text{velocity } v\text{, then}$$

$$e = B\ell v \tag{5.2}$$

The above equation is only true for the case when the conductor is moving at right angles to the magnetic field. If the conductor moves through the field at some angle less than 90°, then the 'cutting' action between the conductor and the flux is reduced. This results in a consequent reduction in the induced emf. Thus, if the conductor is moved horizontally through the field, the 'cutting' action is zero, and so no emf is induced. To be more precise, we can say that *only the component of the velocity at 90° to the flux* is responsible for the induced emf. In general, therefore, the induced emf is given by:

$$e = B\ell v \sin\theta \tag{5.3}$$

where $v \sin\theta$ is the component of velocity at 90° to the field, as illustrated in Figure 5.8.

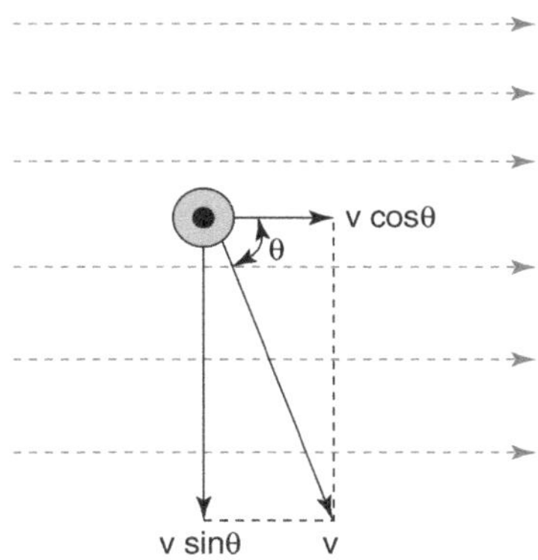

Figure 5.8 Induced emf with only the component of the velocity at 90° to the flux

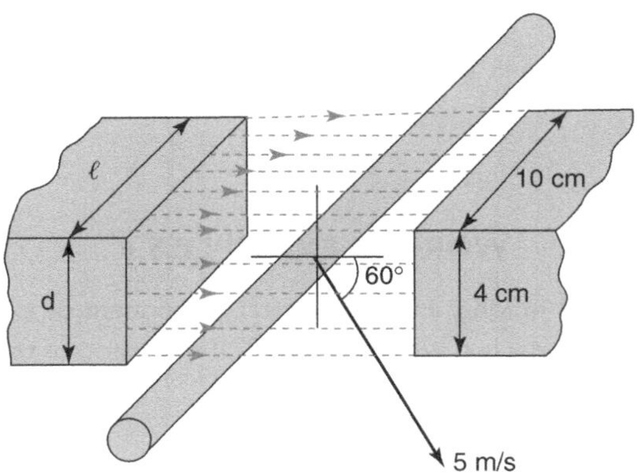

Figure 5.9 Movement of a conductor for Worked Example 5.4

This equation is simply confirmed by considering the previous two extremes; i.e. when conductor moves parallel to the flux, $\theta = 0°$; $\sin\theta = 0$; so $e = 0$. When it moves at right angles to the flux, $\theta = 90°$; $\sin\theta = 1$; so we are back to Equation (5.2).

Note: ℓ is known as the *effective* length of the conductor, since it is that portion of the conductor that actually links with the flux. The total length of the conductor may be considerably greater than this, but those portions that may extend beyond the field at either end will not have any emf induced.

WORKED EXAMPLE 5.4

Q A conductor is moved at a velocity of 5 m/s at an angle of 60° to a uniform magnetic field of 1. 6 mWb. The field is produced by a pair of pole pieces, the faces of which measure 10 cm by 4 cm. If the conductor length is parallel to the longer side of the field, calculate the emf induced; see Figure 5.9.

$v = 5$ m/s; $\theta = 60°$; $\Phi = 1.6 \times 10^{-3}$ Wb; $\ell = 0.1$ m; $d = 0.04$ m

$$B = \frac{\Phi}{A} = \frac{1.6 \times 10^{-3}\ \text{Wb}}{0.1\ \text{m} \times 0.04\ \text{m}} = 0.4\ \text{T}$$

$$e = Blv\sin\theta = 0.4\ \text{m} \times 0.1\ \text{m} \times 5\ \text{m/s} \times \sin 60^\circ = 0.173\ \text{V}$$

WORKED EXAMPLE 5.5

Q A conductor of effective length 15 cm, when moved at a velocity of 8 m/s at an angle of 55° to a uniform magnetic field, generates an emf of 2.5 V. Determine the density of the field.

$\ell = 0.15$ m; $v = 8$ m/s; $\theta = 55^\circ$; $e = 2.5$ V

$$e = B\ell v \sin\theta$$

$$B = \frac{e}{\ell v \sin\theta}$$

$$B = \frac{2.5}{0.15 \times 8 \times \sin 55^\circ} = 2.543\ \text{T}$$

WORKED EXAMPLE 5.6

Q The axle of a lorry is 2.2 m long, and the vertical component of the Earth's magnetic field density, through which the lorry is travelling, is 38 μT. If the speed of the lorry is 80 km/h, then calculate the emf induced in the axle.

$$\ell = 2.2\ \text{m};\ v = \frac{80 \times 10^3}{60 \times 60}\ \text{m/s};\ B = 38 \times 10^{-6}\ \text{T};\ \sin\theta = 1,\ \text{since}\ \theta = 90^\circ$$

$$e = B\ell v\sin\theta = \frac{38 \times 10^{-6} \times 2.2 \times 80 \times 10^3}{3600} = 1.86\ \text{mV}$$

This section, covering the induction or generation of an emf in a conductor moving through a magnetic field, forms the basis of the generator principle. However, most electrical generators are rotating machines, and we have so far considered only linear motion of the conductor.

Consider the conductor now formed into the shape of a rectangular loop, mounted on to an axle. This arrangement is then rotated between the poles of a permanent magnet. We now have the basis of a simple generator as illustrated in Figure 5.10.

The two sides of the loop that are parallel to the pole faces will each have an effective length ℓ metre. At any instant of time, these sides are passing through the field in opposite directions. Applying the right-hand rule at the instant shown in Figure 5.10, the directions of the induced emfs will be as marked, i.e. of opposite polarities. However, if we trace the path around the loop, it will be seen that both emfs are causing current to flow in the same direction around the loop. This is equivalent to two cells connected in series as shown in Figure 5.11.

The situation shown in Figure 5.10 applies only to one instant in one revolution of the loop (it is equivalent to a 'snapshot' at that instant).

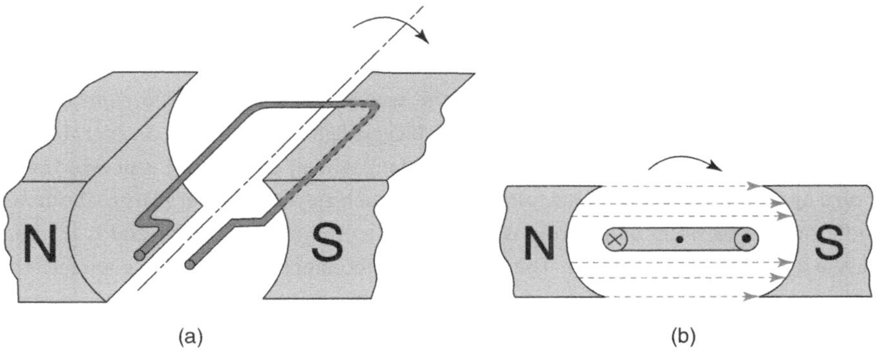

Figure 5.10 A simple generator

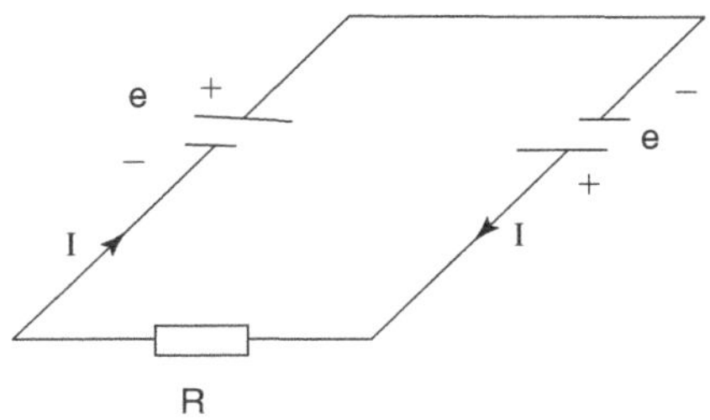

Figure 5.11 Two cells connected in series

If we were to plot a graph of the total emf generated in the loop, for one complete revolution, it would be found to be one cycle of a sine wave, i.e. an alternating voltage. This result should not come as any surprise though, since the equation for the emf generated in each side of the loop is $e = Blv \sin\theta$ volt. This very simple arrangement therefore is the basis of a simple form of a.c. generator or alternator. Exactly the same principles apply to a d.c. generator, but the way in which the inherent a.c. voltage is converted into d.c. automatically by the machine is dealt with in detail in *Further Electrical and Electronic Principles*.

Note: The 'ends' of the loop attached to the axle do not have emf induced in them, since they do not 'cut' the flux. Additionally, current can only flow around the loop provided that it forms part of a closed circuit.

A practical example of an a.c. generator is a bicycle dynamo. An elongated magnet is moved between a number of stationary coils of wire. The magnet rotates and is called the rotor. The coils of wire are fixed and are called the stator. So it is the magnet that rotates and the magnetic field moves over these coils of wire. The magnitude and direction of the induced voltage depend on the position and direction of the rotating magnet. As a result, an alternating current will flow through the externally connected circuit. However, it can also be the other way around. The permanent magnet is fixed (and is therefore called the stator now) and as a result of the mechanical movement the armature (the rotor) is rotated with the various conductors. The end of each conductor (which is actually only rotated once around the rotor) is connected to one copper slip ring and the other end to the other copper slip ring (both insulated from each other). Two brushes bring the generated alternating current to the externally connected circuit. Carbon is used for the brushes, because this is a soft material. Due to wear, this conforms to the slip rings and thus always makes good contact. Generators are constructed in such a way that the carbon brushes can be replaced quickly and easily.

Such a bicycle dynamo is clamped against the wheel. When cycling the central axis of the dynamo rotates internally and thus lights up the bicycle lights. There are considerable mechanical losses because the ribbed drive wheel (the running wheel) on the axle is pressed against the outer tyre by spring force. The transmission ratio in such a dynamo is very large: about 30 times. One pole of the electrical connection is connected to the metal housing and is connected to the bicycle light through the frame. The other pole is connected with a single wire to the front light and with another single wire to the rear light. A normal bicycle dynamo can deliver about 0.5 A alternating current at a voltage of 6 V (or another 3 W of power). The front light usually consumes 400 mA and in parallel the rear light consumes 100 mA.

Although not optimised, a bicycle dynamo can also be used as an a.c. motor: voltage and current are transformed into motion.

5.4 FORCE ON A CURRENT-CARRYING CONDUCTOR

Figure 5.12(a) shows the field patterns produced by two pole pieces, and the current flowing through the conductor. Since the lines of flux obey the rule that they will not intersect, the flux pattern from the poles will be distorted as illustrated in Figure 5.12(b). Also, since the lines of flux tend to act as if elastic, they will try to straighten themselves. This results in a force being exerted on the conductor, in the direction shown.

The direction of this force may be more simply obtained by applying Fleming's *left-hand* rule. This rule is similar to the right-hand rule. The major difference is of course that the fingers and thumb of the left hand are now used. In this case, the **F**irst finger indicates the direction of the main *F* lux (from the poles). The se**C**ond finger indicates the direction of **C**urrent flow. The thu**M**b shows the direction of the resulting force and hence consequent **M**otion. This is shown in Figure 5.13.

Simple experiments can be used to confirm that the force exerted on the conductor is directly proportional to the flux density produced by the pole pieces, the value of current flowing through the conductor and the length of conductor lying inside the field. This yields the following equation for the force *F*, expressed in newton or N:

$$F = BI\ell \tag{5.4}$$

The determination of the effective length ℓ of the conductor is exactly the same as that for the generator principle previously considered. So any conductor extending beyond the main

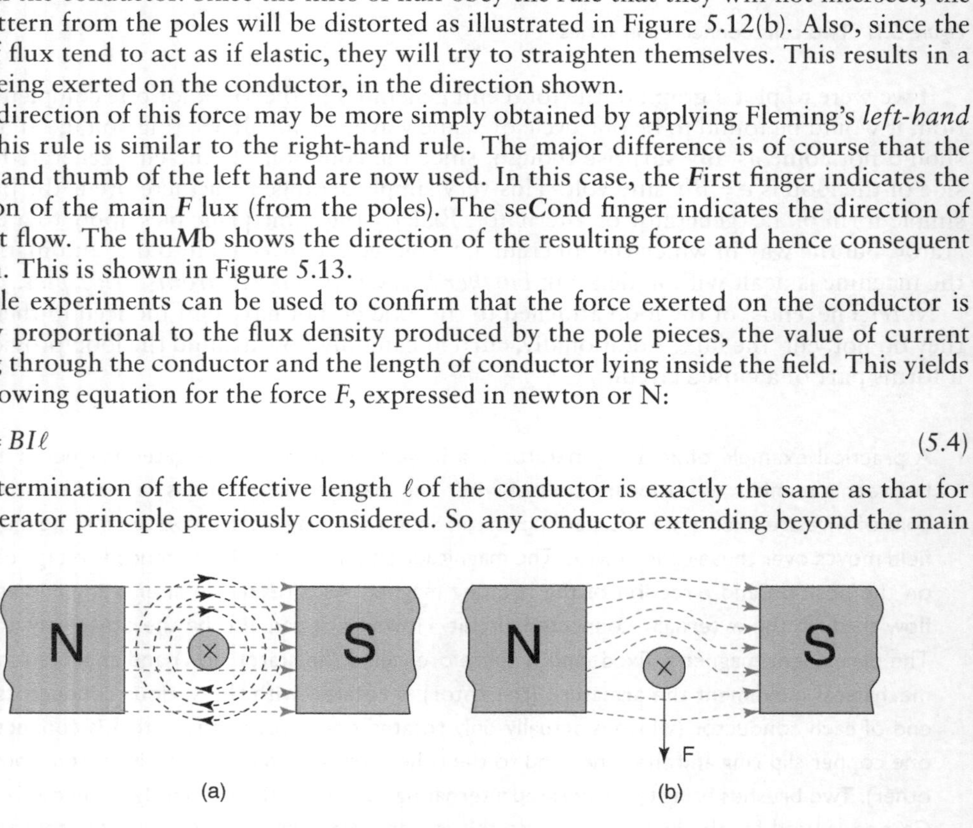

Figure 5.12 Force on a current-carrying conductor

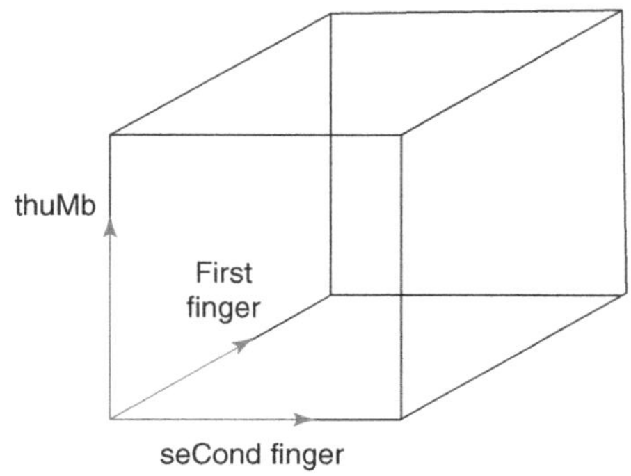

Figure 5.13 Fleming's left-hand rule: the cube

field does not contribute to the force exerted. Equation (5.4) also only applies to the condition when the conductor is perpendicular to the main flux. If it lies at some angle less than 90°, then the force exerted on it will be reduced. Thus, in general, the force exerted is given by

$$F = BI\ell \sin\theta \tag{5.5}$$

WORKED EXAMPLE 5.7

Q A conductor of effective length 22 cm lies at right angles to a magnetic field of density 0.35 T. Calculate the force exerted on the conductor when carrying a current of 3 A.

$\ell = 0.22$ m; $B = 0.35$ T; $I = 3$ A; $\theta = 90^\circ$

$F = BI\ell\ \sin\theta = 0.35 \times 3 \times 0.22 \times 1 = 0.231$ N

WORKED EXAMPLE 5.8

Q A pair of pole pieces 5 cm by 3 cm produce a flux of 2.5 mWb. A conductor is placed in this field with its length parallel to the longer dimension of the poles. When a current is passed through the conductor, a force of 1.25 N is exerted on it. Determine the value of the current. If the conductor was placed at 45° to the field, what then would be the force exerted?

$\Phi = 2.5 \times 10^{-3}$ Wb; $\ell = 0.05$ m; $d = 0.03$ m; $F = 1.25$ N

cross-sectional area of the field, $A = 0.05\ \text{m} \times 0.03\ \text{m} = 1.5 \times 10^{-3}\ \text{m}^2$

$$\text{flux density, } B = \frac{\Phi}{A} = \frac{2.5 \times 10^{-3}}{1.5 \times 10^{-3}} = 1.667\ \text{T}$$

and since $\theta = 90^\circ$, then $\sin\theta = 1$

$$F = BI\ell\ \sin\theta$$

$$I = \frac{F}{B\ell}$$

$$\text{therefore } I = \frac{1.25}{1.667 \times 0.05} = 15\ \text{A}$$

$$F = BI\ell \sin\theta, \text{ where } \theta = 45^\circ$$
$$F = 1.667 \times 15 \times 0.05 \times 0.707 = 0.884 \text{ N}$$

The principle of a force exerted on a current-carrying conductor as described above forms the basis of operation of a linear motor. However, since most electric motors are rotating machines, the above system must be modified.

5.5 THE MOTOR PRINCIPLE

Once more, consider the conductor formed into the shape of a rectangular loop, placed between two poles, and current passed through it. A cross-sectional view of this arrangement, together with the flux patterns produced, is shown in Figure 5.14.

The flux patterns for the two sides of the loop will be in opposite directions because of the direction of current flow through it. The result is that the main flux from the poles is twisted as shown in Figure 5.15. This produces forces on the two sides of the loop in opposite directions. Thus, there will be a turning moment exerted on the loop, in a counterclockwise direction. The distance from the axle (the pivotal point) is r metre, so the torque exerted on each side of the loop is expressed in newton-metre or Nm and is given by

$$T = Fr$$
$$F = BI\ell \sin\theta, \text{ and } \sin\theta = 1$$

so torque on each side = $BI\ell r$. Since the torque on each side is exerting a counterclockwise turning effect, the total torque exerted on the loop will be

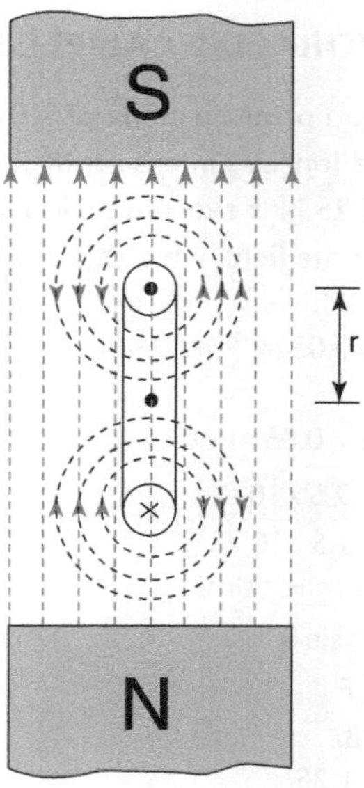

Figure 5.14 The motor principle

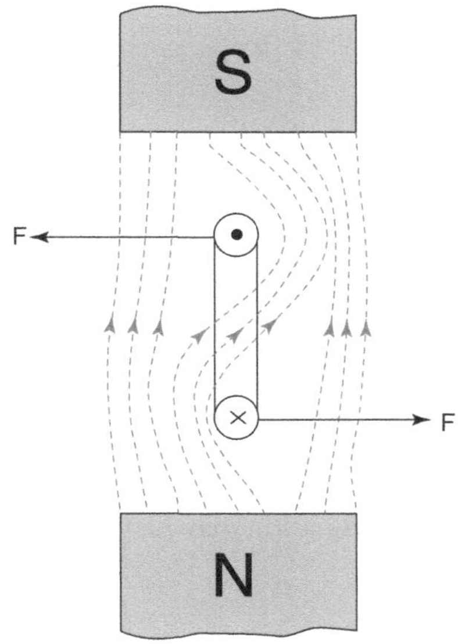

Figure 5.15 Twist of the main flux from the poles

$$T = 2BI\ell r \tag{5.6}$$

WORKED EXAMPLE 5.9

Q A rectangular single-turn loop coil 1.5 cm by 0.6 cm is mounted between two poles, which produce a flux density of 1.2 T, such that the longer sides of the coil are parallel to the pole faces. Determine the torque exerted on the coil when a current of 10 mA is passed through it.

$\ell = 0.015$ m; $B = 1.2$ T; $I = 10^{-2}$ A

radius of rotation, $r = 0.006 \text{ m} / 2 = 0.003$ m

$T = 2BI\ell r = 2 \times 1.2 \times 10^{-2} \times 0.015 \times 0.003 = 1.08\ \mu\text{Nm}$

From the above example, it may be seen that a single-turn loop produces a very small amount of torque. It is acknowledged that the dimensions of the coil specified and the current flowing through it are also small. However, even if the coil dimensions were increased by a factor oftentimes, and the current increased by a factor of a thousand times (to 10 A), the torque would still be only a very modest 0.108 Nm.

The practical solution to this problem is to use a multi-turn coil, as illustrated in Figure 5.16. If the coil now has N turns, then each side has an effective length of $N \times \ell$. The resulting torque will be increased by the same factor. So, for a multi-turn coil, the torque is given by

$$T = 2NBI\ell r$$

The term $2\ell r$ in the above expression is equal to the area 'enclosed' by the coil dimensions, so this is the effective cross-sectional area A of the field affecting the coil.

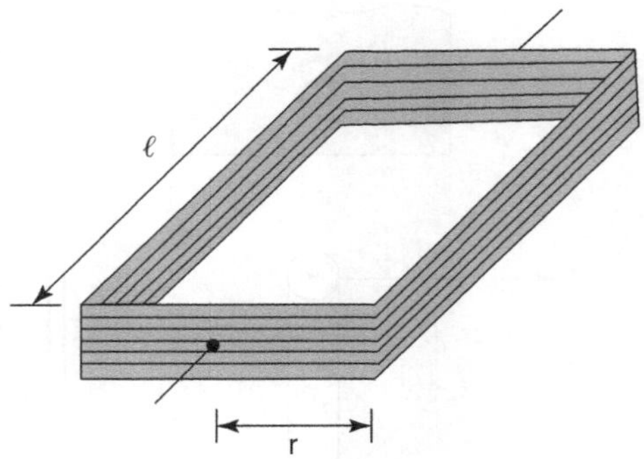

Figure 5.16 The torque for a multi-turn coil

Thus, $2\ell r = A$ m^2, and the above equation may be written

$$T = BANI \tag{5.7}$$

The principle of using a multi-turn current-carrying coil in a magnetic field is therefore used for rotary electric motors. However, the same principles apply to the operation of analogue instruments known as moving coil metres.

WORKED EXAMPLE 5.10

Q The coil of a moving-coil metre consists of 80 turns of wire wound on a former of length 2 cm and radius 1.2 cm. When a current of 45 μA is passed through the coil, the movement comes to rest when the springs exert a restoring torque of 1.4 μNm. Calculate the flux density produced by the pole pieces.

$N = 80$; $\ell = 0.02$ m; $r = 0.012$ m; $I = 45 \times 10^{-6}$ A; $T = 1.4 \times 10^{-6}$ Nm

The metre movement comes to rest when the deflecting torque exerted on the coil is balanced by the restoring torque of the springs.

$$T = BANI$$

$$B = \frac{T}{ANI} = \frac{1.4 \times 10^{-6}}{0.02 \times 2 \times 0.012 \times 80 \times 45 \times 10^{-6}} = 0.81\text{ T}$$

5.6 FORCE BETWEEN PARALLEL CONDUCTORS

When two parallel conductors are both carrying current, their magnetic fields will interact to produce a force of attraction or repulsion between them. This is illustrated in Figure 5.17.

In order to determine the value of such a force, consider first a single conductor carrying a current of I ampere. The magnetic field produced at some distance d from its centre is shown in Figure 5.18.

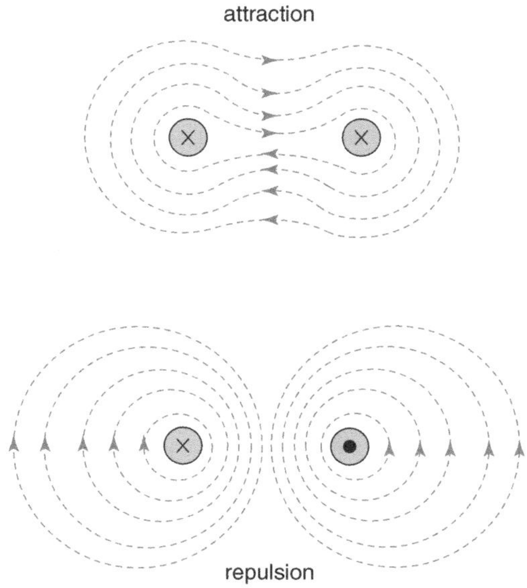

Figure 5.17 Force between parallel conductors

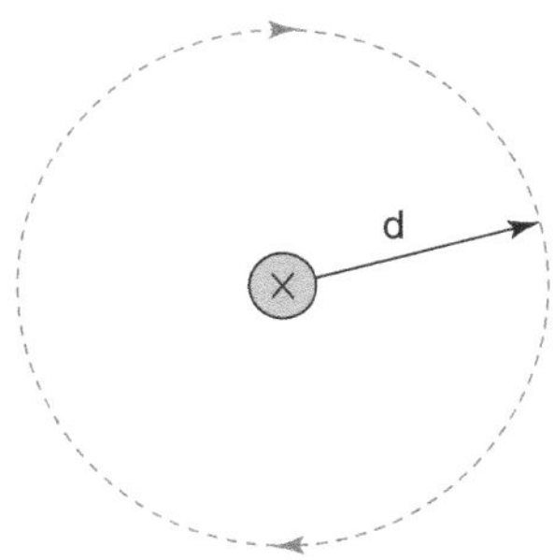

Figure 5.18 The magnetic field produced at some distance *d* from its centre

In general, $H = \frac{NI}{\ell}$ (expressed in ampere turn/metre), but in this case $N = 1$ (one conductor) and $\ell = 2\pi d$ metre (the circumference of the dotted circle), so

$$H = \frac{NI}{2\pi d}$$

Now, the flux density $B = \mu_0 \mu_r H$, and as the field exists in air, then $\mu_r = 1$. Thus, the flux density (expressed in tesla) at distance *d* from the centre is given by

$$B = \frac{I\mu_o}{2\pi d} \quad [1]$$

Consider now two conductors Y and Z carrying currents I_1 and I_2 respectively, at a distance of *d* metres between their centres as in Figure 5.19.

Using Equation [1] we can say that the flux density acting on Z due to current I_1 flowing in Y is:

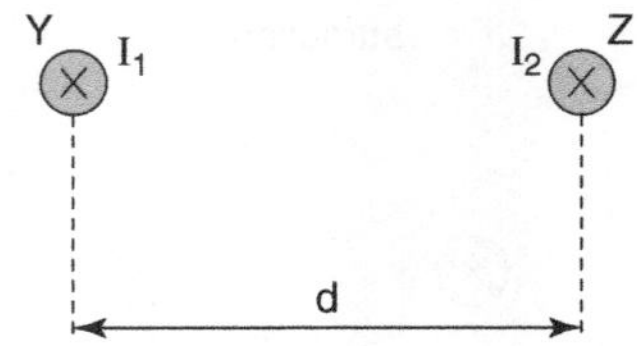

Figure 5.19 Two conductors each carrying currents

$$B_1 = \frac{I_1\mu_0}{2\pi d}$$

and the force exerted on Z, expressed in newton, equals $B_1I_2\ell$, or B_1I_2, expressed in newton per metre length of Z.

Hence, force/metre length acting on Z is expressed in newton and equals

$$= \frac{\mu_0 I_1 I_2}{2\pi d}$$

$$= \frac{4\pi \times 10^{-7} \times I_1 I_2}{2\pi d}$$

So force/metre length acting on Z is expressed in newton and equals

$$= \frac{2 \times 10^{-7} \times I_1 I_2}{d} \qquad (5.8)$$

Now, the current I_2 flowing in Z also produces a magnetic field which will exert a force on Y. Using the same reasoning as above, it can be shown that force/metre length acting on Y equals

$$= \frac{2 \times 10^{-7}}{d} I_1 I_2$$

so if $I_1 = I_2 = 1$ A, and $d = 1$ m, then

force exerted on *each* conductor $= 2 \times 10^{-7}$ N

This value of force forms the basis for the definition of the ampere, namely: that current, when maintained in each of two infinitely long parallel conductors situated *in vacuo*, and separated one metre between centres, produces a force of 2×10^{-7} newton per metre length on each conductor.

WORKED EXAMPLE 5.11

Q Two long parallel conductors are spaced 35 mm between centres. Calculate the force exerted between them when the currents carried are 50 A and 40 A respectively.

$d = 0.035$ m; $I_1 = 50$; $I_2 = 40$

$$F = \frac{2 \times 10^{-7} I_1 I_2}{d} = \frac{2 \times 10^{-7} \times 50 \text{ A} \times 40 \text{ A}}{0.035} = 11.4 \text{ mN}$$

WORKED EXAMPLE 5.12

Q Calculate the flux density at a distance of 2 m from the centre of a conductor carrying a current of 1000 A. If the centre of a second conductor, carrying 300 A, was placed at this same distance, what would be the force exerted?

$$d = 2\text{ m};\ I_1 = 1000\text{ A};\ I_2 = 300\text{ A}$$

$$B = \frac{\mu_0 I_1}{d} = \frac{4\pi \times 10^{-7} \times 1000}{2} = 0.628\text{ mT}$$

$$F = \frac{2 \times 10^{-7} I_1 I_2}{d} = \frac{2 \times 10^{-7} \times 300 \times 1000}{2} = 30\text{ mN}$$

5.7 EDDY CURRENTS

Consider an iron-cored solenoid, as shown in cross-section in Figure 5.20. Let the coil be connected to a source of emf via a switch. When the switch is closed, the coil current will increase rapidly to some steady value. This steady value will depend upon the resistance of the coil. The coil current will, in turn, produce a magnetic field. Thus, this flux pattern will increase from zero to some steady value. This changing flux therefore expands outwards from the centre of the iron core. This movement of the flux pattern is shown by the arrowed lines pointing outwards from the core.

Since there is a changing flux linking with the core, an emf will be induced in the core. As the core is a conductor of electricity, the induced emf will cause a current to be circulated around it. This is known as an eddy current (also called Foucault's current), since it traces out a circular path similar to the pattern created by an eddy of water. The direction of the induced emf and eddy current will be as shown in Figure 5.20. This has been determined by applying Fleming's right-hand rule. Please note, that to apply this rule, we need to consider the movement of the *conductor relative to* the flux. Thus, the *effective* movements of the left

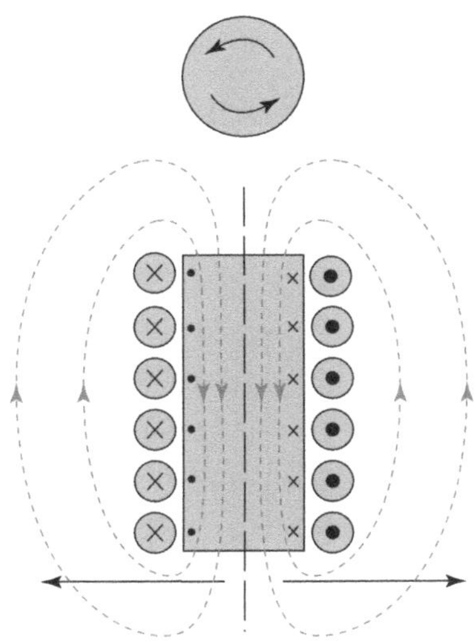

Figure 5.20 Eddy currents

and right halves of the core are *opposite* to the arrows showing the expansion of the flux pattern.

Jean Foucault (1819–1868) was a French physicist and is best known for the Foucault pendulum, an instrument demonstrating the rotation of the earth. He also made one of the first accurate measurements of the speed of light, he explained the principle of the gyroscope and discovered the power of eddy currents, also called Foucault currents.

As the eddy current circulates in the core, it will produce a heating effect. This is normally an undesirable effect. The energy thus dissipated is therefore referred to as the eddy current loss. If the solenoid forms part of a d.c. circuit, this loss is negligible. This is because the eddy current will flow only momentarily – when the circuit is first connected, and again when it is disconnected. However, if an a.c. supply is connected to the coil, the eddy current will be flowing continuously in alternate directions. Under these conditions, the core is also being taken through repeated magnetisation cycles. This will also result in a hysteresis loss.

In order to minimise the eddy current loss, the resistance of the core needs to be increased. On the other hand, the low reluctance needs to be retained. It would therefore be pointless to use an insulator for the core material, since we might just as well use an air core! The technique used for devices such as transformers, used at mains frequency, is to make the core from laminations of iron. This means that the core is made up of thin sheets (laminations) of steel, each lamination being insulated from the next. This is illustrated in Figure 5.21. Each lamination, being thin, will have a relatively high resistance. Each lamination will have an eddy current, the circulation of which is confined to that lamination. If the values of these individual eddy currents are added together, it will be found to be less than that for the solid core.

The hysteresis loss is proportional to the frequency f of the a.c. supply. The eddy current loss is proportional to f^2. Thus, at higher frequencies (e.g. radio frequencies), the eddy current loss is predominant. Under these conditions, the use of laminations is not adequate, and the eddy current loss can be unacceptably high. For this type of application, iron dust cores or ferrite cores are used. With this type of material, the eddy currents are confined to individual 'grains', so the eddy current loss is considerably reduced.

5.8 SELF- AND MUTUAL INDUCTANCE

The effects of self- and mutual inductance can be demonstrated by another simple experiment. Consider two coils, as shown in Figure 5.22. Coil 1 is connected to a battery via a

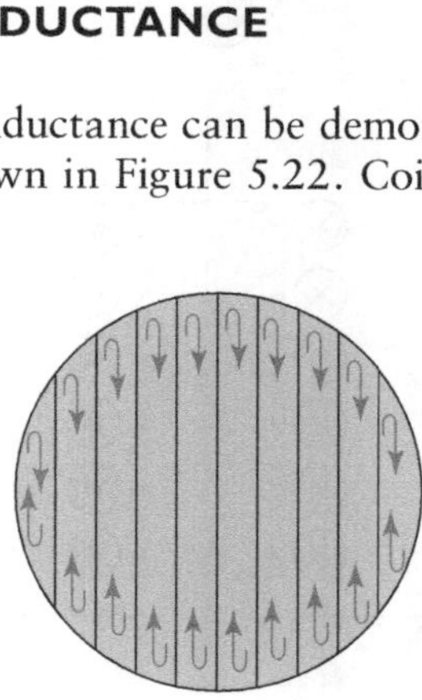

Figure 5.21 Lamination minimising Eddy currents

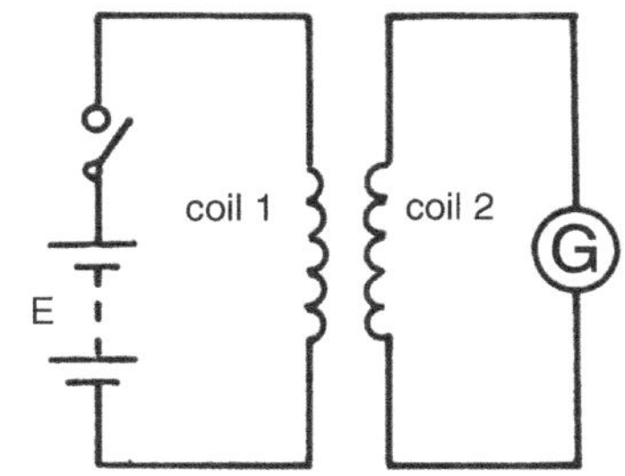

Figure 5.22 Self- and mutual inductance

switch. Coil 2 is placed close to coil 1, but is not electrically connected to it. Coil 2 has a galvo connected to its terminals.

When the switch is closed, the current in coil 1 will rapidly increase from zero to some steady value. Hence, the flux produced by coil 1 will also increase from zero to a steady value. This changing flux links with the turns of coil 2, and therefore induces an emf into it. This will be indicated by a momentary deflection of the galvo pointer.

Similarly, when the switch is subsequently opened, the flux produced by coil 1 will collapse to zero. The galvo will again indicate that a momentary emf is induced in coil 2, but of the opposite polarity to the first case. Thus, an emf has been induced into coil 2, by a changing current (and flux) in coil 1. This is known as a mutually induced emf.

If the changing flux can link with coil 2, then it must also link with the turns of coil 1. Thus, there must also be a momentary emf induced in this coil. This is known as a self-induced emf. Any induced emf obeys Lenz's law. This self-induced emf must therefore be of the opposite polarity to the battery emf. For this reason, it is also referred to as a back emf. Unfortunately, it is extremely difficult to demonstrate the existence of this back emf. If a voltmeter was connected across coil 1, it would merely indicate the terminal voltage of the battery.

5.8.1 Self-Inductance

Self-inductance is that property of a circuit or component which causes a self-induced emf to be produced, when the current through it changes. The unit of self-inductance is the henry, which is defined as follows: A circuit has a self-inductance of one henry (1 H) if an emf of one volt (1 V) is induced in it, when the circuit current changes at the rate of one ampere per second (1 A/s). The quantity symbol for self-inductance is L. From the above definition, we can state the following equation:

$$L = \frac{-e}{\mathrm{d}i/\mathrm{d}t}$$

$$\text{or, self-induced emf, } e = \frac{-L\mathrm{d}i}{\mathrm{d}t} \qquad (5.9)$$

Notes:

1 The minus sign again indicates that Lenz's law applies.
2 The emf symbol is e, because it is only a momentary emf.

3 The current symbol is *i*, because it is the *change* of current that is important.
4 The term d*i*/d*t* is the rate of change of current.

WORKED EXAMPLE 5.13

Q A coil has a self-inductance of 0.25 H. Calculate the value of emf induced, if the current through it changes from 100 mA to 350 mA, in a time of 25 ms.

$$L = 0.25\ \text{H};\ di = (350 - 100) \times 10^{-3}\ \text{A};\ dt = 25 \times 10^{-3}\ \text{s}$$

$$e = \frac{-Ldi}{dt} = \frac{0.25 \times 250 \times 10^{-3}}{25 \times 10^{-3}} = 2.5\ \text{V}$$

WORKED EXAMPLE 5.14

Q Calculate the inductance of a circuit in which an emf of 30 V is induced, when the circuit current changes at the rate of 200 A/s.

$$e = 30\ \text{V};\ \frac{di}{dt} = 200\ \text{A/s}$$

$$L = \frac{-e}{di/dt} = \frac{30}{200} = 0.15\ \text{H}$$

WORKED EXAMPLE 5.15

Q A circuit of self-inductance 50 mH has an emf of 8 V induced into it. Calculate the rate of change of the circuit current that induced this emf.

$$L = 50 \times 10^{-3}\ \text{H};\ e = 8\ \text{V}$$

$$e = \frac{-Ldi}{dt}$$

$$\frac{di}{dt} = \frac{e}{L} = \frac{8}{50 \times 10^{-3}} = 160\ \text{A/s}$$

5.8.2 Flux Linkages

Consider a coil of *N* turns, carrying a current of *I* amp. Let us assume that this current produces a flux of Φ weber. If the current now changes at a uniform rate of d*i*/d*t* ampere per second, it will cause a corresponding change of flux of dφ/d*t* weber per second. Let us also assume that the coil has a self-inductance of *L* henry.

The self-induced emf may be determined from Equation (5.9):

$$e = \frac{-Ldi}{dt}$$

However, the induced emf is basically due to the rate of change of flux linkages. Thus, the emf may also be calculated by using Equation (5.1), namely:

$$e = \frac{-Nd\Phi}{dt}$$

Since both equations above represent the same induced emf, then they must be equal. Thus

$$\frac{Ldi}{dt} = \frac{Nd\Phi}{dt} \left(\text{the minus signs cancel out}\right)$$

$$L = \frac{Nd\Phi}{di} \qquad (5.10)$$

A coil which is designed to have a specific value of self-inductance is known as an inductor. An inductor is the third of the main passive electrical components. The other two are the resistor and the capacitor.

A passive component is one which (a) requires an external source of emf in order to serve a useful function, and (b) does not provide any amplification of current or voltage.

Now, a resistor will have a specific value of resistance, regardless of whether it is in a circuit or not. Similarly, an inductor will have some value of self-inductance, even when the current through it is constant. In other words, an inductor does not have to have an emf induced in it, in order to possess the property of self-inductance. For this reason, Equation (5.10) may be slightly modified as follows. If the current I through an N-turn coil produces a flux of Φ weber, then its self-inductance is given by the equation:

$$L = \frac{N\Phi}{I} \qquad (5.11)$$

In other words, although no *change* of current and flux is specified, the coil will still have some value of inductance. Strictly speaking, Equation (5.11) applies only to an inductor with a non-magnetic core. The reason is that, in this case, the flux produced is directly proportional to the coil current. However, it is a very close approximation to the true value of inductance for an iron-cored inductor which contains an air gap in it.

WORKED EXAMPLE 5.16

Q A coil of 150 turns carries a current of 10 A. This current produces a magnetic flux of 0.01 Wb. Calculate (a) the inductance of the coil, and (b) the emf induced when the current is uniformly reversed in a time of 0.1 s.

$N = 150$; $I = 10$ A; $\Phi = 0.01$ Wb; $dt = 0.1$ s

(a) $L = \frac{N\Phi}{I} = \frac{150 \times 0.10}{10} = 1.5\text{ H}$

(b) Since current is reversed, it will change from 10 A to −10 A, i.e. a *change* of 10−(−10). So, $dI = 20$ A.

$$e = \frac{-Ldi}{dt} = \frac{1.5\,\text{H} \times 20}{0.1} = 300\ \text{V}$$

WORKED EXAMPLE 5.17

Q A current of 8 A, when flowing through a 3000-turn coil, produces a flux of 4 mWb. If the current is reduced to 2 A in a time of 100 ms, calculate the emf thus induced in the coil. Assume that the flux is directly proportional to the current.

$$I_1 = 8\ \text{A};\ N = 3000;\ \Phi_1 = 4 \times 10^{-3}\ \text{Wb};\ I_2 = 2\ \text{A};\ dt = 0.1\ \text{s}$$

This problem may be solved in either of two ways. Both methods will be demonstrated.

$$e = \frac{-Ldi}{dt},\ \text{where } L = \frac{N\Phi}{I_1}$$

$$L = \frac{3000 \times 4 \times 10^{-3}}{8} = 1.5\ \text{H};\ di = 8 - 2 = 6\ \text{A}$$

$$e = \frac{1.5\,\text{H} \times 6\ \text{A}}{0.1\ \text{s}} = 90\ \text{V}$$

Alternatively, $\Phi \propto I$ so $\Phi_1 \propto I_1$ and $\Phi_2 \propto I_2$

$$\frac{\Phi_2}{\Phi_2} = \frac{I_2}{I_1} \text{ and } \Phi_2 = \frac{\Phi_1 I_2}{I_2}$$

$$\text{hence, } \Phi_2 = \frac{2\ \text{A} \times 4 \times 10^{-3}}{8} = 1 \times 10^{-3}\ \text{Wb};$$

$$\text{and } d\Phi = (4 - 1) \times 10^{-3}$$

$$e = \frac{-Nd\Phi}{dt} = \frac{3000 \times 3 \times 10^{-3}}{0.1} = 90\ \text{V}$$

5.8.3 Factors Affecting Inductance

Consider a coil of N turns wound on to a non-magnetic core, of uniform cross-sectional area A metre² and mean length ℓ metre. The coil carries a current of I amp, which produces a flux of Φ weber. From Equation (5.11), we know that the inductance will be

$$L = \frac{N\Phi}{I},\ \text{but } \Phi = BA$$

$$L = \frac{NBA}{I}$$

Also, magnetic field strength,

$$H = \frac{NI}{\ell}; I = \frac{H\ell}{N}$$

and substituting this expression for I into the equation above

$$L = \frac{NBA}{H\ell / N} = \frac{BAN^2}{H\ell}$$

Now, this equation contains the term $\frac{B}{H}$, which equals $\mu_o\mu_r$

$$L = \frac{\mu_0\mu_r N^2 A}{\ell} \quad (5.12)$$

We also know that

$$\frac{\ell}{\mu_0\mu_r A} = \text{reluctance, } S$$

$$L = \frac{N^2}{S} \quad (5.13)$$

NOTES:

1 Equation (5.12) compares with $C = \frac{\varepsilon_o\varepsilon_r A(N-1)}{d}$ for a capacitor.
2 If the number of turns is doubled, then the inductance is quadrupled, i.e. $L \propto N^2$.
3 The terms A and ℓ in Equation (5.12) refer to the dimensions of the core, and *not* the coil.

WORKED EXAMPLE 5.18

Q A 600-turn coil is wound on to a non-magnetic core of effective length 45 mm and cross-sectional area of 4 cm². (a) Calculate the inductance. (b) The number of turns is increased to 900. Calculate the inductance value now produced. (c) The core of the 900-turn coil is now replaced by an iron core having a relative permeability of 75, and of the same dimensions as the original. Calculate the inductance in this case.

$N_1 = 600$; $l = 45 \times 10^{-3}$ m; $A = 4 \times 10^{-4}$ m²; $\mu_{r1} = 1$
$N_2 = 900$; $\mu_{r2} = 1$; $N_3 = 900$; $\mu_{r3} = 75$

(a) $$L_1 = \frac{\mu_0\mu_{r1}N_1^2 A}{\ell} = \frac{4\pi \times 10^{-7} \times 1 \times 600^2 \times 4 \times 10^{-4}}{45 \times 10^{-3}} = 4.02 \text{ mH}$$

(b) Since $L_1 \propto N_1^2$ and $L_2 \propto N_2^2$, then

$$\frac{L_2}{L_1} = \frac{N_2^2}{N_1^2} L_2 = \frac{L_1 N_2^2}{N_1^2} = \frac{4.02 \times 10^{-3} \times 900^2}{600^2} = 9.045 \text{ mH}$$

(c) Since $\mu_{r3} = 75 \times \mu_{r2}$, and there are no other changes, then $L_3 = 75 \times L_2 = 0.678$ H.

5.8.4 Mutual Inductance

When a changing current in one circuit induces an emf in another separate circuit, then the two circuits are said to possess mutual inductance. The unit of mutual inductance is the henry and is defined as follows. Two circuits have a mutual inductance of one henry (1 H), if the emf induced in one circuit is one volt (1 V), when the current in the other is changing at the rate of one ampere per second (1 A/s). The quantity symbol for mutual inductance is M, and expressing the above definition in henry as an equation we have

$$M = \frac{\text{induced emf in coil 2}}{\text{rate of change of current in coil 1}}$$

$$= \frac{-e_2}{\mathrm{d}i_1 / \mathrm{d}t}$$

and transposing this equation for emf e_2

$$e_2 = \frac{-M\mathrm{d}i_1}{\mathrm{d}t} \qquad (5.14)$$

This emf may also be expressed in terms of the flux linking coil 2. If *all of* the flux from coil 1 links with coil 2, then we have what is called 100% flux linkage. In practice, it is more usual for only a proportion of the flux from coil 1 to link with coil 2. Thus, the flux linkage is usually less than 100%. This is indicated by a factor, known as the coupling factor, k. Coupling at 100% is indicated by $k = 1$. If there is no flux linkage with coil 2, then k will have a value of zero. So if zero emf is induced in coil 2, the mutual inductance will also be zero. Thus, the possible values for the coupling factor k lie between zero and 1. Expressed mathematically, this is written as

$$0 \le k \le 1$$

Consider two coils possessing mutual inductance, and with a coupling factor < 1. Let a change of current $\mathrm{d}i_1/\mathrm{d}t$ amp/s in coil 1 produce a change of flux $\mathrm{d}\Phi/\mathrm{d}t$ weber/s. The proportion of this flux change linking coil 2 will be $\mathrm{d}\Phi_2/\mathrm{d}t$ weber/s. If the number of turns on coil 2 is N_2, then

$$e_2 = \frac{-N_2\mathrm{d}\Phi_2}{\mathrm{d}t} \qquad (5.15)$$

However, Equations (5.14) and (5.15) both refer to the same induced emf. Therefore, we can equate the two expressions

$$\frac{M\mathrm{d}i_1}{\mathrm{d}t} = \frac{N_2\mathrm{d}\Phi_2}{\mathrm{d}t}$$

and transposing for M, we have

$$M = \frac{N_2\mathrm{d}\Phi_2}{\mathrm{d}i_1} \qquad (5.16)$$

As with self-inductance for a single coil, mutual inductance is a property of a pair of coils. They therefore retain this property, regardless of whether or not an emf is induced. Hence, Equation (5.16) may be modified to

$$M = \frac{N_2 \Phi_2}{I_1} \tag{5.17}$$

WORKED EXAMPLE 5.19

Q Two coils, A and B, have 2000 turns and 1500 turns respectively. A current of 0.5 A, flowing in A, produces a flux of 60 μWb. The flux linking with B is 83% of this value. Determine (a) the self-inductance of coil A, and (b) the mutual inductance of the two coils.

$N_A = 2000$; $N_B = 1500$; $I_A = 0.5$ A; $\Phi_A = 60 \times 10^{-6}$ Wb

(a) $$L_A = \frac{N_A \Phi_A}{I_A} = \frac{2000 \times 60 \times 10^{-6}}{0.5} = 0.24\text{ H}$$

(b) $$M = \frac{N_B \Phi_B}{I_A} = \frac{1500 \times 0.83 \times 60 \times 10^{-6}}{0.5} = 0.149\text{ H}$$

5.8.5 Relationship between Self- and Mutual Inductance

Consider two coils of N_1 and N_2 turns, respectively, wound on to a common non-magnetic core. If the reluctance of the core is S ampere turns/weber, and the coupling coefficient is unity, then

$$L_1 = \frac{N_1^2}{S} \text{ and } L_2 = \frac{N_2^2}{S}$$

therefore,

$$L_1 L_2 = \frac{N_1^2 N_2^2}{S^2}$$

$$M = \frac{N_2 \Phi}{I_1}\text{, and multiplying by } \frac{N_1}{N_1}$$

$$M = \frac{N_1 N_2 \Phi}{N_1 I_1}$$

The above expression contains the term

$$\frac{\Phi}{N_1 I_1} = \frac{1}{S}$$

$$M = \frac{N_1 N_2}{S}$$

$$M^2 = \frac{N_1^2 N_2^2}{S^2}$$

and comparing both equations results in

$$M^2 = L_1 L_2$$

$$M = \sqrt{L_1 L_2} \tag{5.18}$$

The above equation is correct only provided that there is 100% coupling between the coils; i.e. $k = 1$. If $k < 1$, then the general form of the equation, shown below, applies.

$$M = k\sqrt{L_1 L_2} \tag{5.19}$$

WORKED EXAMPLE 5.20

Q A 400-turn coil is wound on to a cast steel toroid having an effective length of 25 cm and cross-sectional area of 4.5 cm^2. If the steel has a relative permeability of 180 under the operating conditions, calculate the self-inductance of the coil.

$$N = 400;\ \ell = 0.25\text{ m};\ A = 4.5 \times 10^{-4}\text{ m}^2;\ \mu_r = 180$$

$$L = \frac{\mu_0 \mu_r N^2 A}{\ell} = \frac{4\pi \times 10^{-7} \times 180 \times 400^2 \times 4.5 \times 10^{-4}}{0.25} = 65\text{ mH}$$

WORKED EXAMPLE 5.21

Q Considering Worked Example 5.20, a second coil of 650 turns is wound over the first, and the current through coil 1 is changed from 2 A to 0.5 A in a time of 3 ms. If 95% of the flux thus produced links with coil 2, then calculate (a) the self-inductance of coil 2, (b) the value of mutual inductance, (c) the self-induced emf in coil 1, and (d) the mutually induced emf in coil 2.

$$L_1 = 65 \times 10^{-3}\text{ H};\ dI_1 = 2 - 0.5 = 1.5\text{ A};\ dt = 3 \times 10^{-3}\text{ s};\ k = 0.95$$

(a) From the equation for inductance we know that $L \propto N^2$, and since all other factors for the two coils are the same, then

$$\frac{L_2}{L_1} = \frac{N_2^2}{N_1^2}$$

$$L_2 = \frac{N_2^2 L_1}{N_1^2} = \frac{650^2 \times 65}{400^2}\text{ mH} = 172\text{ mH}$$

It is left to the student to confirm this answer by using the equation

$$L = \frac{\mu_0 \mu_r N^2 A}{\ell}$$

(b) $M = k\sqrt{L_1L_2} = 0.95\sqrt{65 \times 175} = 100$ mH

(c) $e_1 = -L_1\dfrac{di}{dt} = \dfrac{65 \times 10^{-3} \times 1.5}{3 \times 10^{-3}} = 32.5$ V

(d) $e_2 = -M\dfrac{di_1}{dt} = \dfrac{100 \times 1.5}{3 \times 10^{-3}} = 50$ V

5.9 ENERGY STORED

As with an electric field, a magnetic field also stores energy. When the current through an inductive circuit is interrupted, by opening a switch, this energy is released. This is the reason why a spark or arc occurs between the contacts of the switch, when it is opened.

Consider an inductor connected in a circuit, in which the current increases uniformly, to some steady value I amp. This current change is illustrated in Figure 5.23. The magnitude of the emf induced by this change of current is expressed in volt and given by

$$e = \frac{LI}{t}$$

The average power input to the coil during this time is:

average power = $e \times$ average current

From the graph, it may be seen that the average current over this time is $I/2$ amp. Therefore,

$$\text{average power expressed in watt} = \frac{1}{2}eI = \frac{LII}{2t} = \frac{LI^2}{2t}$$

but, energy stored, expressed in joule = average power × time

$$= \frac{LI^2t}{2t}$$

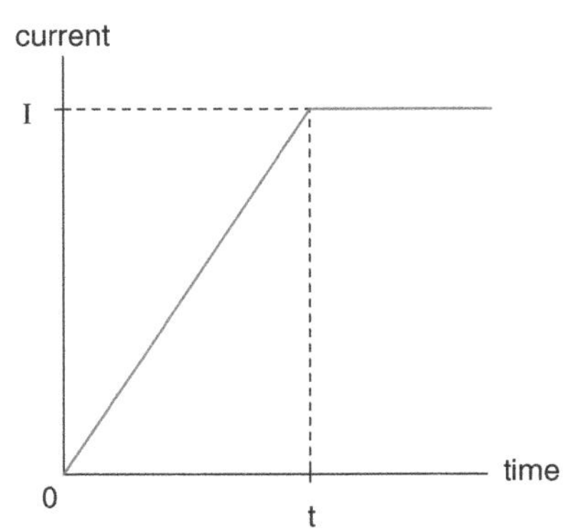

Figure 5.23 The current as function of the time

thus, energy stored, expressed in joule, $W = \frac{1}{2}LI^2$ (5.20)

Equation (5.20) applies to a single inductor. When two coils possess mutual inductance, and are connected in series, both will store energy. In this situation, the total energy stored is given by the equation

$$W = \frac{1}{2}L_1I_1^2 + \frac{1}{2}L_2I_2^2 \pm MI_1I_2 \quad (5.21)$$

Consider a parallel connection of a light bulb and a coil, which can be turned on and off with a switch. Without the coil, the light bulb would just give light. But with the coil, the behaviour changes completely. The light bulb can be thought of as a resistor (here one that creates heat to make the wire glow). The wire of the coil is also a resistor, but with a much smaller resistance value. When the switch connects, we expect the light bulb to give light very softly. After all, most current follows the path of least resistance (i.e. through the coil). However, what happens when making the connection is that the lamp first shines very hard and then softer. When you open the switch again, the light bulb also lights up very brightly and then goes out completely. We can explain this by looking at the coil. Once current starts flowing through the coil, a magnetic field is built up and energy is stored in the coil. As a result, the coil inhibits the current (and therefore more current flows through the light bulb). When the magnetic field has built up completely, the current flows normally (and therefore less through the light bulb). When switched off, the magnetic field ensures that the current continues to flow through the coil until the magnetic field completely disappears. Note that the magnetic field can be influenced by bringing a ferromagnetic material or a magnet close to the coil.

WORKED EXAMPLE 5.22

Q Calculate the energy stored in a 50 mH inductor when it is carrying a current of 0.75 A.

$$L = 50 \times 10^{-3}\ \text{H};\ I = 0.75\ \text{A}$$

$$W = \frac{1}{2}LI^2 = \frac{50 \times 10^{-3} \times 0.75^2}{2} = 14.1\ \text{mJ}$$

WORKED EXAMPLE 5.23

Q Two inductors of inductance 25 mH and 40 mH, respectively, are wound on a common ferromagnetic core, and are connected in series with each other. The coupling coefficient, k, between them is 0.8. When the current flowing through the two coils is 0.25 A, calculate (a) the energy stored in each and (b) the total energy stored when the coils are connected (i) in series aiding and (ii) in series opposition.

$L_1 = 25 \times 10^{-3}\,\text{H}; L_2 = 40 \times 10^{-3}\,\text{H}; I_2 = 0.25\,\text{A}; k = 1.8$

(a)
$$W_1 = \frac{1}{2}L_1 I_1^2 = 0.5 \times 25 \times 10^{-3} \times 0.25^2 = 0.78\,\text{mJ}$$
$$W_2 = \frac{1}{2}L_2 I_2^2 = 0.5 \times 40 \times 10^{-3} \times 0.25^2 = 1.25\,\text{mJ}$$

(b) The general equation for the energy stored by two inductors with flux linkage between them is:

$$W = \frac{1}{2}L_1 I_1^2 + \frac{1}{2}L_1 I_2^2 \pm M I_1 I_2 \text{ joule}$$

When the coils are connected in series such that the two fluxes produced act in the same direction, the total flux is increased and the coils are said to be connected in series aiding. In this case the total energy stored in the system will be increased, so the last term in the above equation is added, i.e. the + sign applies. If, however, the connections to one of the coils are reversed, then the two fluxes will oppose each other, the total flux will be reduced, and the coils are said to be in series opposition. In this case, the minus sign is used. These two connections are shown in Figure 5.24.

$$M = k\sqrt{L_1 L_2} = 0.8\sqrt{25 \times 40} = 25.3\,\text{mH}$$

The values for $\frac{1}{2}L_1 I_1^2$ and $\frac{1}{2}L_2 I_2^2$ have been calculated in part (a) and $M I_1 I_2 = 25.3 \times 10^{-3}\,\text{H} \times 0.25^2\,\text{A}^2 = 1.58\,\text{mJ}$

(i) For series aiding:

$$W = 0.78 + 1.25 + 1.58 = 3.6\,\text{mJ}$$

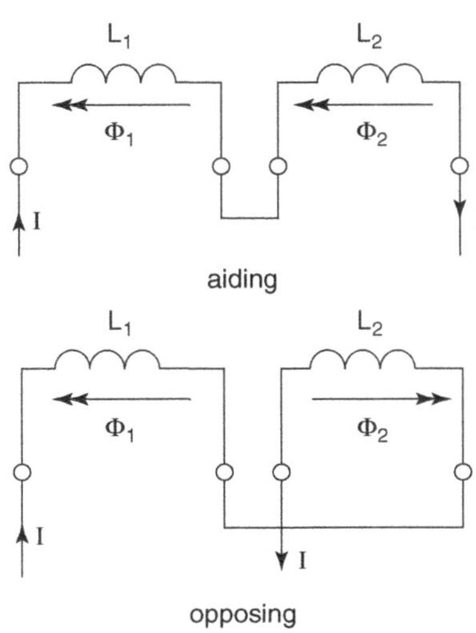

Figure 5.24 Coils connected in series

(ii) For series opposition:

$W = 0.78 + 1.25 - 1.58 = 0.45$ mJ

5.10 THE TRANSFORMER PRINCIPLE

A transformer is an a.c. machine, which utilises the mutual inductance between two coils, or windings. The two windings are wound on to a common iron core, but are not electrically connected to each other. The purpose of the iron core is to reduce the reluctance of the magnetic circuit. This ensures that the flux linkage between the coils is almost 100%.

a.c. means alternating current, i.e. one which flows alternately, first in one direction, then in the opposite direction. It is normally a sinewave

Since it is an a.c. machine, an alternating flux is produced in the core. The core is therefore laminated to minimise the eddy current loss. Indeed, the transformer is probably the most efficient of all machines. Efficiencies of 98% to 99% are typical. This high efficiency is due mainly to the fact that there are no moving parts.

The general arrangement is shown in Figure 5.25. One winding, called the primary, is connected to an a.c. supply. The other winding, the secondary, is connected to a load. The primary will draw an alternating current I_1 from the supply. The flux, Φ, produced by this winding, will therefore also be alternating; i.e. it will be continuously changing. Assuming 100% flux linkage, then this flux is the only common factor linking the two windings. Thus, a mutually induced emf, E_2, will be developed across the secondary. Also, there will be a back emf, E_1, induced across the primary. If the secondary is connected to a load, then it will cause the secondary current I_2 to flow. This results in a secondary terminal voltage, V_2. Figure 5.26 shows the circuit symbol for a transformer.

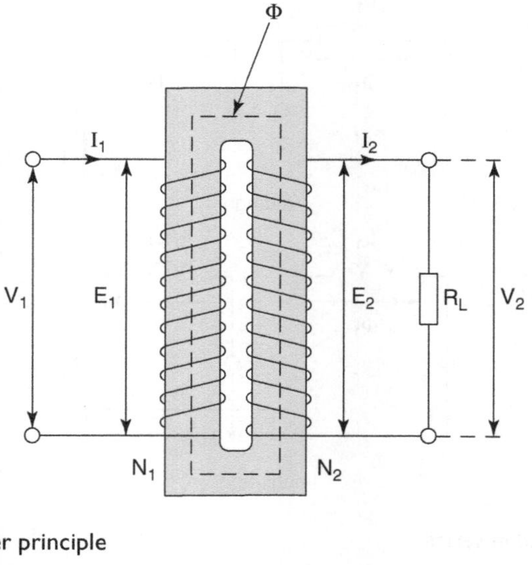

Figure 5.25 The transformer principle

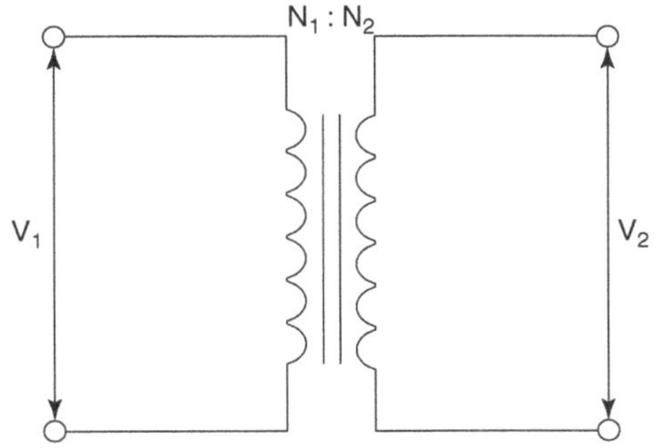

Figure 5.26 The circuit symbol for a transformer

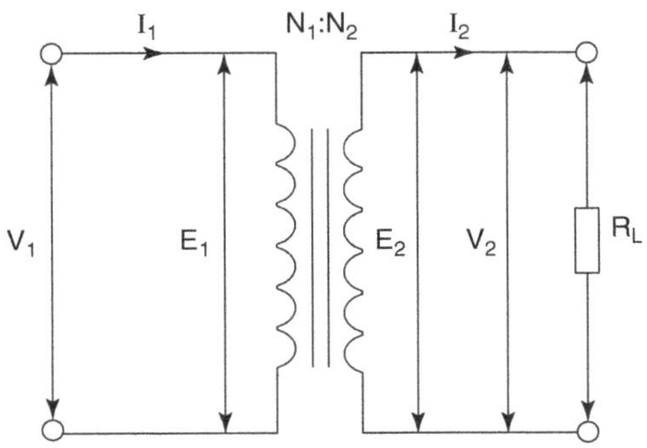

Figure 5.27 An ideal transformer connected to purely resistive load

A load is any device or circuit connected to some source of emf. Thus, a load will draw current from the source. The term 'load' is also loosely used to refer to the actual current drawn from a source

Let us consider an ideal transformer. This means that the resistance of the windings is negligible, and there are no core losses due to hysteresis and eddy currents. Also, let the secondary be connected to a purely resistive load, as shown in Figure 5.27.

Under these conditions, the primary back emf, E_1, will be of the same magnitude as the primary applied voltage, V_1. The secondary terminal voltage, V_2, will be of the same magnitude as the secondary induced emf, E_2. Finally, the output power will be the same as the input power. The two emfs are given by

$$E_1 = \frac{-N\mathrm{d}\Phi}{\mathrm{d}t} \text{ and } E_2 = \frac{-N_2\mathrm{d}\Phi}{\mathrm{d}t}$$

$$\frac{\mathrm{d}\Phi}{\mathrm{d}t} = \frac{E_1}{N_1} \cdots\cdots\cdots\cdots [1]$$

$$\text{and } \frac{\mathrm{d}\Phi}{\mathrm{d}t} = \frac{E_2}{N_2} \cdots\cdots\cdots\cdots [2]$$

Since both Equations [1] and [2] refer to the same rate of change of flux in the core, then [1] = [2]:

$$\frac{E_1}{N_1} = \frac{E_2}{N_2}$$

$$\frac{E_1}{E_2} = \frac{N_1}{N_2}$$

and since $E_1 = V_1$, and $E_2 = V_2$, then

$$\frac{V_1}{V_2} = \frac{N_1}{N_2} \tag{5.22}$$

From this equation, it may be seen that the voltage ratio is the same as the turns ratio. This is perfectly logical, since the same flux links both windings, and each induced emf is directly proportional to its respective number of turns. This is the main purpose of the transformer. It can therefore be used to 'step-up' or 'step-down' a.c. voltages, depending upon the turns ratio selected.

Wireless charging is a process in which electrical energy is transported between two objects via a magnetic field, so that no cabling is required for the energy transfer. A coil is built into each object, so that the transformer principle plays and the voltage and current in one coil is converted to a voltage and current in the other coil. This process is possible through induction (inductive coupling) and is mainly used to charge batteries. The standard for energy transfer of low power (up to 5 W) over a distance of up to 4 cm is Q_i and is used for wireless charging of an electric toothbrush. General Motors and Toyota also use wireless charging for a number of electric cars. And a number of smartphone and smartwatch models are equipped with this technology.

Transformers are used in public electricity networks, because it is more efficient (and therefore with less power loss) to allow high voltages and low currents to travel a long distance than low voltages and high currents. The public electricity network contains so-called high-voltage lines with 'step-up' and 'step-down' transformers to bridge these great distances.

Two or more voltages can be tapped from the same transformer by also providing a connection halfway to the conductor of the secondary coil. For example, a midway connection divides the secondary coil into two smaller coils. We can then choose whether to use the voltage across one part (or the other part) of the coil or take the voltage across the entire secondary coil (the two small coils together).

When determining the transformer ratio, take possible losses into account: not only in the windings themselves but also in any subsequent circuit. It is therefore better to obtain a slightly higher secondary voltage, which can then possibly be weakened further on (reducing a voltage is always easier than increasing it). Also even though no secondary current is consumed, the secondary voltage is and remains generated by the magnetic field in and around the soft iron core. In other words, the larger the core, the more power it consumes regardless of the load at the output. Leaving a transformer unused in the socket is really energy-consuming. You better get him out!

WORKED EXAMPLE 5.24

Q A transformer is to be used to provide a 60 V output from a 240 V a.c. supply. Calculate (a) the turns ratio required and (b) the number of primary turns, if the secondary is wound with 500 turns.

$V_2 = 60$ V; $V_1 = 240$ V; $N_2 = 500$

(a) $$\frac{V_1}{V_2} = \frac{N_1}{N_2} = \frac{240}{60}$$

so, turns ratio, $\frac{N_1}{N_2} = \frac{4}{1}$ or 4 : 1

(b) $$\frac{N_1}{500} = \frac{4}{1}$$

$$N_1 = 2000$$

Since the load is purely resistive, the output power, P_2, is given by V_2I_2 and the input power, $P_1 = V_1I_1$. Also since the transformer has been considered to be 100% efficient (no losses), then

$$P_2 = P_1$$

$$V_2I_2 = V_1I_2$$

$$\frac{I_1}{I_2} = \frac{V_2}{V_1} \text{ but } \frac{V_2}{V_1} = \frac{N_2}{N_1}$$

hence, $$\frac{I_1}{I_2} = \frac{N_2}{N_1} \qquad (5.23)$$

i.e. the current ratio is the *inverse* of the turns ratio.

This result is also logical. For example, if the voltage was 'stepped up' by the ratio N_2/N_1, then the current must be 'stepped down' by the same ratio. If this was not the case, then we would get more power out than was put in! Although this result would be very welcome, it is a physical impossibility. It would require the machine to be *more* than 100% efficient.

WORKED EXAMPLE 5.25

Q A 15 Ω resistive load is connected to the secondary of a transformer. The terminal p.d. at the secondary is 240 V. If the primary is connected to a 600 V a.c. supply, calculate (a) the transformer turns ratio, (b) the current and power drawn by the load and (c) the current drawn from the supply. Assume an ideal transformer.

$R_L = 15\ \Omega;\ V_2 = 240\ \text{V};\ V_1 = 600\ \text{V}$

The appropriate circuit diagram is shown in Figure 5.27.

(a)
$$\frac{N_1}{N_2} = \frac{V_1}{V_2} = \frac{600}{240}$$

turns ratio, $N_1 / N_2 = 2.5:1$

(b)
$$I_2 = \frac{V_2}{R_L} = \frac{240\ \text{V}}{15\ \Omega} = 16\ A$$

$$P_2 = V_2 I_2 = 240\ \text{V} \times 16\ \text{A} = 3.84\ \text{kW}$$

(c)
$$P_1 = P_2 = 3.84\ \text{kW}$$

$$P_1 = V_1 I_1$$

$$I_1 = \frac{P_1}{V_1} = \frac{3840}{600} = 6.4\ \text{A}$$

Alternatively, using the inverse of the turns ratio:

$$I_1 = I_2 \times \frac{N_2}{N_1} = \frac{16}{2.5}$$

$$I_1 = 6.4\ \text{A}$$

SUMMARY OF EQUATIONS

Self-induced emf: $e = -N\dfrac{d\Pi}{dt}$

Emf in a straight conductor: $e = B\ell v \sin\theta$

Force on a current-carrying conductor: $F = BI\ell \sin\theta$

Motor principle: $T = BANI$

Force between current-carrying conductors: $F = \dfrac{2\times10^{-7} I_1 I_2}{d}$

Voltmeter figure of merit: $\dfrac{I}{I_{fsd}}$

Self-inductance: self-induced emf,

$$e = -L\frac{di}{dt}$$

$$L = N\frac{d\varphi}{di} = \frac{N\Phi}{I}$$

$$L = \frac{N^2}{S} = \frac{\mu_0 \mu_r N^2 A}{\ell}$$

Energy stored: $W = 0.5\,LI^2$

Mutual inductance: mutually induced emf,

$$e_2 = -M\frac{di_1}{dt}$$

$$M = N_2\frac{d\varphi_2}{di_1} = \frac{N_2\Phi_2}{I_1}$$

$$M = k\sqrt{L_1L_2}$$

Energy stored: $W = 0.5\,L_1I_1^2 + 0.5\,L_2I_2^2 \pm MI_1I_2$

Transformer:

$$\text{Voltage ratio, } \frac{V_2}{V_1} = \frac{N_2}{N_1}$$

$$\text{Current ratio, } \frac{I_2}{I_1} = \frac{N_2}{N_2}$$

ASSIGNMENT QUESTIONS

1 The flux linking a 600-turn coil changes uniformly from 100 mWb to 50 mWb in a time of 85 ms. Calculate the average emf induced in the coil.

2 An average emf of 350 V is induced in a 1000-turn coil when the flux linking it changes by 200 μWb. Determine the time taken for the flux to change.

3 A flux of 1.5 mWb, linking with a 250-turn coil, is uniformly reversed in a time of 0.02 s. Calculate the value of the emf so induced.

4 A coil of 2000 turns is linked by a magnetic flux of 400 μWb. Determine the emf induced in the coil when (a) this flux is reversed in 0.05 s and (b) the flux is reduced to zero in 0.15 s.

5 When a magnetic flux linking a coil changes, an emf is induced in the coil. Explain the factors that determine (a) the magnitude of the emf and (b) the direction of the emf.

6 State Lenz's law, and hence explain the term 'back emf'.

7 A coil of 15 000 turns is required to produce an emf of 15 kV. Determine the rate of change of flux that must link with the coil in order to provide this emf.

8 A straight conductor, 8 cm long, is moved with a constant velocity at right angles to a magnetic field. If the emf induced in the conductor is 40 mV and its velocity is 10 m/s, calculate the flux density of the field.

9 A conductor of effective length 0.25 m is moved at a constant velocity of 5 m/s, through a magnetic field of density 0.4 T. Calculate the emf induced when the direction of movement relative to the magnetic field is (a) 90°, (b) 60° and (c) 45°.

10 Figure 5.28 represents two of the armature conductors of a d.c. generator, rotating in a clockwise direction.

An armature is the rotating part of a d.c. machine. If the machine is used as a generator, it contains the coils into which the emf is induced. In the case of a motor, it contains the coils through which current must be passed, to produce the torque.

Copy this diagram and hence:

a Indicate the direction of the field pattern of the magnetic poles.

b Indicate the direction of induced emf in each side of the coil.

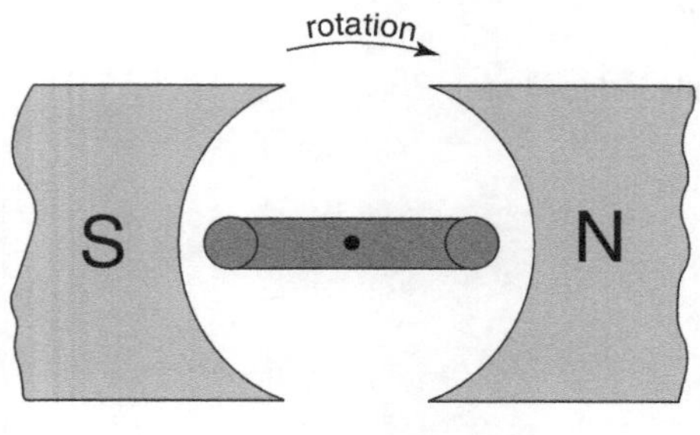

Figure 5.28 Two armature conductors for Assignment Question 10

c If this arrangement was to be used as a motor, with the direction of rotation as shown, indicate the direction of current flow required through the coil.

11 A conductor of effective length 0.5 m is placed at right angles to a magnetic field of density 0.45 T. Calculate the force exerted on the conductor if it carries a current of 5 A.

12 A conductor of effective length 1.2 m is placed inside a magnetic field of density 250 mT. Determine the value of current flowing through the conductor if a force of 0.75 N is exerted on the conductor.

13 A conductor, when placed at right angles to a magnetic field of density 700 mT, experiences a force of 20 mN, when carrying a current of 200 mA. Calculate the effective length of the conductor.

14 A conductor, 0.4 m long, lies between two pole pieces, with its length parallel to the pole faces. Determine the force exerted on the conductor if it carries a current of 30 A and the flux density is 0.25 T.

15 Two long parallel conductors are spaced 12 cm between centres. If they carry 100 A and 75 A, respectively, calculate the force per metre length acting on them. If the currents are flowing in opposite directions, will this be a force of attraction or repulsion? Justify your answer by means of a sketch of the magnetic field pattern produced.

16 The magnetic flux density at a distance of 1.4 m from the centre of a current-carrying conductor is 0.25 mT. Determine the value of the current.

17 Calculate the self-inductance of a 700-turn coil, if a current of 5 A flowing through it produces a flux of 8 mWb.

18 A coil of 500 turns has an inductance of 2.5 H. What value of current must flow through it in order to produce a flux of 20 mWb?

19 When a current of 2.5 A flows through a 0.5 H inductor, the flux produced is 80 μWb. Determine the number of turns.

20 A 1000-turn coil has a flux of 20 mWb linking it when carrying a current of 4 A. Calculate the coil inductance, and the emf induced when the current is reduced to zero in a time of 25 ms.

21 A coil has 300 turns and an inductance of 5 mH. How many turns would be required to produce an inductance of 0.8 mH if the same core material were used?

22 If an emf of 4.5 V is induced in a coil having an inductance of 200 mH, calculate the rate of change of current.

23 An iron ring having a mean diameter of 300 mm and cross-sectional area of 500 mm^2 is wound with a 150-turn coil. Calculate the inductance if the relative permeability of the ring is 50.

24 An iron ring of mean length 50 cm and cross-sectional area of 0.8 cm^2 is wound with a coil of 350 turns. A current of 0.5 A through the coil produces a flux density of 0.6

T in the ring. Calculate (a) the relative permeability of the ring, (b) the inductance of the coil and (c) the value of the induced emf if the current decays to 20% of its original value in 0.01 s, when the current is switched off.

25 When the current in a coil changes from 2 A to 12 A in a time of 150 ms, the emf induced into an adjacent coil is 8 V. Calculate the mutual inductance between the two coils.

26 The mutual inductance between two coils is 0.15 H. Determine the emf induced in one coil when the current in the other decreases uniformly from 5 A to 3 A, in a time of 10 ms.

27 A coil of 5000 turns is wound on to a non-magnetic toroid of cross-sectional area of 100 cm^2 and mean circumference of 0.5 m. A second coil of 1000 turns is wound over the first coil. If a current of 10 A flows through the first coil, determine (a) the self-inductance of the first coil, (b) the mutual inductance, assuming a coupling factor of 0.45 and (c) the average emf induced in the second coil if interruption of the current causes the flux to decay to zero in 0.05 s.

28 Two air-cored coils, A and B, are wound with 100 and 500 turns, respectively. A current of 5 A in A produces a flux of 15 μWb. Calculate (a) the self-inductance of coil A, (b) the mutual inductance if 75% of the flux links with B and (c) the emf induced in each of the coils when the current in A is reversed in a time of 10 ms.

29 Two coils, of self-inductance 50 mH and 85 mH respectively, are placed parallel to each other. If the coupling coefficient is 0.9, calculate their mutual inductance.

30 The mutual inductance between two coils is 250 mH. If the current in one coil changes from 14 A to 5 A in 15 ms, calculate (a) the emf induced in the other coil, and (b) the change of flux linked with this coil if it is wound with 400 turns.

31 The mutual inductance between the two windings of a car ignition coil is 5 H. Calculate the average emf induced in the high-tension winding, when a current of 2.5 A, in the low-tension winding, is reduced to zero in 1 ms. You may assume 100% flux linkage between the two windings.

32 Sketch the circuit symbol for a transformer, and explain its principle of operation. Why is the core made from laminations? Is the core material a 'hard' or a 'soft' magnetic material? Give the reason for this.

33 A transformer with a turns ratio of 20:1 has 240 V applied to its primary. Calculate the secondary voltage.

34 A 4:1 voltage 'step-down' transformer is connected to a 110 V a.c. supply. If the current drawn from this supply is 100 mA, calculate the secondary voltage, current and power.

35 A transformer has 450 primary turns and 80 secondary turns. It is connected to a 240 V a.c. supply. Calculate (a) the secondary voltage and (b) the primary current when the transformer is supplying a 20 A load.

36 A coil of self-inductance 0.04 H has a resistance of 15 Ω. Calculate the energy stored when it is connected to a 24 V d.c. supply.

37 The energy stored in the magnetic field of an inductor is 68 mJ, when it carries a current of 1.5 A. Calculate the value of self-inductance.

38 What value of current must flow through a 20 H inductor if the energy stored in its magnetic field, under this condition, is 60 J?

SUGGESTED PRACTICAL ASSIGNMENTS

Note: The majority of these assignments are only qualitative in nature.

Assignment 1

To investigate Faraday's laws of electromagnetic induction.

Apparatus

Several coils, having different numbers of turns

2 × permanent bar magnets
1 × multimeter

Method

1 Carry out the procedures outlined in Section 5.1 at the beginning of this chapter.
2 Write an assignment report, explaining the procedures carried out, and stating the conclusions that you could draw from the observed results.

Assignment 2

Force on a current-carrying conductor.

Apparatus

1 × current balance
1 × variable d.c. psu
1 × multimeter

Method

1 Assemble the current balance apparatus.
2 Adjust the balance weight to obtain the balanced condition, prior to connecting the psu.
3 With maximum length of conductor, and all the magnets in place vary the conductor current in steps. For each current setting, re-balance the apparatus and note the setting of the balance weight.
4 Repeat the balancing procedure with a constant current, and maximum magnets, but varying the effective length of the conductor.
5 Repeat once more, this time varying the number of magnets. The current must be maintained constant, as must the conductor length.
6 Tabulate all results obtained, and plot the three resulting graphs.
7 Write an assignment report. This should include a description of the procedures carried out, and conclusions drawn, regarding the relationships between the force produced and I, ℓ and B.

Assignment 3

To determine the relationship between turns ratio and voltage ratio for a simple transformer.

Apparatus

Either 1 × single-phase transformer with tappings on both windings *or* Several different coils with a ferromagnetic core

Either a low voltage a.c. supply *or* 1 × a.c. signal generator
1 × multimeter

Method

1 Connect the primary to the a.c. source.
2 Measure both primary and secondary voltages, and note the corresponding number of turns on each winding.
3 Vary the number of turns on each winding, and note the corresponding values of the primary and secondary voltages.
4 Tabulate all results. Write a brief report, explaining your findings.

Chapter 6

Semiconductor Theory and Diodes

LEARNING OUTCOMES

This chapter explains the behaviour of semiconductors and the way in which they are employed in diodes.

On completion of this chapter you should be able to:

1. Understand the way in which conduction takes place in semiconductor materials.
2. Understand how these materials are employed to form devices such as diodes.
3. Understand the action of a zener diode and perform basic calculations involving a simple regulator circuit.
4. Understand the action of a LED and a solar cell and perform basic calculations on the use of it.

6.1 ATOMIC STRUCTURE

In Chapter 1 it was stated that an atom consists of a central nucleus containing positively charged protons, and neutrons, the latter being electrically neutral, surrounded by negatively charged electrons orbiting in layers or shells. Electrons in the inner orbits or shells have the least energy and are tightly bound into their orbits due to the electrostatic force of attraction between them and the nucleus. Electrons in the outermost shell experience a much weaker binding force, and are known as valence electrons.

In conductors, like copper or silver, it is these valence electrons that can gain sufficient energy to break free from their parent atoms. They have a hard time keeping that electron and therefore copper and silver are good conductors: when these valence electrons transfer to another atom, they create a changing electric field. These 'free' electrons are available to drift through the material under the influence of an emf and hence are mobile charge carriers which produce current flow.

The shells are identified by letters of the alphabet, beginning with the letter K for the innermost shell, L for the next and so on. Each shell represents a certain energy level, and each shell can contain only up to a certain maximum number of electrons. This maximum possible number of electrons contained in a given shell is governed by the relationship $2n^2$, where n is the number of the shell. Thus the *maximum* number of electrons in the first four shells will be as shown in Table 6.1.

All things in nature tend to stabilise at their lowest possible energy level, and atoms and electrons are no exception. This results in the lowest energy levels (shells) being filled first until all the electrons belonging to that atom are accommodated. Another feature of the

DOI: 10.1201/9781003308294-6

Table 6.1 The **maximum** number of electrons in the first four shells

Shell	*n*	*Max. no. of electrons*
K	1	$2 \times 1^2 = 2$
L	2	$2 \times 2^2 = 8$
M	3	$2 \times 3^2 = 18$
N	4	$2 \times 4^2 = 32$

Table 6.2 Maximum number of electrons for the subshells

Shell	*L*			*M*			*N*		
Subshell	2s	2p	3s	3p	3d	4s	4p	4d	4f
Max. no.	2	6	2	6	10	2	6	10	14
Total	8		18			32			

system is that if the outermost shell of an atom is completely full (contains its maximum permitted number of electrons) then the binding force on these valence electrons is very strong and the atom is very stable. To illustrate this consider the inert gas neon. The term 'inert' is used because it is very difficult to make it react to external influences. A neon atom has a total of ten electrons, two of which are in the K shell and the remaining eight completely fill the L shell. Having a full valence shell is the reason why neon, krypton and xenon are inert gases. In contrast, a hydrogen atom has only one electron, so its valence shell is almost empty and it is a highly reactive element. One further point to bear in mind is that the electrons in the shells (from L onwards) may exist at slightly different energy levels known as subshells. These subshells may also contain only up to a certain maximum number of electrons. This is shown, for the L, M and N shells, in Table 6.2.

6.2 INTRINSIC (PURE) SEMICONDUCTORS

Semiconductors are group 4 elements, which means they have four valence electrons. For this reason they are also known as tetravalent elements. Among this group of elements are carbon (C), silicon (Si), germanium (Ge) and tin (Sn). Of these only silicon and germanium are used as intrinsic semiconductors, with silicon being the most commonly used. Carbon is not normally considered as a semiconductor because it can exist in many different forms, from diamond to graphite. Similarly, tin is not used because at normal ambient temperatures it acts as a good conductor. The following descriptions of the behaviour of semiconductor materials will be confined to silicon although the general properties and behaviour of germanium are almost the same. The arrangement of electrons in the shells and subshells of silicon is shown in Table 6.3.

From Table 6.3 it may be seen that the four valence electrons are contained in the M shell, where the 3s subshell is full but the 3p subshell contains only two electrons. However, from Table 6.2 it can be seen that a 3p subshell is capable of containing up to a maximum of six electrons before it is full, so in the silicon atom there is space for a further four electrons to be accommodated in this outermost shell.

Silicon has an atomic bonding system known as covalent bonding whereby each of the valence electrons orbits not only its 'parent' atom, but also orbits its closest neighbouring atom. This effect is illustrated in Figure 6.1, where the five large, shaded circles represent

Table 6.3 The arrangement of electrons in the shells and subshells of silicon

K	*L*		*M*		
1s	2s	2p	3s	3p	3d
2	2	6	2	2	–

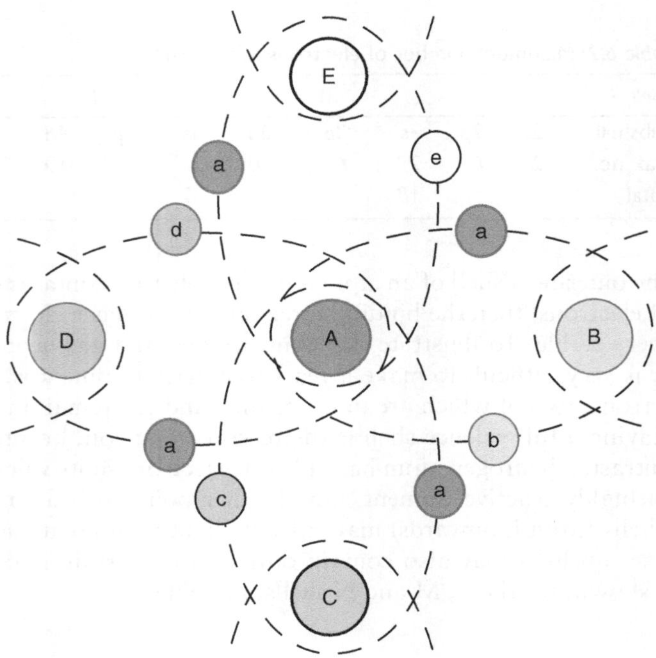

Figure 6.1 Atomic bonding system for silicon

the nucleus and shells K and L of five adjacent atoms (identified as A, B, C, D and E) and the small circles represent their valence electrons, where the letters a, b, c, etc. identify their 'parent' atoms.

Concentrating on the immediate space surrounding atom A, it may be seen that there are actually eight valence electrons orbiting this atom: four of its own plus one from each of its four nearest neighbours. This figure is only two-dimensional and is centred on atom A. However, the same arrangement would be found if the picture was centred on any given atom in the crystal lattice. In addition, the actual lattice is of course three-dimensional. In this case imagine atom A being located at the centre of an imaginary cube with the other four neighbouring atoms being at four of the corners of the cube. Each of these 'corner' atoms is in turn at the centre of another imaginary cube, and so on throughout the whole crystal lattice. The result is what is known as the diamond crystal lattice.

From the above description it may be seen that each silicon atom has an *apparent* valency of 8, which is the same as for the inert gases such as neon. The covalent bonding system is a very strong one so the valence electrons are quite tightly bound into it. It is for this reason that intrinsic silicon is a relatively poor conductor of electrical current, and is called a semiconductor.

In contrast to conductors (having 'free' electrons conducting electric current), in insulator material electric current does not flow freely. The atoms have tightly bound electrons which cannot readily move, resulting in higher resistivity compared to semiconductors and conductors. The most common examples are non-metals, like glass, paper, rubber-like polymers and most plastics. Insulators are used in electrical equipment to support and separate electrical conductors without allowing current through themselves.

6.3 ELECTRON–HOLE PAIR GENERATION AND RECOMBINATION

Although the covalent bond is strong, it is not perfect. Thus, when a sample of silicon is at normal ambient temperature, a few valence electrons will gain sufficient energy to break free from the bond and so become free electrons available as mobile charge carriers. Whenever such an electron breaks free and drifts away from its parent atom it leaves behind a space in the covalent bond, and this space is referred to as a hole. Thus, whenever a bond is broken, an electron–hole pair is generated. This effect is illustrated in Figure 6.2, where the short straight lines represent electrons and the small circle represents a corresponding hole. The large circles again represent the silicon atoms complete with their inner shells of electrons.

The atom which now has a hole in its valence band is effectively a positive ion because it has lost an electron which would normally occupy that space. On the atomic scale, the ion is very massive, is locked into the crystal lattice and so cannot move. However, electron–hole pair generation will be taking place in a random manner throughout the crystal lattice, and a generated free electron will at some stage drift into the vicinity of one of these positive ions, and be captured, i.e. the hole will once more be filled by an electron. This process is known as recombination, and when it occurs the normal charge balance of that atom is restored.

The hole-pair generation and recombination processes occur continuously and, since heat is a form of energy, will increase as the temperature increases. This results in more mobile charge carriers being available, and accounts for the fact that semiconductors have a negative temperature coefficient of resistance, i.e. as they get hotter they conduct more easily. It

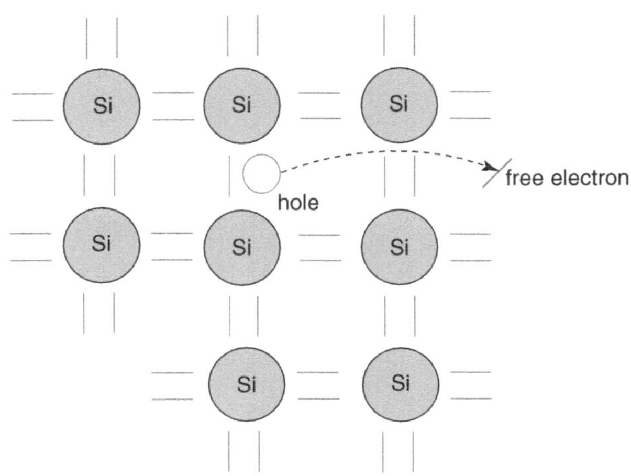

Figure 6.2 Electron-hole pair generation and recombination

must be borne in mind that although these thermally generated mobile charge carriers are being produced, the sample of material *as a whole* still remains electrically neutral. In other words, if a 'head count' of all the positive and negative charged particles could be made, there would still be a balance between positive and negative, i.e. for every free electron there will be a corresponding hole.

The concept of the drift of free electrons through the material may be readily understood, but the concept of hole mobility is more difficult to appreciate. In fact the holes themselves cannot move – they are merely generated and filled. However, when a bond breaks down, the electron that drifts away will at some point fill a hole elsewhere in the lattice. Thus the hole that has been filled is replaced elsewhere by the newly generated hole, and will *appear* to have drifted to a new location. In order to simplify the description of conduction in a semiconductor, the holes are considered to be mobile positive charge carriers whilst the free electrons are of course mobile negative charge carriers.

6.4 CONDUCTION IN INTRINSIC AND EXTRINSIC (IMPURE) SEMICONDUCTORS

Figure 6.3 illustrates the effect when a source of emf is connected across a sample of pure silicon. The electric field produced by the battery will attract free electrons towards the positive plate and the corresponding holes towards the negative plate. Since the external circuit is completed by conductors, and holes exist only in semiconductors, then how does current actually flow around the circuit without producing an excess of positive charge (the holes) at the left-hand end of the silicon? The answer is quite simple. For every electron that leaves the right-hand end and travels to the positive plate of the battery, another is released from the negative plate and enters the silicon at the left-hand end, where a recombination can occur. This recombination will be balanced by fresh electron–hole pair generation. Thus, within the silicon there will be a continuous drift of electrons in one direction with a drift of a corresponding number of holes in the opposite direction. In the external circuit the current flow is of course due only to the drift of electrons.

Although pure silicon and germanium will conduct, as explained earlier, their characteristics are still closer to insulators than to conductors. In order to improve their conduction very small quantities (in the order of 1 part in 10^8) of certain other elements are added. This process is known as doping. The impurity elements that are added are either pentavalent (have five valence electrons) or are trivalent (have three valence electrons) atoms. Depending upon which type is used in the doping process determines which one of the two types of

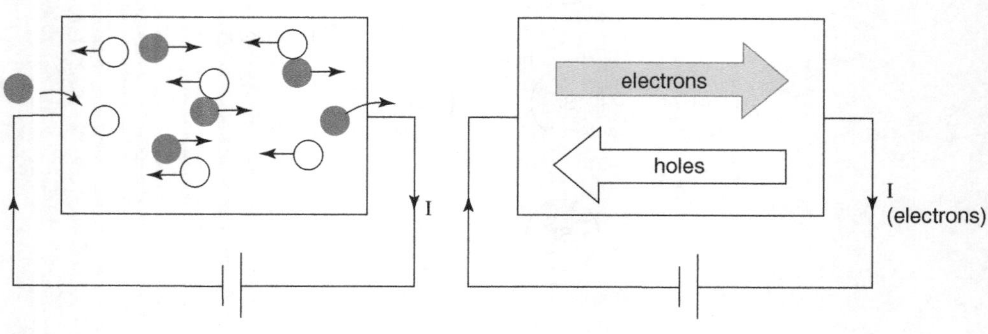

Figure 6.3 Conduction in pure silicon

extrinsic semiconductor is produced: an n-type or a p-type semiconductor, explained in the following sections.

6.5 N-TYPE SEMICONDUCTOR

To produce this type of semiconductor, pentavalent impurities are employed. The most commonly used are arsenic (As), phosphorus (P) and antimony (Sb). When atoms of such an element are added to the silicon a bonding process takes place such that each impurity atom joins the covalent bonding system of the silicon. However, since each impurity atom has five valence electrons, one of these cannot find a place in a covalent bond. These 'extra' electrons then tend to drift away from their parent atoms and become additional free electrons in the lattice. Since these impurities donate an extra free electron to the material they are also known as donor impurities.

As a consequence of each donor atom losing one of its valence electrons, they become positive ions locked into the crystal lattice. Note that free electrons introduced by this process *do not* leave a corresponding hole, although thermally generated electron–hole pairs will still be created in the silicon. The effect of the doping process is illustrated in Figure 6.4.

Since the extra charge carriers introduced by the impurity atom are negatively charged electrons, and these will be in addition to the electron–hole pairs, there will be more mobile negative charge carriers than positive, which is why the material is known as an n-type semiconductor. In this case the electrons are the majority charge carriers and the holes are the minority charge carriers. It should again be noted that the material as a whole still remains electrically neutral since for every extra donated free electron there will be a fixed positive ion in the lattice. Thus a sample of n-type semiconductor may be represented as consisting of a number of fixed positive ions with a corresponding number of free electrons, in addition to the thermally generated electron–hole pairs. This is shown in Figure 6.5.

The circuit action when a battery is connected across the material is illustrated in Figure 6.6. Once more, only electrons flow around the external circuit, whilst within the semiconductor there will be movement of majority carriers in one direction and minority carriers in the opposite direction.

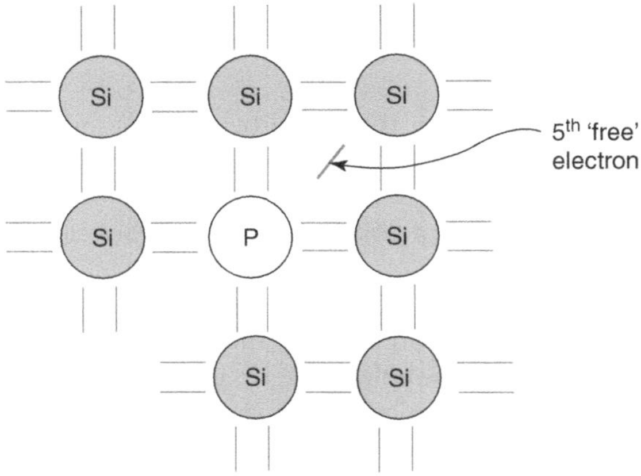

Figure 6.4 Doping in n-type semiconductor

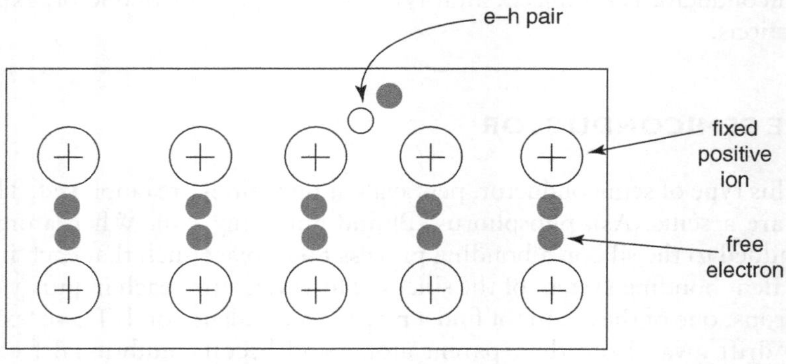

Figure 6.5 The effect of the doping process

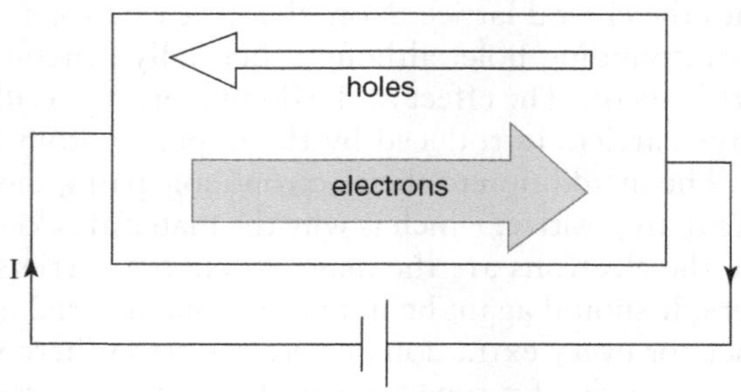

Figure 6.6 An n-type semiconductor with a number of free electrons

6.6 P-TYPE SEMICONDUCTOR

In this case a trivalent impurity such as aluminium (Al), gallium (Ga) or indium (In) is introduced. These impurity atoms also join the covalent bonding system, but since they have only three valence electrons there will be a gap or hole in the bond where an electron would normally be required. Due to electron–hole pair generation in the lattice, this hole will soon become filled, and hence the hole will have effectively drifted off elsewhere in the lattice. Since each impurity atom will have accepted an extra electron into its valence band they are known as acceptor impurities, and become fixed negative ions. The result of the doping process is illustrated in Figure 6.7.

We now have the situation whereby there will be more mobile holes than there are free electrons. Since holes are positive charge carriers, and they will be in the majority, the doped material is called a p-type semiconductor, and it may be considered as consisting of a number of fixed negative ions and a corresponding number of mobile holes as shown in Figure 6.8.

The circuit action when a battery is connected across the material is shown in Figure 6.9. As the holes approach the left-hand end they are filled by incoming electrons from the battery. At the same time, fresh electron–hole pairs are generated, the electrons being swept to and out of the right-hand end, and the holes drifting to the left-hand end to be filled. Once more, the current flow in the semiconductor is due to the movement of holes and electrons in opposite directions, and only electrons in the external circuit. As with the n-type material, p-type is also electrically neutral.

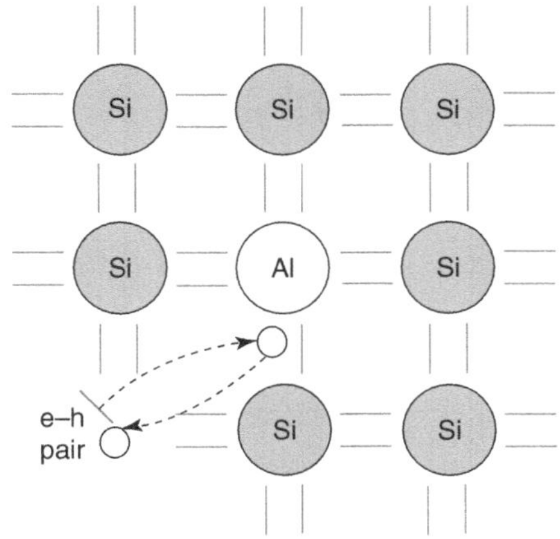

Figure 6.7 Doping in p-type semiconductor

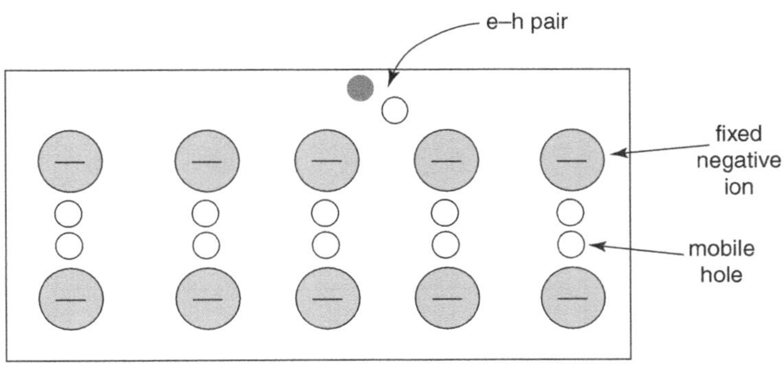

Figure 6.8 The effect of the doping process

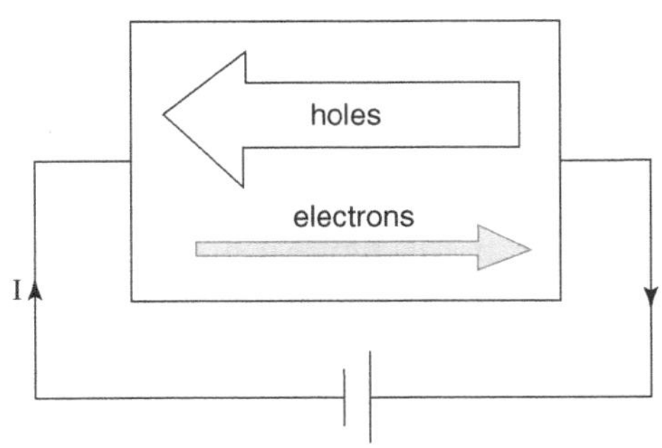

Figure 6.9 An p-type semiconductor with a number of free holes

6.7 THE P-N JUNCTION

When a sample of silicon is doped with both donor and acceptor impurities so as to form a region of p-type and a second region of n-type material in the *same crystal lattice*, the boundary where the two regions meet is called a p-n junction. This is illustrated in Figure 6.10.

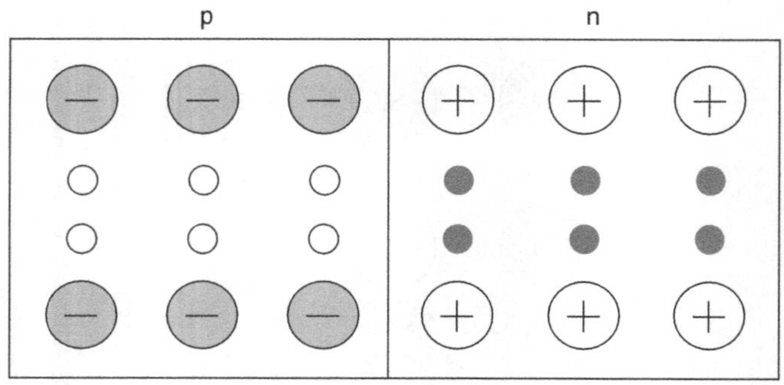

Figure 6.10 The p-n junction in the same crystal lattice

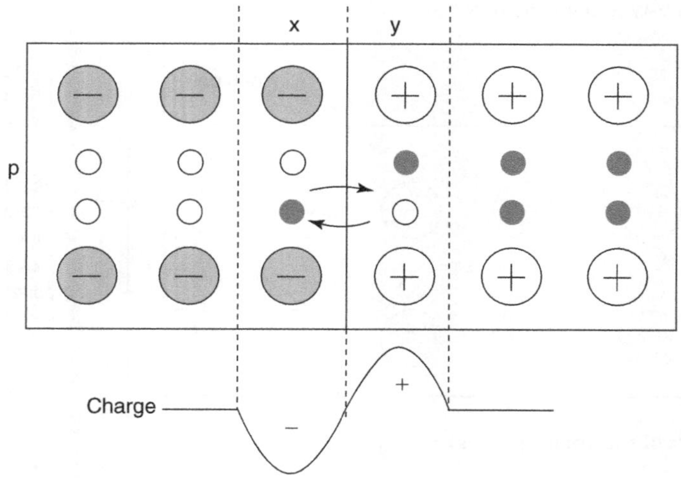

Figure 6.11 Diffusion of electrons and holes

Due to their random movement some of the electrons will diffuse across the junction into the p-type, and similarly some of the holes will diffuse across into the n-type. This effect is illustrated in Figure 6.11, and from this figure it may be seen that region x acquires a net negative charge whilst region y acquires an equal but positive net charge.

The region between the dotted lines is only about 1 μm wide, and the negative charge on x prevents further diffusion on electrons from the n-type. Similarly the positive charge on y prevents further diffusion of holes from the p-type. This redistribution of charge results in a potential barrier across the junction. In the case of silicon this barrier potential will be in the order of 0.6 to 0.7 V, and for germanium about 0.2 to 0.3 V. Once again note that although there has been some redistribution of charge, the sample of material *as a whole* is still electrically neutral (count up the numbers of positive and negative charges shown in Figure 6.11).

A diode is a semiconductor device, based on a p-n junction, allowing current to flow through it in one direction only. It is the electronic equivalent of a mechanical valve, for example the valve in a car tyre. This device allows air to be pumped into the tyre, but prevents the air from escaping. A diode is so called because it has two terminals: the anode,

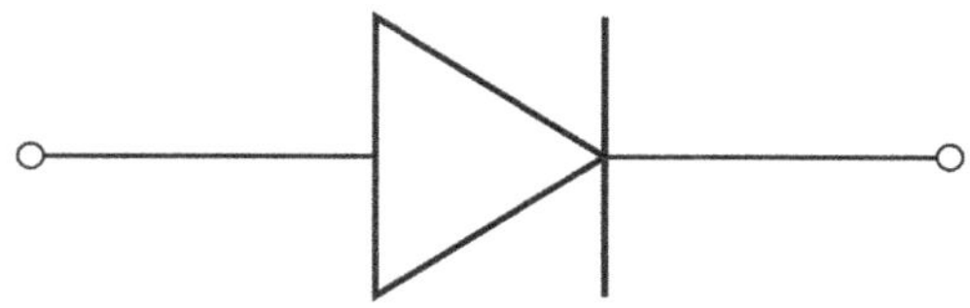

Figure 6.12 The circuit symbol for a diode

which is the positive terminal, and the cathode, which is the negative terminal. In the case of a p-n junction diode the anode is the p-type and the cathode is the n-type.

The circuit symbol for a diode is shown in Figure 6.12. The 'arrow head' part of the symbol is known as the anode. This indicates the direction in which *conventional* current can flow through it. The 'plate' part of the symbol is the cathode, and indicates that conventional current is prevented from entering at this terminal. Thus, provided that the anode is more positive than the cathode, the diode will conduct. This is known as the forward bias condition. If the cathode is more positive than the anode, the diode is in its blocking mode, and does not conduct. This is known as reverse bias.

Note: The potentials at anode and cathode do not have to be positive and negative. Provided that the anode is *more* positive than the cathode, the diode will conduct. So if the anode potential is (say) + 10 V and the cathode potential is + 8 V, then the diode will conduct. Similarly, if these potentials are reversed, the diode will not conduct.

You can measure a diode (which is not included in a circuit) with a multimeter in the mode to test diodes or in the resistance measurement option otherwise. With a properly functioning diode, the multimeter then indicates the voltage across the diode. In the forward direction, for a silicon diode it will be in the order of 0.6 V to 0.7 V (which is in fact the barrier potential) and for a germanium diode about 0.2 V to 0.3 V. In the reverse direction there is a greater voltage across the diode: depending on the type of diode between 2.5 V and 3.5 V or 'O.L.' (short for Over Length) appears on the screen. When the diode is not functioning and thus broken, two options exist. When it is open, in both forward and reverse direction 'O.L.' appears. In the case of a shorted diode, the 0 V reading is shown in the forward and reverse direction.

6.8 FORWARD-BIASED AND REVERSE-BIASED DIODE

Figure 6.13 shows a battery connected across a diode such that the positive terminal is connected to the anode and the negative terminal to the cathode.

The electric field produced by the battery will cause holes and electrons to be swept towards the junction, where recombinations will take place. For each of these an electron from the battery will enter the cathode. This would have the effect of disturbing the charge balance within the semiconductor, so to counterbalance this a fresh electron–hole pair will be created in the p-type. This newly freed electron will then be attracted to the positive plate of the battery, whilst the hole will be swept towards the junction. Thus the circuit is complete, with electrons moving through the external circuit, and a movement of holes and electrons in the semiconductor. Hence, when the anode of the diode is made positive with respect to the cathode it will conduct, and it is said to be forward biased.

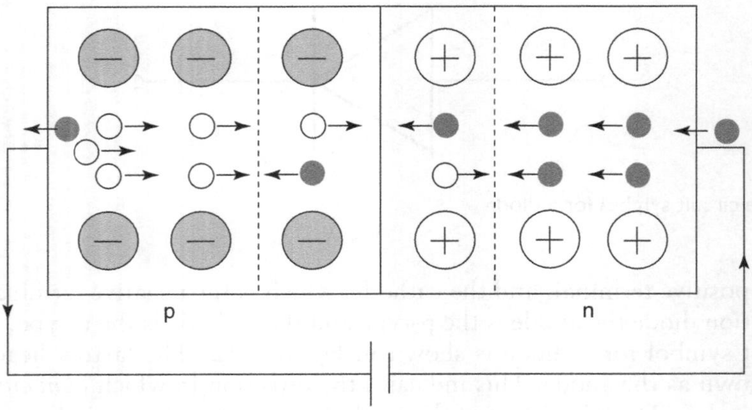

Figure 6.13 Forward-biased diode

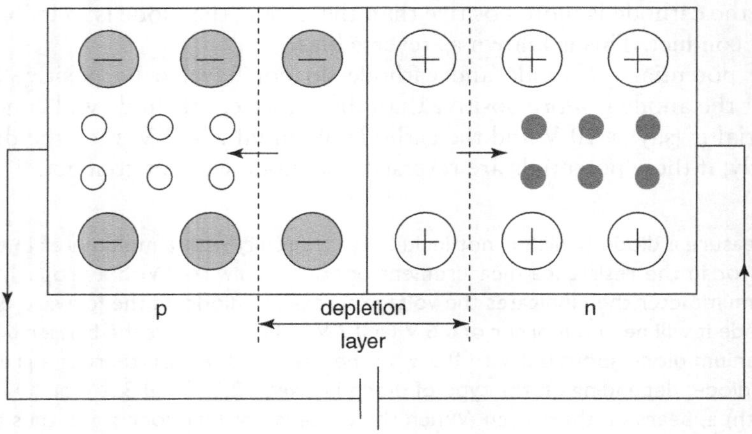

Figure 6.14 Reverse-biased diode

Consider what now happens when the battery connections are reversed (Figure 6.14). The electric field of the battery will now sweep all the mobile holes into the p-type and all the free electrons into the n-type. This leaves a region on either side of the junction which has been depleted of all of its mobile charge carriers. This layer thus acts as an insulator, and is called the depletion layer. There has been a redistribution of charge within the semiconductor, but since the circuit has an insulating layer in it, current cannot flow. The diode is said to be in its blocking mode.

However, there is no such thing as a perfect insulator, and the depletion layer is no exception. Although all the mobile charge carriers provided by the doping process have been swept to opposite ends of the semiconductor, there will still be some thermally generated electron–hole pairs. If such a pair is generated in the p-type region, the electron will be swept across the junction by the electric field of the battery. Similarly, if the pair is generated in the n-type, the hole will be swept across the junction. Thus a very small reverse current (in the order of microamps) will flow, and is known as the reverse leakage current. Since this leakage current is the result of thermally generated electron–hole pairs, as the temperature increases so too will the leakage current.

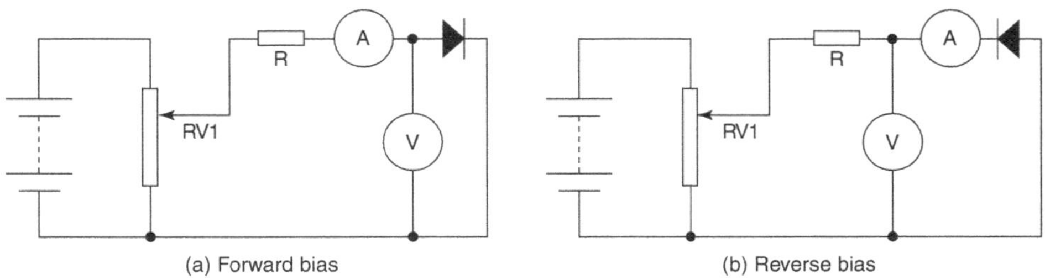

Figure 6.15 Circuits for determining forward and reverse characteristics

6.9 DIODE CHARACTERISTICS

The characteristics of a device such as a diode can be best illustrated by means of a graph (or graphs) of the current flow through it versus applied voltage. Circuits for determining both the forward and reverse characteristics are shown in Figure 6.15.

Note the change of position of the voltmeter for the two different tests. In (a) the voltmeter measures only the small p.d. across the diode itself, and not any p.d. across the ammeter. In (b) the ammeter measures only the leakage current of the diode, and does not include any current drawn by the voltmeter. The procedure in each case is to vary the applied voltage, in steps, by means of the variable resistor RV1 and record the corresponding current values. When these results are plotted, for both silicon and germanium diodes, the graphs will typically be as shown in Figure 6.16. The very different scales for both current and voltage for the forward and reverse bias conditions should be noted. Also, the actual values shown for the forward current scale and the reverse voltage scale can vary considerably from those shown, depending upon the type of diode being tested, i.e. whether it be a small signal diode or a power rectifying diode. In the case of the latter, the forward current would usually be in amperes rather than milliamps.

The sudden increase in reverse current occurs at a reverse voltage known as the reverse breakdown voltage. The effect occurs because the intensity of the applied electric field

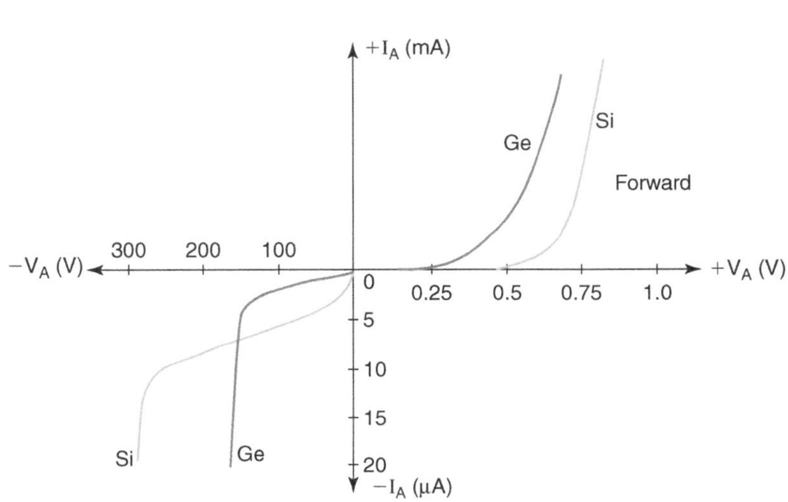

Figure 6.16 Diode characteristics

causes an increase in electron–hole pair generation. These electron–hole pairs are not due to temperature, but the result of electrons being torn from bonds by the electric field. This same field will rapidly accelerate the resulting charge carriers and as they cross the junction they will collide with atoms. These collisions will free more charge carriers, and the whole process builds up very rapidly. For this reason the effect is known as avalanche breakdown, and will usually result in the destruction of the diode.

When the impurity doping of the semiconductor is heavier than 'normal', the depletion layer produced is much thinner. In this case, when breakdown occurs, the charge carriers can pass through the depletion with very little chance of collisions taking place. This type of breakdown is known as zener breakdown and such diodes are called zener diodes.

The current flow through a resistor as function of the applied voltage is a straight line through the graph origin. The slope of the line equals the resistor value, according to Ohm's law. By comparing this straight line with the diode characteristics, the differences can be explained by the semiconductor technology.

Diodes are most often used as rectifiers, because current is only allowed to flow in one direction and not in the other direction. Different types of rectifier diodes exist, with different breakdown voltages. Some of them are listed in Table 6.4 and they all can pass currents up to 1 A without any problems. The forward voltage drop is approximately 0.7 V for each. For each type you can find a datasheet on the web, describing the entire behaviour of the component.

Another type of diode is the signal diode. This one is actually designed for much smaller currents than the rectifier diode described above. Typically that is about 100 mA to 200 mA. The two most common diodes are the 1N4148 and 1N914, which both have similar properties: the forward voltage drop is 0.7 V and the breakdown voltage is 100 V. Furthermore, a maximum of 200 mA current can flow through. These diodes are more compact than rectifier diodes and are usually made of glass with a black band designation for the cathode. These signal diodes are widely used in audio and radio frequency circuits because they respond quickly to high frequencies. That is why they are also called high-speed diodes or switching diodes.

Table 6.4 Different types of rectifier diodes with their breakdown voltage

Type	*Breakdown voltage*
1N4001	50 V
1N4002	100 V
1N4003	200 V
1N4004	400 V
1N4005	600 V
1N4006	800 V
1N4007	1000 V

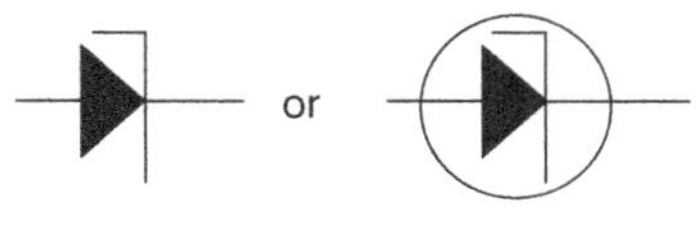

Figure 6.17 The circuit symbol for a zener diode

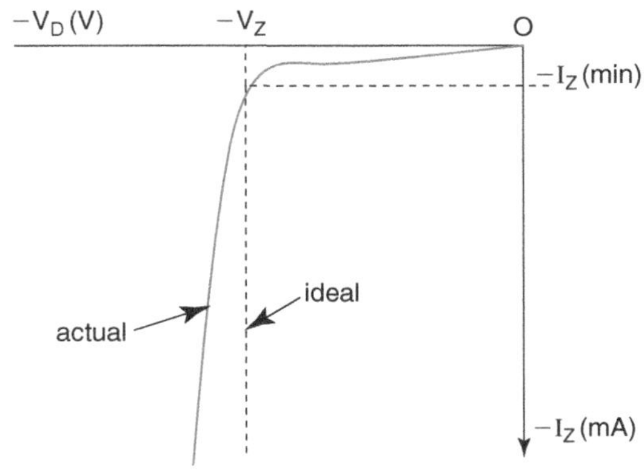

Figure 6.18 Zener diode characteristics

6.10 THE ZENER DIODE

The main feature of the zener diode is its ability to operate in the reverse breakdown mode without sustaining permanent damage. In addition, during manufacture, the precise breakdown voltage (zener voltage) for a given zener diode can be predetermined. For this reason they are also known as voltage reference diodes. The major application for these devices is to limit or stabilise a voltage between two points in a circuit. Zener diodes are available with zener voltages from 2.6 V to about 200 V. The circuit symbol for a zener diode is shown in Figure 6.17.

The forward characteristic for a zener diode will be the same as for any other p-n junction diode, and also, since the device is always used in its reverse bias mode, only its reverse characteristic need be considered. Such a reverse characteristic is shown in Figure 6.18.

Clarence Zener (1905–1993) was an American theoretical physicist and studied superconductivity, metallurgy and geometric programming. He was the first to describe the property concerning the breakdown of electrical insulators. This principle was exploited by the Bell Labs to invent a special-purpose diode, named after Zener.

In Figure 6.18, V_Z represents the zener breakdown voltage, and if it were an ideal device, this p.d. across it would remain constant, regardless of the value of current, I_Z, flowing through it. In practice the graph will have a fairly steep slope as shown. The inverse of the slope of the graph is defined as the zener diode slope resistance, r_Z, expressed in ohm as follows

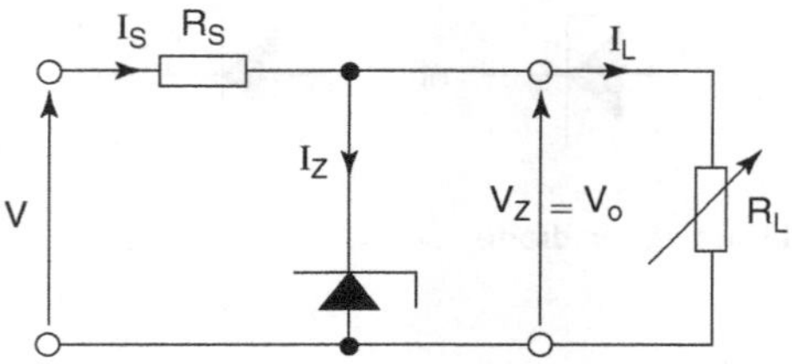

Figure 6.19 Circuit with zener diode to provide simple voltage stabilisation

$$r_Z = \frac{\delta V_D}{\delta I_Z} \tag{6.1}$$

Typical values for r_z range from 0.5 Ω to about 150 Ω. For satisfactory operation the current through the zener diode must be at least equal to $I_{z(min)}$. Due to the zener diode slope resistance, the p.d. across the zener diode will vary by a small amount from the ideal of V_z volt as the diode current changes. For example, if r_z = 1 Ω and V_z = –15 V, a change in diode current of 30 mA would cause only a 0.02% change in the zener diode p.d. This figure may be verified by applying Equation (6.1).

The value of current that may be allowed to flow through the device must be limited so as not to exceed the diode power rating. This power rating is always quoted by the manufacturer, and zener diodes are available with power ratings up to about 75 W.

Consider now the application of a zener diode to provide simple voltage stabilisation to a load. A circuit is shown in Figure 6.19.

In order for satisfactory operation the supply voltage, V_s, needs to be considerably greater than the voltage required at the load. The purpose of the series resistor R_S is to limit the maximum diode current to a safe value, bearing in mind the diode's power rating. Considering Figure 6.19, the diode current will be at its maximum when the load is disconnected, because under this condition all of the current from the supply will flow through the diode, i.e. $I_Z = I_S$. When the load is connected it will draw a current I_L, and since $I_Z = I_S - I_L$, then under this condition the diode current will decrease, since it must divert current to the load. The output voltage, however, will remain virtually unchanged. Knowing the diode power rating, a suitable value for R_S may be calculated as shown in the following worked example. This example also demonstrates the stabilising action of the circuit.

WORKED EXAMPLE 6.1

Q A 9.1 V, 500 mW zener diode is used in the circuit of Figure 6.19 to supply a 2.5 kΩ load. The diode has a slope resistance of 1.5 Ω and the input supply has a nominal value of 12 V.

(a) Calculate a suitable value for the series resistor R_s.
(b) Calculate the value of diode current when the load resistor is connected to the circuit.
(c) If the input supply voltage decreases by 10%, calculate the percentage change in the p.d. across the load.

$V_Z = 9.1$ V; $P_Z = 0.5$ W; $r_Z = 1.5\ \Omega$; $V = 12$ V; $R_L = 2500\ \Omega$

(a)

$$P_Z = V_Z I_Z$$

$$I_Z = \frac{P_Z}{V_Z} = \frac{0.5}{9.1} = 54.95 \text{ mA}$$

($I_S = I_Z$ because for this condition the load is disconnected)

$$V_S = V - V_Z = 12 - 9.1 = 2.9 \text{ V}$$

$$R_S = \frac{V_S}{I_S} = \frac{2.9}{54.95 \times 10^{-3}} = 52.78 \ \Omega$$

A resistor of this precise value would not be readily available, so the nearest preferred value resistor would be chosen. However, to ensure that the zener diode power rating cannot be exceeded, the nearest preferred value *greater than* 52.78 Ω would be chosen. Thus a 56 Ω resistor would be chosen. In order to protect the resistor, its own power rating must be taken into account. In this circuit, the maximum power dissipated by R_S is:

$$P_{max} = I_S^2 R_S = \left(54.95 \times 10^{-3}\right)^2 \times 56 = 0.169 \text{ W}$$

so a 0.25 W resistor would be chosen, and the complete answer to part (a) is:
R_S should be a 56 Ω, 0.25 W resistor

With $R_L = 2500 \ \Omega$ and $V_O = 9.1$ V

(b) $$I_L = \frac{V_O}{R_L} = \frac{9.1}{2.5 \times 10^3} = 3.64 \text{ mA}$$

$$I_Z = I_s - I_L = 55 - 3.64 = 51.36 \text{ mA}$$

(c) When V falls by 10% from its nominal value, then

$$V = 12 - (0.1 \times 12) = 12 - 1.2 = 10.8 \text{ V}$$

$$I_S = \frac{V - V_Z}{R_S} = \frac{10.8 - 9.1}{56} = 30.36 \text{ mA}$$

The current for the load must still be diverted from the zener diode, so

$$I_Z = I_S - I_L = 30.36 - 3.64 = 26.72 \text{ mA}$$

therefore, $\delta I_Z = 51.36 \text{ mA} - 26.72 \text{ mA} = 24.64 \text{ mA}$, and from equation (6.1):

$$\delta V_Z = \delta I_Z r_Z = 24.64 \times 10^{-3} \text{ mA} \times 1.5 \ \Omega$$

$$\delta V_Z = 0.037 \text{ V}$$

Thus the voltage applied to the load changes by 0.037 V, which expressed as a percentage change is:

$$\text{change} = \frac{0.037}{9.1} \times 100$$

$$\text{change} = 0.41\% \text{ (compare with a 10\% change in supply)}$$

WORKED EXAMPLE 6.2

Q A d.c. voltage of 15 V ± 5% is required to be supplied from a 24 V unstabilised source. This is to be achieved by the simple regulator circuit of Figure 6.20.

The available diodes and resistors are listed below. For each zener diode listed determine the appropriate resistor required and hence determine the total unit cost for each circuit.

Diode no.	V_Z *(V)*	*Slope resistance (Ω)*	*Max power (W)*	*Unit cost (£)*
1	15	30	0.5	0.07
2	15	15	1.3	0.20
3	15	2.5	5.0	0.67

Resistors are available in the following values and unit costs:
18 Ω, 27 Ω, 56 Ω, 100 Ω, 120 Ω, 150 Ω, 220 Ω, 270 Ω and 330 Ω.

0.25 W	£0.026
0.5 W	£0.038
1.0 W	£0.055
2.5 W	£0.260
7.5 W	£0.280

For all three diodes:

$$V_s = V - V_Z = 24 - 15 = 9\ \text{V}$$

Diode 1:

$$I_Z = \frac{P_Z}{V_Z} = \frac{0.5}{15} = 33.3\ \text{mA}$$

$$R_S = \frac{V_S}{I_Z} = \frac{9}{33.3 \times 10^{-3}} = 270\ \Omega$$

$$P_S = \frac{V_S^2}{R_s} = \frac{81}{270} = 0.3\ \text{W}$$

and, $R_S = 270\ \Omega$, 0.5 W

total unit cost = 0.07 + 0.038 = £0.045

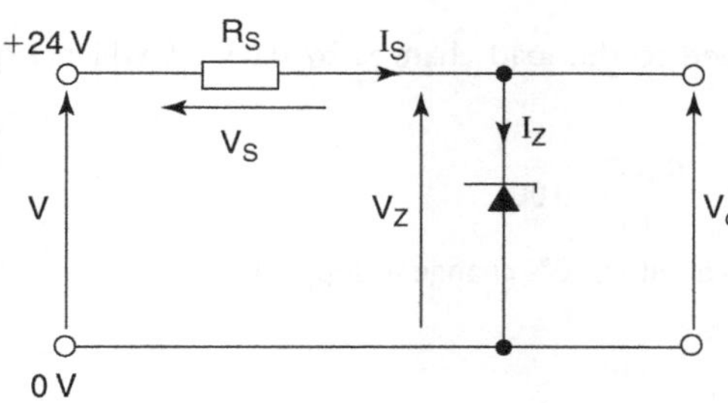

Figure 6.20 The circuit diagram for Worked Example 6.2

Diode 2:

$$I_Z = \frac{1.3}{15} = 86.7 \text{ mA}$$

$$R_S = \frac{9}{86.7 \times 10^{-3}} = 103.85\ \Omega\text{, so choose the } 120\ \Omega \text{ resistor}$$

$$P_s = \frac{81}{120} = 0.675 \text{ W so choose } 1.0 \text{ W rating}$$

$$\text{hence, } R_S = 120\ \Omega, 1.0 \text{ W}$$

$$\text{total unit cost} = 0.20 + 0.055 = £0.26$$

Diode 3:

$$I_Z = \frac{5}{15} = 333.3 \text{ mA}$$

$$R_S = \frac{9}{333.3 \times 10^{-3}} = 27\ \Omega$$

$$P_s = \frac{81}{27} = 3 \text{ W so choose } 7.5 \text{ W resistor}$$

$$\text{hence, } R_S = 27\ \Omega, 7.5 \text{ W}$$

$$\text{total unit cost} = 0.67 + 0.28 = £0.95$$

6.11 THE LED

A LED is short for light emitting diode. It is a special diode that emits light. The circuit symbol for a LED is shown in Figure 6.21. You will clearly recognise the symbol of the diode in it, as discussed earlier. The only difference is in the arrows pointing away from the diode, indicating that the diode is emitting light.

The characteristic of a LED is the same as an ordinary diode: the current flow through it versus the applied voltage. However, only the most important part of the characteristic is shown, being the positive part of the characteristic and in which the LED will emit light. The p.d. of a LED is denoted as V_F and the current as I_F. For infrared LEDs the V_F is typically 1.5 V and goes up to 3 V for blue LEDs. Red, yellow and green LEDs are usually in the middle with values of about 2 V. For the same voltage, the current values differ according to the semiconductor material used and thus according to the colour of the light emitted by the LED. The nominal current of a LED is usually 20 mA. As the current increases, light output generally also increases. These are all parameters that should be checked in the datasheets of the component that will eventually be used.

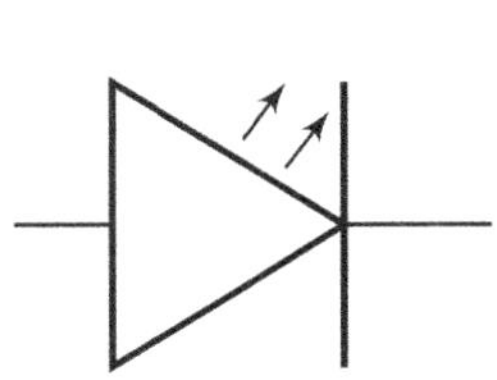

Figure 6.21 The circuit symbol for a led

A typical LED is the UR5366X, where the characteristic can be found online in the datasheet. That same datasheet also shows the relative intensity as a function of the current. When the LED starts to emit light, the current gradually increases. To prevent the current from exceeding the maximum allowable value (as mentioned in that datasheet), a resistor R_V is placed in series with the LED. It can be calculated by using Equation 6.2, with V the external applied voltage.

$$R_V = \frac{V - V_F}{I_F} \tag{6.2}$$

It is best to round this value up for an available resistance value. This is to prevent the final current from becoming too large. Note that the unavoidable heat development in the resistor can be calculated with the formula P = V.I. The resistance must be able to handle this; otherwise it will overheat.

The 7-segment display is a common way to visualise digits, as is used in displays like, for instance, clock radios. Each segment is a LED and in this way we can easily visualise all numbers and somewhat less easily all letters. In some the anode is common, in others it is the cathode. So watch out!

A photodiode is a component designed to detect light. The circuit symbol for a photodiode is shown in Figure 6.22, where the two arrows represent the incoming light.
If there is enough light, the photodiode can be used in two ways: normal polarised or reverse polarised. With normal polarisation, the light generates an electrical voltage through which a limited current in the forward direction flows. With reverse polarisation (and therefore a reverse voltage), this photodiode has a very high resistance and a reverse current flows, virtually independent of the reverse voltage, but dependent on the amount of light. A photodiode is usually used with reverse polarity. Although not optimised for it, LEDs also can convert light into electricity. The LED then works the other way around, but according to the same principle as with a photodiode. All photons of a specific LED actually all have almost the same energy content (which also determines the colour of the LED). From the forward voltage drop it can be deduced that the photons from red LEDs contain less energy than those from green LEDs. However, if a LED is used as a photodiode, then that LED only detects those photons that contain sufficient energy. This means that a red LED can be illuminated as a photodiode with a green LED (after all, the photons have enough energy). But a green LED as a photodiode cannot be illuminated with a red LED (after all, the photons do not have enough energy). This can also be measured by measuring the forward voltage drop across that LED as a photodiode (and thus detecting photons).

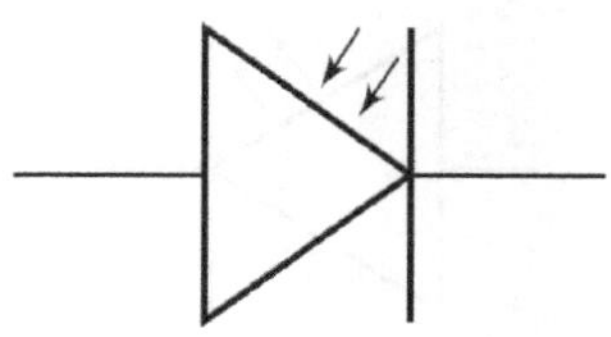

Figure 6.22 The circuit symbol for a photodiode

WORKED EXAMPLE 6.3

Q For lighting a LED, the datasheet indicates a forward current of 10 mA with a forward voltage of 2.1 V. Calculate the required resistor for a power supply of 5 V. Determine the dissipated power in the resistor.

$$I_F = 10\text{ mA};\ V_F = 2.1\text{ V};\ V = 5\text{ V}$$

$$R_V = \frac{V - V_F}{I_F} = \frac{3 - 2.1}{10} = 90\ \Omega$$

$$P = R_V.I^2 = 90.10.10 = 9\text{ mW}$$

6.12 THE SOLAR CELL

There is enough solar energy available on earth every hour to meet the global demand for electricity for an entire year. From an ecological point of view, people are increasingly talking and making use of the possibility to convert that solar energy into electrical energy. New developments in semiconductor material also make it possible to convert this energy very efficiently. The physical principle is called the photovoltaic effect. A photovoltaic cell (or Photovoltaic Cell or PV cell) or solar cell converts the solar energy into electrical energy. Sunlight contains energy packets or photons that can generate an electron–hole pair. The electrons end up in the n-type of the p-n junction, leaving a hole in the p-type. More and more electrons accumulate in the n-type and the holes in the p-type. In this way they realise a voltage across the photovoltaic cell. If a load is then externally connected, a current flows.

Although other types of photovoltaic cells or solar cells exist and there are many new developments, the crystalline silicon solar cell is the most common. The bottom (on the p-type side) gets a flat metal plate. On top of the solar cell, a conductive grid is placed, so that the electrons from the n-type can easily leave the silicon via the grid and form a current. To absorb as much light as possible and thus avoid reflection of sunlight, an anti-reflective layer is applied on top of that conductive grid. The last layer is a glass (or sometimes plexiglass) plate to protect the photovoltaic cell against the weather conditions. The result is available in different sizes, shapes and colours.

Solar cells are standard 100 cm² to 225 cm² in size, with the obtained voltage being approximately 0.5 to 0.6 V. The light intensity only influences this voltage to a limited extent, but it strongly influences the current: the more sunlight, the more current. For a solar cell of 100 cm² and with a light intensity of 1000 W/m², a maximum current of approximately 2 A can be achieved. In addition, the load resistance must be adjusted to maintain the same voltage at a different light intensity and therefore current, which is done, among other things, in the charge controller. By placing different solar cells in series, you can simply add up the voltage of each solar cell. Due to this series connection, however, the current remains identical compared to a single solar cell. Therefore, several series-connected solar cells are in turn connected in parallel. Several solar cells together and for a specific power are called solar panels. In practice, more solar cells are connected together to compensate for all kinds of losses.

Although electrical energy can also be generated on a cloudy day, the yield is still less. At night even nothing can be generated. It can hence be useful to include a battery in the system to compensate for the lower energy yield. The aforementioned charge controller ensures that the batteries are not overcharged. After all, overcharging batteries shortens their lifespan and can also damage them. To prevent current flowing back to the solar cell when the battery voltage exceeds the voltage of a solar cell, the charge controller also contains a correctly polarised diode.

If the solar cells or solar panels are used for alternating current and voltage, we still have to take additional steps. After all, a solar panel supplies direct current. In the converter (or inverter), the direct current is switched on and off electronically and filtered with a low-pass filter to obtain an alternating current. This is usually followed by a transformer (see also Chapter 5) to realise the correct alternating voltage. If the solar panel is connected to the public electricity network, it can lose any surplus of power there. The interaction between own yield and the return of surpluses to the grid is much more cost-effective than storage on a battery. The battery is therefore often omitted in these systems.

ASSIGNMENT QUESTIONS

1 A simple voltage stabiliser circuit is shown in Figure 6.23. The zener diode is a 7.5 V, 500 mW device and the supply voltage is 12 V. Calculate (a) a suitable value for R_S, and (b) the value of zener current when $R_L = 470\ \Omega$.

2 Using the circuit of Figure 6.23 with a 10 V, 1.3 W zener diode having a slope resistance of 2.5 Ω, and a supply voltage of 24 V, calculate (a) a suitable value for R_S, (b) the value of zener current when on no-load and (c) the variation of zener voltage when R_L is changed from 500 Ω to 200 Ω.

3 How many solar cells are needed in order to have 3 V and 4 A?

SUGGESTED PRACTICAL ASSIGNMENTS

Assignment 1

To obtain the forward and reverse characteristics for silicon and germanium p-n junction diodes. An investigation into the effects of increased temperature on the reverse leakage current could also be undertaken.

Assignment 2

To investigate the operation of the zener diode.

Apparatus

1 × 5.6 V, 400 mW zener diode
1 × 9.1 V, 400 mW zener diode

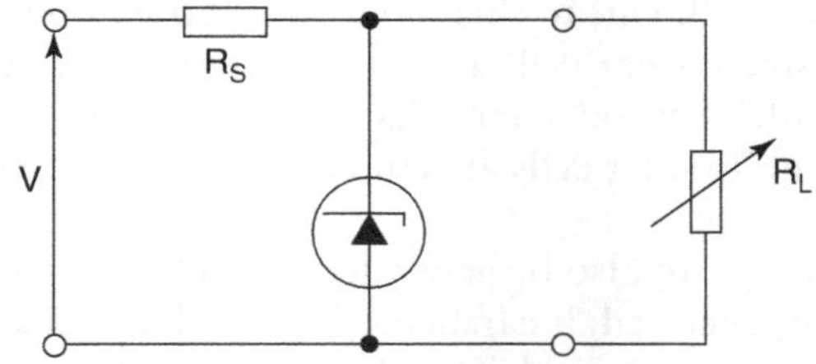

Figure 6.23 The circuit diagram for Assignment Question 1

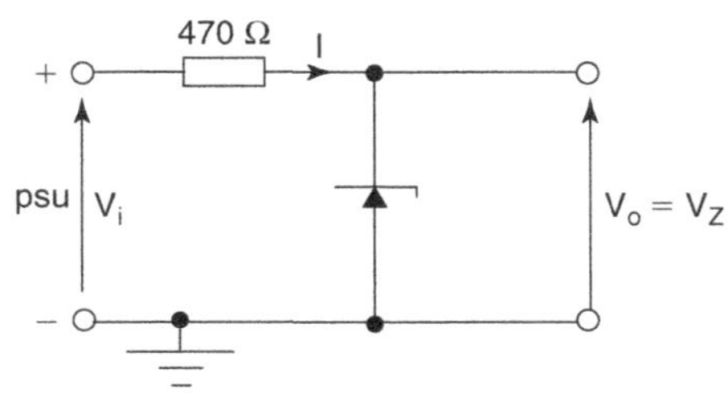

Figure 6.24 The circuit diagram for Practical Assignment 2

1 × 470 Ω resistor
1 × variable d.c. power supply unit (psu)
1 × voltmeter
1 × ammeter

Method

1 Connect the circuit of Figure 6.24 using the 5.6 V diode.
2 Vary the input voltage in 1 V steps from 0 V to + 15 V, and note the corresponding values of V_Z and I.
3 Tabulate your results and plot the reverse characteristic for the diode.
4 Repeat steps 1 to 3 above for the 9.1 V diode.
5 From the plotted characteristics, determine the diode slope resistance in each case.

Chapter 7

Transistors

LEARNING OUTCOMES

This chapter explains the behaviour of transistors and the way in which they are used in circuits. On completion of this chapter you should be able to:

1 Understand the way in which n-p-n junctions and p-n-p junctions result in the transistor effect within a bipolar transistor.
2 Understand how bipolar transistors are employed to amplify signals.
3 Describe the operation of a MOSFET and a JFET.
4 Explain the principle of inverting and non-inverting amplifiers with opamps.

7.1 BIPOLAR TRANSISTOR

In Chapter 6, a diode as a semiconductor with one single p-n junction was discussed. The diode conducts current when forward biased and blocks current (no current is conducting) when reverse biased. A bipolar transistor is basically a diode with an extra third layer of either p-type or n-type. Such a bipolar transistor is therefore a semiconductor with two different p-n junctions. It can be built like an n-p-n semiconductor (with one p-type sandwiched in between the two n-types at the outside) or like a p-n-p semiconductor (with one n-type in the middle and two p-types at the outside). The basic idea of the transistor is that it has the ability to vary the amount of current driven by a much smaller current. You can therefore consider it as an electronic leverage: with a very small effort at the input, you can exert a large effort at the output with this leverage. The word 'transistor' is actually a contraction of 'transfer' and 'resistor', because during the operation of the transistor, the input resistance is high and the output resistance is low. So the transistor is actually a component that not only amplifies the signals, but also changes the resistance from high to low. In the early days, it was mainly used to convert the change of a small signal (such as a sound wave) into a larger signal (with therefore much larger changes). Nowadays, it has been a basic part of every integrated circuit of every electronic device. At AT&T's Bell Labs, three engineers Brattain, Bardeen and Shockley were able to demonstrate the first working transistor, consisting of a thin wafer of germanium with two adjacent gold contacts on one side and a larger contact on the other. The amplification factor at that time of this point-contact transistor was 18 times.

 DOI: 10.1201/9781003308294-7

Walter Brattain (1902–1987) was an American physicist, best known as the co-inventor (with John Bardeen and William Shockley) of the transistor. In 1956, he was awarded the Nobel Prize in Physics for this. At AT&T's Bell Labs, his research interest was the surface properties of solids, in particular of semiconductors. It was already known that semiconductors could rectify an alternating current and that this effect was a surface property of semiconductors.

John Bardeen (1908–1991) was an American physicist and two-time Nobel laureate: the first time in 1956 along with Walter Brattain and William Shockley for the invention of the transistor; the second time in 1972 with Leon Cooper and Robert Schrieffer for the theoretical explanation of superconductivity. He wanted, together with Brattain, to find a solid-state alternative to the fragile vacuum tube amplifiers, based on Shockley's ideas to use an external electric field to influence the conductivity of semiconductors.

William Shockley (1910–1989) was an American physicist and co-inventor of the transistor with John Bardeen and Walter Brattain. Shockley was supervising Bardeen and Brattain. While he was on a holiday in December 1947, Bardeen and Brattain worked on a new circuit and they succeeded in demonstrating the first working point-contact transistor. It was only after this that Shockley was informed, who was dismayed that he had been kept out of this major breakthrough.

Packaging of transistors exists in all kinds of sizes and colours. However, there is only one constant: it has three contact pins. Each of the three semiconductor parts has its own contact pin. The collector (shortened to C) connects to the largest part of semiconductor material on one outside. The emitter (E) connects to the second largest part of semiconductor material on the other outside. The base (B) connects to the central part of semiconductor material and serves to determine how much current is allowed to flow between the collector and emitter. Figure 7.1 gives the circuit symbols for both types of bipolar transistors: the n-p-n transistor is shown on the left and the p-n-p transistor on the right. The symbol differs only in the direction of the arrowhead that is always drawn on the emitter. This arrowhead also indicates the current direction in the emitter with a correctly set transistor.

There are two main current paths: on the one hand, the collector-emitter path, where the voltage is often referred to as V_{CE} and the corresponding current with I_C as the collector

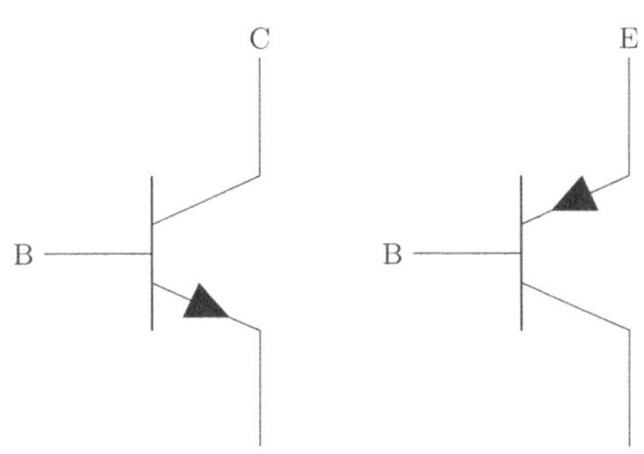

Figure 7.1 The circuit symbols for an n-p-n transistor (left) and a p-n-p transistor (right)

current, and on the other hand, the base–emitter path, where the bias voltage is referred to as V_{BE} and the corresponding current with I_B as the base current. In an n-p-n transistor, the emitter is the negative side, while in a p-n-p transistor it is the positive side. In the circuit representation, the positive side is always on top and in circuits all n-p-n transistors can always be replaced by p-n-p transistors if the polarity of the power lines is also reversed.

You can find transistors in different packaging. Which connections are the emitter, the base or the collector depends on the type of housing. The manufacturers of transistors indicate in their documentation or datasheet for each type how the connectors are located on the housing. On the housing of the transistor, you will usually find the type number of the transistor according to one of the following coding systems:

1. The European coding consists of two or three letters followed by a number: for instance AC117, BC557, BD137, BU108 or BDY20. The first letter indicates whether the transistor is made of germanium (A) or silicon (B). So AC117 is a germanium transistor and BC547 is a silicon transistor. The second letter indicates the application area of the transistor. For example, the U in BU108 indicates that this transistor can be used as a power switch. A possible third letter is the type number and indicates that the transistor is used in a specific application.
2. With American coding, the type number starts with the code 2N followed by a number: 2N2222.
3. The Japanese coding starts with the code 2S (the S for semiconductor) followed by a second letter indicating the application of the transistor and a number: 2SD1406.

7.2 TRANSISTOR EFFECT

When analysing an n-p-n bipolar transistor, one notices that it consists of two different p-n junctions: the p-n junction between the base and the emitter and the p-n junction between the base and the collector. In Chapter 6, we said that a p-n junction actually forms a diode, where the connection to the p-type is called anode and the connection to the n-type is called cathode. So apparently one can think of an n-p-n bipolar transistor as two diodes with their anodes internally connected, as shown on the left in Figure 7.2. This can be seen as a replacement for an n-p-n bipolar transistor, but this does not at all mean that you can make a transistor by connecting two diodes. You only get a bipolar transistor if the p-n junctions are close to each other and together form a whole. The replacement scheme for a p-n-p bipolar transistor is shown on the right in Figure 7.2.

Taking into account this replacement circuit, a bipolar transistor that is not part of a circuit can be tested with a multimeter in the resistance measurement mode (or in the diode test mode). In order to do this, measure the diodes between the base and emitter and between the base and collector in both forward and reverse direction. Just like with diodes, for a properly working transistor in forward bias you get a small resistance value when in resistance measurement mode or the p.d. across the diode when in diode test mode. Measured in reverse bias for a properly working transistor, a large resistance value is obtained when in resistance measurement mode or O.L. (short for 'Over Length') when in diode test mode. As can be noticed in Figure 7.2, measuring between the collector and emitter, one of the two internal diodes is always off, resulting in a large resistance value when both forward and

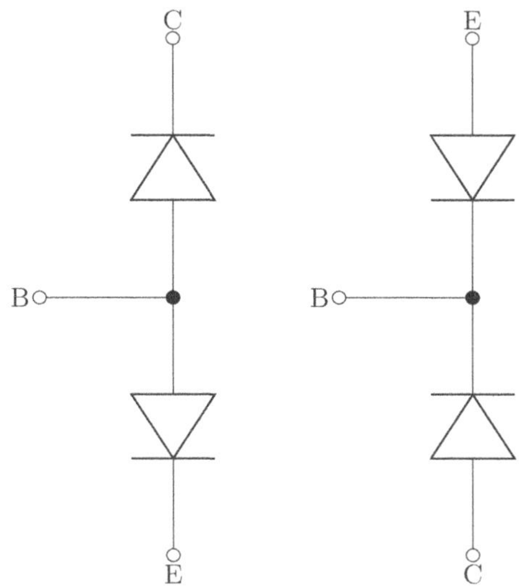

Figure 7.2 Equivalent model for an n-p-n transistor (left) and a p-n-p transistor (right)

reverse bias is measured. If these measurements are performed correctly, you can be sure that the transistor is not damaged or broken.

A bipolar transistor can only amplify if both the base and collector are connected to a supply voltage. These voltages must be connected in such a way that the base–emitter diode is forward biased and the base–collector diode is reversed biased. If the base–emitter voltage V_{BE} is greater than the p.d. of the corresponding diode (usually 0.7 V), this diode will be forward biased. If the base–collector diode remains in the reverse state, a current will still flow through the blocking base–collector diode. This is called the transistor effect and will be explained further. Also note that the collector current I_C is greater than the base current I_B. The ratio between the collector current I_C and the base current I_B is the current gain factor h_{FE} of the transistor. It is defined as follows:

$$h_{FE} = \beta = \frac{I_C}{I_B} \tag{7.1}$$

Typical values for h_{FE} range from 100 to 300. This current gain factor is one of the most important parameters of that bipolar transistor and clearly indicates that it is a current amplifier.

The transistor effect can be explained by recalling that a current consists of electrons. However, the direction of the electron current is opposite to the direction of the conventional current. In an n-p-n transistor, the electrons hence flow from the emitter to the base (through the forward-biased base–emitter diode). When a large positive voltage is applied to the collector of the reverse polarised base–collector diode, it exerts a large attraction on those negative electrons travelling from the emitter to the base. As already stated, this base semiconductor material is very thin (less than 1 µm), so almost all electrons coming from the emitter (even up to 99%) are attracted by the large positive voltage of the collector and sucked right through the thin base to the electron-attracting collector. The remaining 1% of the electrons end up in the base itself, following their normal route. Translated into a conventional current direction, this means that the collector current I_C is much larger than the base current I_B. In conventional current direction, the emitter current I_E is the sum of the base current I_B and the collector current I_C. Thus, despite the base–collector diode being reverse biased, significant current flow from the collector is possible when the base–emitter diode is made conductive. The principle is explained here by relying on the electron current direction, while we almost always work with the conventional current direction.

WORKED EXAMPLE 7.1

Q The current gain factor of a bipolar transistor can be found in the datasheet and equals 250. The base current equals 1 mA. Determine the collector current and the emitter current.

$$h_{FE} = 250;\ I_B = 1\,\text{mA}$$
$$I_C = h_{FE} \cdot I_B = 250 \cdot 1\,\text{mA} = 250\,\text{mA}$$
$$I_E = I_B + I_C = 1 + 250 = 251\,\text{mA}$$

WORKED EXAMPLE 7.2

Q Determine the current gain h_{FE} and the emitter current I_E for a bipolar transistor where I_B = 60 μA and I_C = 3 mA.

$$I_B = 60\,\mu\text{A};\ I_C = 3\,\text{mA}$$
$$h_{FE} = \frac{I_C}{I_B} = \frac{3}{60} = 50$$
$$I_E = I_B + I_C = 60 + 3 = 3.06\,\text{mA}$$

7.3 TRANSISTOR AS SWITCH

Bipolar transistors can be used as switches. They switch millions of times faster, compared to the slow operation of traditional, mechanical switches. You can therefore use a transistor as an electronic switch that switches a certain load on or off at a certain input voltage from, for example, a temperature or light sensor. The big advantage here is that the base current may be very limited, while a large collector current can still flow. This is conveniently used with a twilight switch. This electronic switch is operated by the action of light on a photosensitive cell (or Light-Dependent Resistor (LDR)). This electronic component behaves as a resistor, the value R_{LDR} of which is influenced by the amount of incident light. The more light, the lower the resistance value. This photosensitive cell also depends on temperature, influencing the resistor value and hence its behaviour. The circuit symbol can be found in Figure 7.3.

The schematic of the twilight switch can be found in Figure 7.4. Thus, when there is sufficient light, the resistance value of R_{LDR} is small and the voltage drop across R_{LDR} and

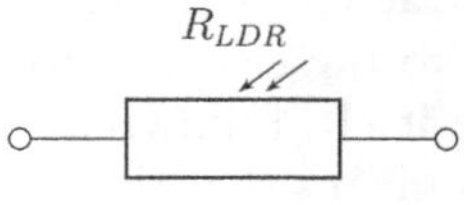

Figure 7.3 The circuit symbol of an LDR

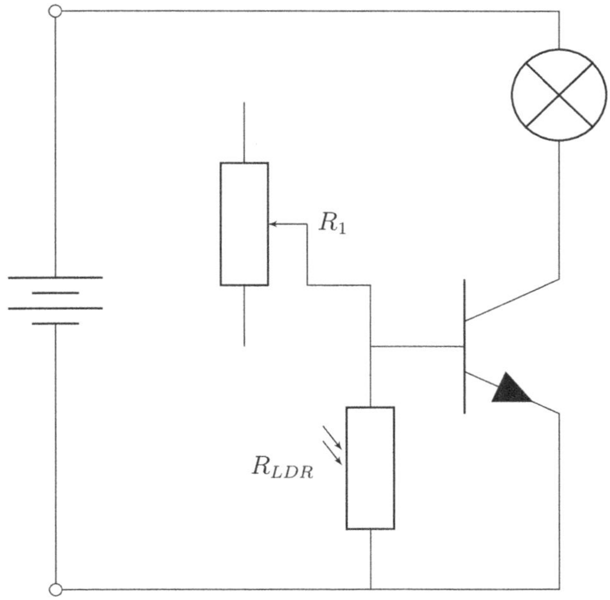

Figure 7.4 The schematic of the twilight switch with an LDR

hence across the base–emitter diode of the transistor is also small. The base–emitter voltage V_{BE} is adjusted with the potentiometer R_1 in such a way that the transistor is then cut off. The light will not blink. When it gets dark, the resistance value of R_{LDR} and therefore also the voltage V_{BE} increase. When the voltage between the base and emitter exceeds the p.d. of 0.7 V, the transistor is conducting and current is flowing in a conventional direction from the collector to emitter. This collector current of the conducting transistor makes the lamp blink.

When R_{LDR} and R_1 are swapped, the behaviour of the circuit also changes. When there is sufficient light, the resistance value of R_{LDR} remains low. In relation to the resistance value R_1, the voltage across the base and emitter is large. Above a p.d. of 0.7 V across the base and emitter the transistor conducts and a current flows. The light is on. On the other hand, the light is switched off with no or limited light. This is reversed behaviour as described above and makes no sense for a twilight lamp.

The bipolar transistor here operates under clipping conditions. Therefore, when the base is saturated, the base–emitter voltage $V_{BE} > 0.7$ V. The collector current flows without any limitation and so the switch is on. On the other hand, if there is no base–emitter voltage ($V_{BE} < 0.7$ V), the switch is off. This is called cutoff.

7.4 TRANSISTOR AS AMPLIFIER

A bipolar transistor can also be used as an amplifier for small voltages of a microphone signal, for example. This has been applied in the electronic diagram of Figure 7.5. The base–emitter is connected in forward bias by the d.c. voltage source V_{BE} with at least 0.7 V. The

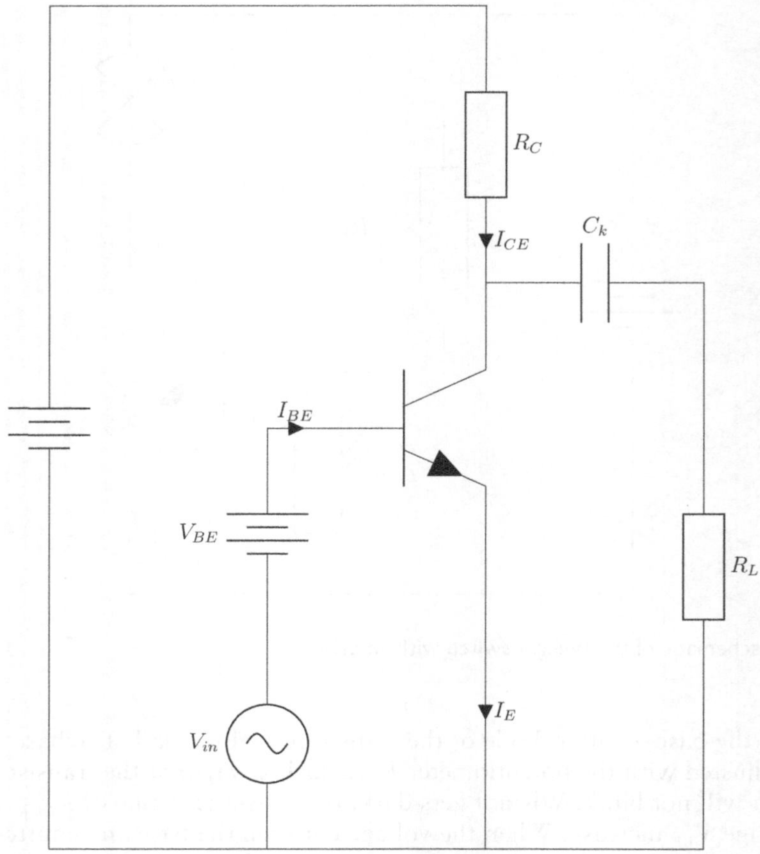

Figure 7.5 The schematic for amplification with a transistor

a.c. voltage source V_{in} is connected in series with this d.c. voltage V_{BE}. This a.c. voltage is the input voltage of the amplifier and originates, for example, from a microphone. Because of this voltage, the total base–emitter voltage varies and the base current I_B also changes. This base current also controls the collector current I_C, as we know that $I_C = I_B.h_{FE}$. In other words, the variations or time dependencies of I_C are identical to those of I_B, but in a wider range thanks to the current gain factor h_{FE}. This changing collector current I_C causes a changing voltage across the collector resistance R_C (with the same variations as the collector current I_C). The supply voltage divides between the collector resistance R_C and the collector-emitter voltage V_{CE} of the transistor. The coupling capacitor C_k only passes a.c. and blocks d.c. This results in an output voltage V_{out} across the load resistor R_L, which is actually a pure a.c. voltage. This V_{out} is the opposite of V_{in}, but in a wider range. So we have amplified the input signal, although phase inverted.

Instead of two separate voltage sources, usually the voltage V_{BE} is realised by branching it from the supply voltage with the use of a voltage divider (by integrating an additional resistor). The output voltage V_{out}, measured across R_L, can be regarded as the input voltage V_{in} of a subsequent amplifier stage. To amplify an audio signal recorded with a microphone, different amplifier stages are cascaded to obtain an amplification that can deliver the desired output power to the loudspeakers.

The output voltage V_{out} of the amplifier is several times greater than the input voltage V_{in}. The ratio of the voltage variations from output voltage to the input voltage is called the voltage gain and in formula form this becomes:

$$\text{voltage gain} = \frac{\Delta V_{out}}{\Delta V_{in}} \tag{7.2}$$

This voltage gain has no unity. Sometimes the voltage gain is expressed in dB (decibels), which can be calculated with the following formula:

$$\text{voltage gain in dB} = 20.\log_{10} \frac{\Delta V_{out}}{\Delta V_{in}} \tag{7.3}$$

with $\log_{10}$ as an abbreviation of the base-10 logarithm.

Please note that this bipolar transistor is biased. This means that some collector current flows even if there is no input signal to be amplified, which is a major drawback in terms of power consumption. However, this biasing is necessary to have the output voltage value at approximately half of the supply voltage value, eventually realised with a voltage divider (where an extra resistor must be added to the emitter). This way the a.c. signal can be higher as well as lower within the predefined limits and omitting clipping the a.c. output signal.

The circuit with a bipolar transistor described here is most commonly used and is also called the common emitter amplifier, because the emitter of the transmitter is used for both the input (together with the base of the transmitter) and the output (together with the collector of the transmitter). We can say that the emitter is used to apply the input voltage and read the output voltage at the same time. A major drawback is the phase inversion: when the input is high, the output is much lower and vice versa. In addition, there are also two other types for circuits with bipolar transistors: the common collector amplifier and the common base amplifier. As the name suggests: in the common collector amplifier is the collector pin used commonly for both the input (together with the base) and the output (together with the emitter). This amplifier is also called an emitter-follower and no phase inversion is involved. The common base amplifier is sometimes called the ground base. The emitter and the base form the input and the collector with the base is used as the output. There is also no phase inversion to be taken into account.

7.5 FIELD EFFECT TRANSISTOR

The field effect transistor (FET) differs fundamentally from the bipolar transistors described above. Compared to bipolar transistors, FETs can be made even more compact and consume far less power. The only disadvantage is that they are incredibly sensitive to static electricity. If you touch one and you feel a (small) shock, you know that you have not taken the proper precautions against static discharge. It is better to throw that component away immediately. Although internally a FET is very different from a bipolar transistor, it behaves very much like a bipolar transistor. They have their own naming convention. Instead of base, emitter and collector, the contact pins are called gate (or G), drain (or D) and source (or S).

There are many different types of field effect transistors, the most common of which is the MOSFET (short for Metal Oxide Semiconductor Field Effect Transistor). This is shown schematically in Figure 7.6: on the left an n-type MOSFET is shown and on the right a p-type MOSFET. As the name itself suggests, it contains a characteristic insulating oxide

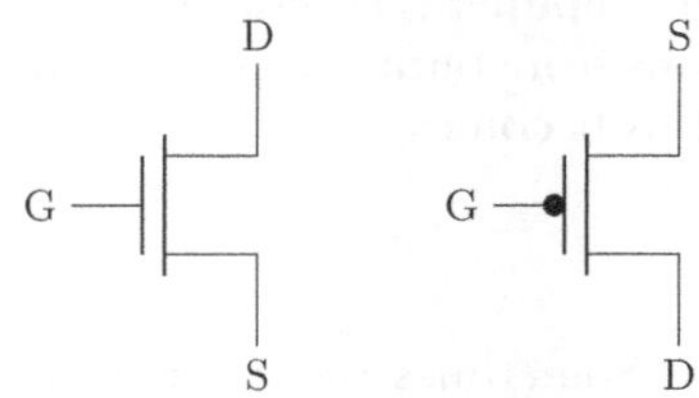

Figure 7.6 The circuit symbols for an n-type MOSFET (left) and a p-type MOSFET (right)

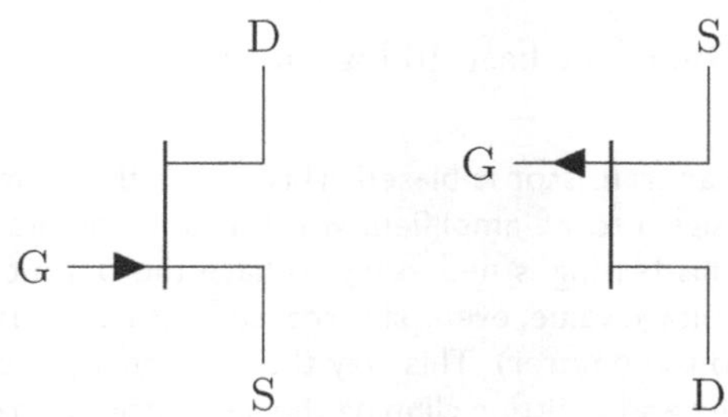

Figure 7.7 The circuit symbols for an n-type JFET (left) and a p-type JFET (right)

with a piece of metal on top that makes contact with the gate G. The control voltage is applied to that contact pin G, which acts as a capacitor across the oxide and is connected to the bulk of the material. The actual current-carrying contact pins are called the source S and the drain D, with the current flowing in an n-type MOSFET through an n-p-n semiconductor and in a p-type MOSFET through a p-n-p semiconductor. By applying a control voltage, a conduction channel is created through which the current flows. That current encounters a certain resistance that can be influenced by an electrical voltage. So we can think of a MOSFET as a voltage-controlled current source.

In addition to the MOSFET, there is another commonly used field effect transistor: the Junction Field Effect Transistor (JFET). This is schematically shown in Figure 7.7: on the left the n-type and on the right the p-type. Here, too, the width of the conductive channel and thus the resistance can be adjusted. The main difference with the MOSFET is that the gate is now not electrically separated from the channel by an insulating layer: there is no oxide. In an n-type JFET, both the bulk and the gate G are p-type which can control the current flowing from drain D to source S. In a p-type JFET, the opposite is true. Here, too, the current encounters a certain resistance that can be influenced by an electrical voltage. Again, we can say that a JFET is a voltage-controlled current source.

Field effect transistors have many advantages compared to bipolar transistors. The power dissipation is noticeably lower, so that less heat has to be dissipated. Furthermore, the spatial structure and manufacturing process are simpler. This results in a greater packing or integration density, because more transistors can be integrated on the same space. This is then called an integrated circuit, which we will cover in detail in the next section.

7.6 INTEGRATED CIRCUITS AND MOORE'S LAW

An integrated circuit (or IC), sometimes called a chip, is actually a complete electronic circuit. It consists of various individual components such as transistors, diodes, resistors, capacitors and conductive connections between all these components, all made from a single piece of silicon. The various components are directly embedded on the silicon itself. It is therefore not an extensive miniaturisation of a printed circuit board or pcb, on which components are soldered. Instead of one p-n junction in a diode or two p-n junctions in a transistor, an IC has thousands of p-n junctions. For an Intel Core i7 computer chip, there are even 731 million on a surface of barely 263 mm^2.

Gordon Moore (1929–2023) was one of the founders of Intel. In 1965, he made an important prediction: every year the number of transistors in an IC will double. In the 1970s, this was adjusted to: every two years the number of transistors in an IC doubles. This ruling has since become known as Moore's Law. We should note that the complexity of electronic technology is increasing exponentially and not incrementally like many other technologies (think of the diesel consumption per 100 km of new cars). On several occasions, the chip manufacturers had the feeling that they were encountering some kind of physical limitation, but each time there were new breakthroughs. Moore's Law has been in effect for over 60 years now and is expected to remain so for some time to come. Because the design and development goals for all engineers are always held against the light of this law, Moore's law has actually become a self-fulfilling prophecy.

To understand how these ICs work, it is important to briefly explain how they are made. The fabrication of such ICs is actually very complex and depends on the type of chip that actually needs to be made. A number of typical steps are successively explained here.

1. A very large, cylindrical piece of silicon crystal is cut into thin slices (or wafers) with a thickness of 0.2 to 0.3 mm. Each of those wafers is used to make up to a thousand finished ICs.
2. A fairly thick oxide is applied on top of each wafer. This will serve as an insulating layer for the following steps. A photosensitive layer is then applied. This is a special varnish that becomes soluble under the influence of light.
3. On top of that there will be a mask that is an image of the circuit to be realised. In addition, some zones are transparent, so that the light can pass through. Other zones are then frosted to shield the oxide from light.
4. The wafer is exposed to UV light so that the photosensitive resist is only dissolved on the transparent areas of the mask. Afterwards, the mask is removed and the oxide not protected by photosensitive lacquer is etched away. Then all photosensitive varnish is removed.
5. The wafer is then exposed to dopant, which creates n-type or p-type zones at the oxide-removed sites.
6. For each layer, the above steps are repeated to arrive at a multi-layer IC.
7. The wafer is cut into several ICs and each IC is assembled in its final packaging (see below).

This entire process is carried out in a cleanroom. This is a very clean working environment, where contamination from unwanted particles in the ambient air is kept to a minimum. This pollution is specified by the number of polluting particles of a certain size per cubic metre.

Manufacturers use different techniques to make ICs. They can choose bipolar transistors, MOSFETS or something else. One can choose a compact chip where many components are close to each other or one can opt for low power consumption or fast switching. ICs are categorised by their design approach. In this context, we speak of logic families. There are

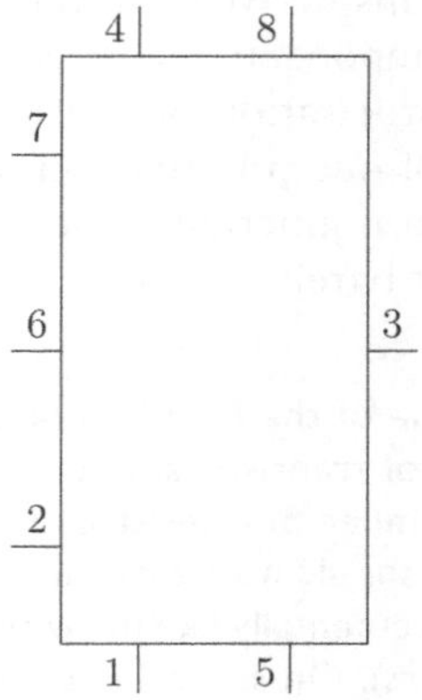

Figure 7.8 The numbering of a DIP

dozens of them, two of which are very common: the TTL and the CMOS. TTL stands for Transistor-Transistor Logic and uses bipolar transistors. They can be manufactured very cheaply, but consume a lot of power and require a specific voltage of 5 V. CMOS, on the other hand, stands for Complementary Metal Oxide Semiconductor and uses MOSFETS. They are slightly more expensive than their TTL equivalent, but they use much less power and operate at a variable supply voltage between 3 V and 15 V. A disadvantage is that they are sensitive to static electricity.

ICs come in different types of packaging. The most common for us is the Dual Inline Package (DIP). It is actually a small plastic or resin box in which the actual IC is packaged. It contains two rows of contact pins along the long side of the rectangle. They squeak out of the package and are then bent 90 degrees, like the legs of a beetle. The pins themselves are spaced 0.1 inch or 2.54 mm apart and the two rows of pins are spaced 0.3 inch or 7.62 mm apart, making it a perfect fit on a breadboard. Each pin of a DIP is uniquely numbered. The reference is indicated by a notch, a groove or an elevation and is placed at the top. Conventionally, numbers are then counterclockwise: from the top left, to the bottom left, to the bottom right and finally to the top right. In an electrical diagram, an IC is usually represented by a rectangle, with connections to the environment being made all around. This can also be seen in Figure 7.8. The numbering at the rectangle on the diagram itself indicates which connection should arrive at which pin. In general, the physical position of the pin on the DIP is not taken into account. Unused pins are even omitted completely. When realising the circuit itself, the connecting wires must be adapted to the pin order on the DIP itself.

If this IC is included in an electronic circuit, you have to watch out for both static electricity and heat. This makes soldering difficult. Therefore, special sockets as placeholders are available. These are first soldered and when that is all finished and tested, the actual IC can be clicked into the socket itself.

7.7 OPAMPS

An opamp or operational amplifier is a supersensitive amplifier circuit, specially designed to amplify the difference between two input voltages. It is represented by a triangle, with two inputs and only one output. The output voltage is sometimes 10 times or even 100 to 1000 times greater than the difference in the input voltages. It was invented to be used as an amplifier in telephone networks. Afterwards, computer engineers discovered that these opamps could easily be modified to also perform mathematical operations, hence the name. With opamps, addition, subtraction, multiplication or division can be performed quite easily, which are the basics of digital electronics. The operation of an opamp in a circuit can be explained by analysing the five connections of the triangle separately, as indicated also in Figure 7.9.

- The supply voltage for an opamp is connected via two pins: $+V$ and $-V$. This is called split voltage sources. For example, the voltage sources +9 V and –9 V can be realised by connecting two 9 V batteries in series. The ground is then the connection between the two batteries. Some opamps do not require a split voltage source and then the $-V$ pin is connected to ground.
- The output of the opamp can be found on V_{out}. This voltage can become positive or negative depending on the voltage difference between the two inputs. The maximum output voltage is usually a little bit lower than the supply voltage.
- The input voltages $V-$ and $V+$ are sometimes also indicated with a minus and a plus sign within the triangle itself: the negative terminal $V-$ and the positive terminal $V+$. An opamp behaves as a differential amplifier: if $V+$ is greater than $V-$, the output will be positive. If $V+$ is lower than $V-$, the output will be negative. If the $V+$ is connected to ground, the output voltage will always be opposite to the input voltage. That is why the $V-$ is also called the inverting input. If, on the other hand, the $V-$ is connected to ground, the output voltage will always have the same sign as the input voltage. That is why the $V+$ is also called the non-inverting input.

In Figure 7.4, a twilight switch is realised with a bipolar transistor. But it is also possible with an opamp as the basis, as is shown in Figure 7.10. When there is enough light, the resistance value of R_{LDR} is very small. Due to the voltage division, there is a very low voltage on the positive terminal of the opamp. The potentiometer is adjusted in such a way that there is virtually no difference with the voltage at the negative terminal, so that there is also virtually no voltage at the output. When it gets dark, the resistance value of R_{LDR} becomes much larger, so that the voltage at the positive terminal rises. The difference between the positive and negative terminal becomes much larger, so that the output voltage also increases enormously. The lamp lights up.

Even though commercial opamps are not ideal, their properties are very close to those of an ideal opamp. We would therefore like to sum up the properties of such an ideal opamp.

- An ideal opamp has infinite gain. This means that any voltage difference at the input results in an infinite voltage at the output. However, with realistic opamps, this voltage is limited to the supply voltage. Since the voltage cannot be infinite, the gain in realistic opamps is also not infinite.

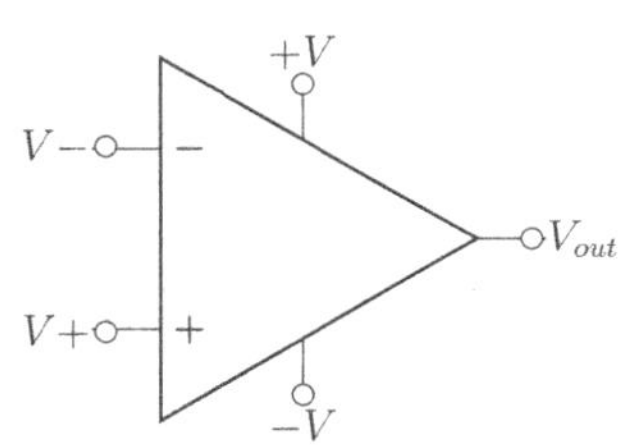

Figure 7.9 The five connections of an opamp

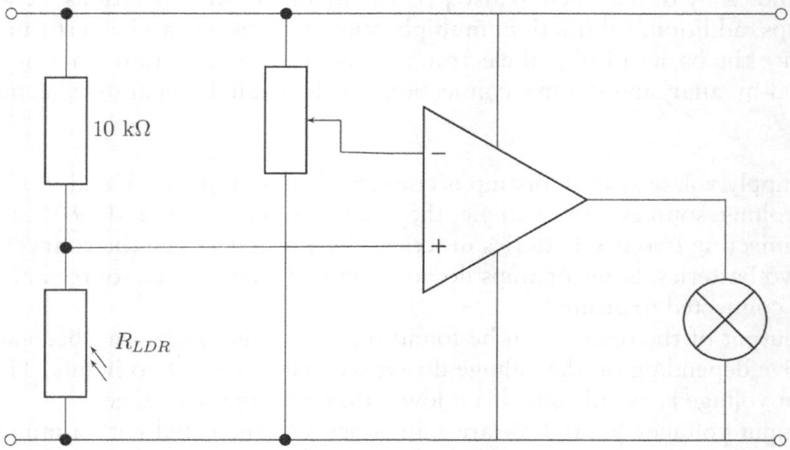

Figure 7.10 The schematic of the twilight switch with an opamp

- In an ideal opamp, no current flows from the positive and negative terminals to the internal circuit. We say that the input impedance is infinite. The internal circuit therefore only 'sees' the value of the voltage, without any current flowing. This means that an ideal opamp has no influence at all on the voltage at the input.
- The output impedance is equal to zero. This means that the output voltage is independent of the value of the load resistance at the output. In practice, this output impedance is not zero, but very small.
- If the two input voltages are equal, the output voltage is 0 V. This is called the offset voltage and is therefore 0 V for an ideal opamp.
- In an ideal opamp, each input signal is amplified, independent of the frequency value of the applied a.c. voltage. In practice, there is a maximum frequency, below which this is the case. Above it, the output signal will no longer be recognisable.

7.8 INVERTING AND NON-INVERTING AMPLIFIERS WITH OPAMPS

In most cases, an opamp is used as an amplifier of signals. If an alternating voltage is applied to the input of an opamp, also at the output an alternating voltage is produced. That output signal is an amplified version of the input signal, because the opamp provides a voltage gain of up to a factor 1000. For the amplification of sound recorded with a microphone, the microphone should be connected to the two inputs of the opamp. One clamp of the microphone and the negative clamp of the opamp are grounded. The other clamp of the microphone is connected to the plus terminal of the opamp. It is possible that the signal from the microphone is too large, causing the amplified output signal to drive the opamp to clipping or saturation. We have lost our dependence on time. Therefore another circuit (as depicted in Figure 7.11) is used to amplify the microphone input signal.

The output voltage V_{out} is actually fed back to the input through the resistor R_2. If the input voltage V_{in} increases, the output voltage V_{out} will become much larger, but negative because the microphone is connected to the negative terminal. Because V_{out} has dropped, the voltage on the negative terminal $V-$ will also drop. As a result, the voltage difference between the positive and negative terminal of the opamp also becomes smaller. The final voltage gain of the signal is reduced in such a way that the output voltage no longer reaches the extreme limits of the supply voltage. The signal gain is determined by the resistors R_1

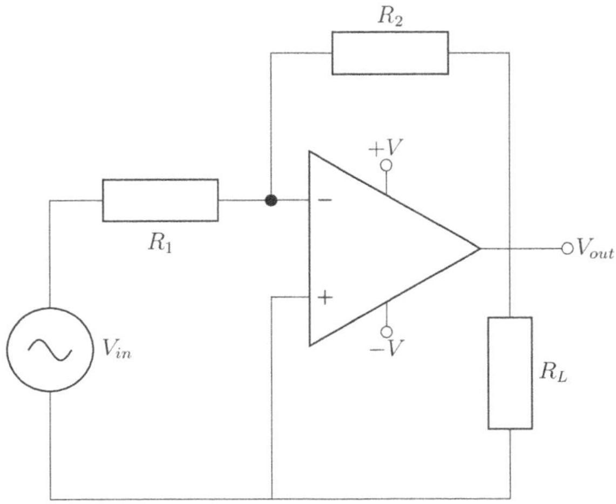

Figure 7.11 The schematic for amplification with an opamp

and R_2 and no longer by the opamp itself. These two resistors must therefore be chosen well. In an ideal opamp, the current to the plus and minus terminal is equal to zero. The voltage division across those two resistors can then be determined as follows:

$$V- = V_{in} + \frac{R_1}{R_1 + R_2}(V_{out} - V_{in}) \quad (7.4)$$

The ideal opamp with its infinite gain tries to avoid saturation at all times. This is only possible if $V- = V+$. Since $V+ = 0$ V (the positive terminal is connected to ground), the voltage gain A can be calculated as a ratio of the output voltage to the input voltage. This results in:

$$A = \frac{V_{out}}{V_{in}} = -\frac{R_2}{R_1} \quad (7.5)$$

This voltage gain formula clearly indicates that the gain only depends on the ratio of the two resistors used. These values must be chosen so that the opamp does not saturate or in other words that the amplified input difference remains lower than the applied supply voltage.

WORKED EXAMPLE 7.3

Q Calculate the voltage gain of an inverting opamp circuit when $R_1 = 0.5$ kΩ and $R_2 = 47$ kΩ.

$R_1 = 0.5$ kΩ; $R_2 = 47$ kΩ

$$A = -\frac{R_2}{R_1} = -\frac{47}{0.5} = -94$$

Instead of connecting the input signal to the negative terminal of the opamp via the resistor R_1, for a non-inverting amplifier, the input signal is connected to the positive terminal, where the negative terminal is grounded via the resistor R_1. It works like a voltage follower circuit because this circuit uses a negative feedback connection, giving a part of the output signal as feedback to the inverting input terminal. The input signal is applied to the positive terminal. The output signal is in phase with the input, hence the name 'non-inverting opamp'.

Combining the voltage divider rule and the ideal opamp assumption, the amplification can be derived. The feedback signal is applied to the inverting input of the opamp. The voltage at the inverting input is equal to the voltage at the non-inverting input because of the high input impedance of the opamp. Therefore, we can assume that the voltage at the inverting input is equal to V_{in}. The output voltage of an ideal opamp is given by $V_{out} = A_{openloop}(V+ - V-)$ where $A_{openloop}$ is the open-loop gain of the opamp and $V+$ and $V-$ are the voltages at the non-inverting and inverting inputs respectively. Since

$$V+ = V_{in} \tag{7.6}$$

$$V- = \frac{V_{out}}{A} = \frac{R_1}{R_1 + R_2} V_{out}, \tag{7.7}$$

we can substitute these values into the equation for V_{out} and solving for the voltage gain A gives us:

$$A = \frac{V_{out}}{V_{in}} = 1 + \frac{R_2}{R_1} \tag{7.8}$$

Also in this case the voltage gain A is only dependent of the used resistors R_1 and R_2 and of the open-loop gain of the opamp itself.

The LM741 is a very common eight-pin DIP, containing one opamp circuit. Of these eight pins, only five are in use. The inverting input $V-$ goes to pin 2, the non-inverting input $V+$ goes to pin 3. The supply voltage $+V$ is connected to pin 7, while the other supply voltage of the split power source goes to pin 4. The output and thus the amplified signal can be tapped from pin 6. If you need more than one opamp circuit, you should better use the LM324 which contains four opamp circuits in one single package. No split power source is needed here; only a positive voltage (on pin 4) and ground (on pin 11). The first opamp has the inverting input on pin 2, the non-inverting input on pin 3 and the output on pin 1. For the second opamp, these are pins 6, 5 and 7 respectively. For the third opamp these will be pins 9, 10 and 8 respectively. For the fourth and last opamp these are pins 13, 12 and 14. Note that all 14 pins are used here.

WORKED EXAMPLE 7.4

Q Calculate the voltage gain of a non-inverting opamp circuit when $R_1 = 0.5\ \text{k}\Omega$ and $R_2 = 47\ \text{k}\Omega$.

$$R_1 = 0.5\ \text{k}\Omega;\ R_2 = 47\ \text{k}\Omega$$

$$A = 1 + \frac{R_2}{R_1} = 1 + \frac{47\ \text{k}\Omega}{0.5\ \text{k}\Omega} = 95$$

ASSIGNMENT QUESTIONS

1 Determine the current gain h_{FE} and the emitter current I_E for a bipolar transistor where $I_B = 40\ \mu A$ and $I_C = 2\ mA$.

2 Calculate the value R_1 for an inverting opamp with an absolute value for the voltage gain of 100 and $R_2 = 63\ k\Omega$.

3 Calculate the value R_1 for a non-inverting opamp with an absolute value for the voltage gain of 100 and $R_2 = 63\ k\Omega$.

Chapter 8

Alternating Quantities

LEARNING OUTCOMES

This chapter deals with the concepts, terms and definitions associated with alternating quantities. The term alternating quantities refers to any quantity (current, voltage, flux, etc.) whose polarity is reversed alternately with time. For convenience, they are commonly referred to as a.c. quantities. Although an a.c. can have any waveshape, the most common waveform is a sinewave. For this reason, unless specified otherwise, you may assume that sinusoidal waveforms are implied.

On completion of this chapter you should be able to:

1 Explain the method of producing an a.c. waveform.
2 Define all of the terms relevant to a.c. waveforms.
3 Obtain values for an a.c., both from graphical information and when expressed in mathematical form.
4 Understand and use the concept of phase angle.
5 Use both graphical and phasor techniques to determine the sum of alternating quantities.

8.1 PRODUCTION OF AN ALTERNATING WAVEFORM

From electromagnetic induction theory, we know that the average emf induced in a conductor, moving through a magnetic field, is expressed in volt and given by

$$e = Blv\sin\theta \qquad [1]$$

where B is the flux density of the field (in tesla)

ℓ is the effective length of conductor (in metre)
v is the velocity of the conductor (in metre/s)
θ is the angle at which the conductor 'cuts' the lines of magnetic flux (in degrees or radians)

i.e. $v\sin\theta$ is the component of velocity at right angles to the flux. Consider a single-turn coil, rotated between a pair of poles, as illustrated in Figure 8.1(a) and (b). Part (a) shows the general arrangement and part (b) shows a cross-section at one instant in time, such that the coil is moving at an angle θ to the flux.

DOI: 10.1201/9781003308294-8

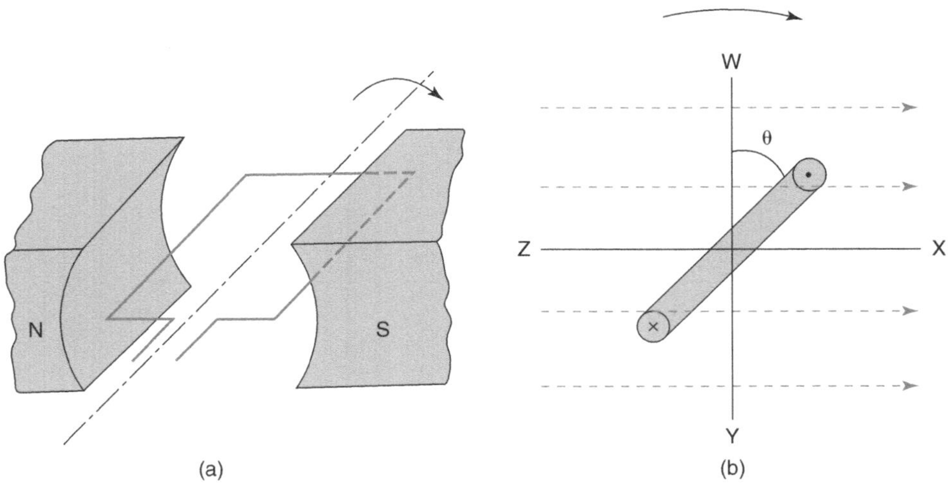

Figure 8.1 A single-turn coil, rotated between a pair of poles

Considering Figure 8.1(b), each side of the coil will have the same value of emf induced, as given by Equation [1] above. The polarities of these emfs will be as shown, according to Fleming's right-hand rule. Although these emfs are of opposite polarities, they both tend to cause current to flow in the same direction around the coil. Thus, the total emf generated is given by:

$$e = 2 \times Blv\sin\theta \qquad [2]$$

Still considering Figure 8.1(b), at the instant the coil is in the plane W–Y, angle $\theta = 0°$. Thus the emf induced is zero. At the instant that it is in the plane X–Z, $\theta = 90°$. Thus, the emf is at its maximum possible value, given by:

$$e = 2 \times Blv \qquad [3]$$

Let us consider just one side of the coil, starting at position W. After 90° rotation (to position X), the emf will have increased from zero to its maximum value. During the next 90° of rotation (to position Y), the emf falls back to zero. During the next 180° rotation (from Y to Z to W), the emf will again increase to its maximum, and reduce once more to zero. However, during this half revolution, the polarity of the emf is reversed.

If the instantaneous emf induced in the coil is plotted, for one complete revolution, the sinewave shown in Figure 8.2 will be produced. For convenience, it has been assumed that the maximum value of the coil emf is 1 V, and that the plot starts with the coil in position W.

When the coil passes through one complete revolution, the waveform returns to its original starting point. The waveform is then said to have completed one cycle. Note that one cycle is the interval between any two *corresponding* points on the waveform. The number of cycles generated per second is called the *frequency*, *f*, of the waveform. The unit for frequency is the hertz (Hz). Thus, one cycle per second is equal to 1 Hz.

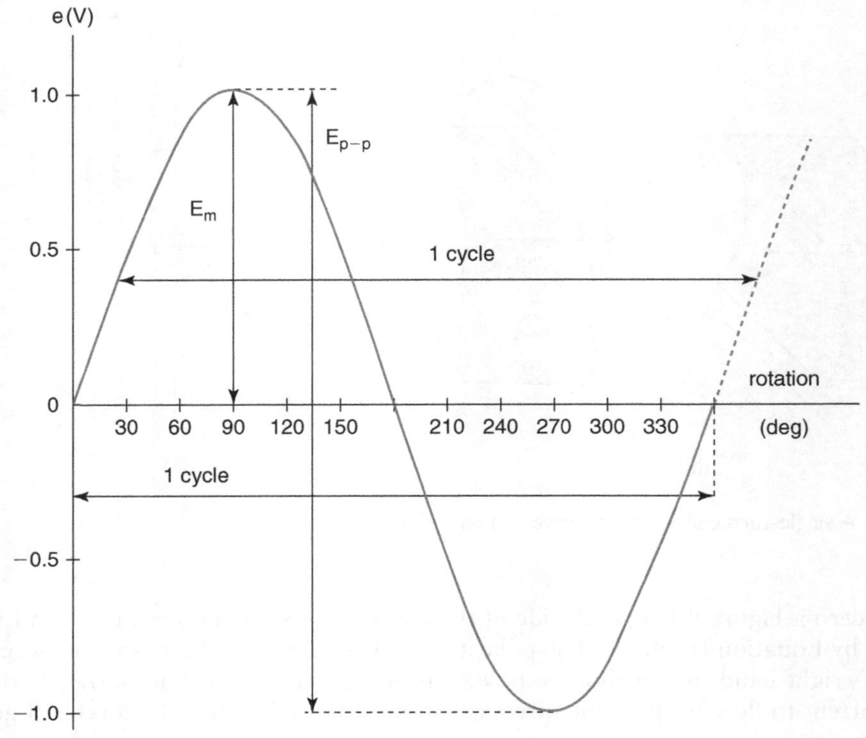

Figure 8.2 The instantaneous emf induced in the coil as function of degrees

Heinrich Hertz (1857–1894) was a German physicist interested in the electromagnetism theory elaborated earlier by James Maxwell. Hertz is best known for the discovery of radio waves, which was a direct confirmation of that electromagnetism theory. He also accidentally discovered the phenomenon of photoelectric effect: by irradiating a metal with ultraviolet light an electric current could be generated.

For the simple two-pole arrangement considered, one cycle of emf is generated in one revolution. The frequency of the waveform is therefore the same as the speed of rotation, measured with the quantity symbol n and expressed in revolutions per second (rev/s). This yields the following equation

$$f = np \tag{8.1}$$

where p = the number of pole *pairs*.

Therefore, if the coil is rotated at 50 rev/s, the frequency will be

$$f = 50 \text{ rev/s} \times 1 \text{ pair of poles} = 50 \text{ Hz}$$

The time taken for the waveform to complete one cycle is called the periodic time, T. Thus, if 50 cycles are generated in one second, then one cycle must be generated in 1/50 of a second. The relationship between frequency and period is therefore

$$T = \frac{1}{f} \text{ or } f = \frac{1}{T} \qquad (8.2)$$

The maximum value of the emf in one cycle is shown by the peaks of the waveform. This value is called either the maximum or peak value, or the *amplitude* of the waveform. The quantity symbol used may be either $\hat{E}$, or E_m. The voltage measured between the positive and negative peaks is called the peak-to-peak value and has the quantity symbol E_{pk-pk}, or E_{p-p}.

8.2 ANGULAR VELOCITY AND FREQUENCY

In SI units, angles are measured in radians, shortened as rad, rather than degrees. Similarly, angular velocity is measured in radians per second, rather than revolutions per second. The quantity symbol for angular velocity is ω (lower-case Greek omega).

A radian is the angle subtended at the centre of a circle, by an arc on the circumference, which has length equal to the radius of the circle. Since the circumference = $2\pi r$, there must be 2π such arcs in the circumference. Hence there are 2π radians in one complete circle; i.e. 2π rad = 360°.

If the coil is rotating at n rev/s, then it is rotating at 360° × n degrees/second. Since there are 2π radians in 360°, the coil must be rotating at $2\pi n$ radians per second.

Thus, angular velocity $\omega = 2\pi n$ and is expressed in rad/s,

but for a 2-pole system, $f = n$

$$\text{therefore, } \omega = 2\pi f \qquad (8.3)$$

$$\text{and, } f = \frac{\omega}{2\pi} \qquad (8.4)$$

If the coil is rotating at ω rad/s, then in a time of t seconds, it will rotate through an angle of ω radian. Hence the waveform diagram may be plotted to a base of degrees, radians or time. In the latter case, the time interval for one cycle is, of course, the periodic time, T. These are shown in Figure 8.3.

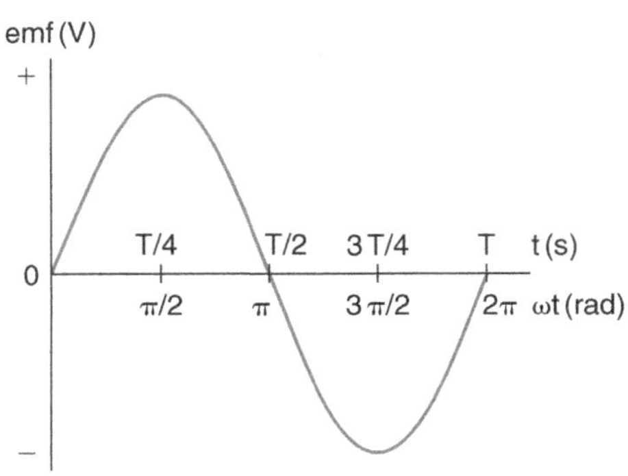

Figure 8.3 The instantaneous emf induced in the coil as function of radians

8.3 STANDARD EXPRESSION FOR AN ALTERNATING QUANTITY

All the information regarding an a.c. can be presented in the form of a graph. The information referred to here is the amplitude, frequency, period and value at any instant. The last is normally called the instantaneous value. However, presenting this information in a graph is not always very convenient. To overcome these difficulties, the a.c. is expressed in a more convenient form. This results in an equation, sometimes referred to as the algebraic form of the a.c. More correctly, it should be called the trigonometric form. Since many students are put off by these terms, we shall refer to it simply as the standard expression for a waveform.

The emf for an N-turn coil is:

$$e = 2 \times NBlv\sin\theta, \text{ where } \theta \text{ is in degrees}$$
$$\text{or, } e = 2 \times NBlv\sin(\omega t), \text{ where } \omega t \text{ is in radians}$$

and the emf is at its maximum value when $\sin(\omega t)$, or $\sin\theta$ is equal to 1. With $E_m = 2 \times NB\ell v$ the expression becomes:

$$e = E_m\sin\theta \tag{8.5}$$
$$\text{or, } e = E_m\sin(\omega t) \tag{8.6}$$
$$\text{or, } e = E_m\sin(2\pi ft) \tag{8.7}$$

All three of the above equations are the so-called standard expressions for this a.c. voltage. Equations (8.6), and (8.7) in particular, are those most commonly used. Using these, all the relevant information concerning the waveform is contained in a neat mathematical expression.

WORKED EXAMPLE 8.1

Q An alternating voltage is represented by the expression $v = 35\sin(314.2t)$ V. Determine, (a) the maximum value, (b) the frequency, (c) the period of the waveform and (d) the value 3.5 ms after it passes through zero, going positive.

(a) $v = 35\sin(314.2t)$ and comparing this to the standard, $v = v_m \sin(2\pi ft)$, we can see that: $v_m = 35$ V

(b) Again, comparing the two expressions:

$$2\pi f = 314.2$$
$$\text{so, } f = \frac{314.2}{2\pi} = 50\,\text{Hz}$$

(c) $T = \frac{1}{f} = \frac{1}{50}$ so, $T = 20\,\text{ms}$

When $t = 3.5\,\text{ms}$; then:

(d) $v = 35\sin\left(2\pi \times 50 \times 3.5 \times 10^{-3}\right) = 35\sin(1.099)^* = 35 \times 0.891$

therefore, $v = 31.19\,\text{V}$

*The term inside the brackets is an angle in *radian*. You must therefore remember to switch your calculator into the *radian mode*.
So far, we have dealt only with an alternating voltage. However, all of the terms and definitions covered are equally applicable to any alternating quantity. Thus, exactly the same techniques apply to a.c. currents, fluxes, etc. The same applies also to mechanical alternating quantities involving oscillations, vibrations, etc.

WORKED EXAMPLE 8.2

Q For a current, $i = 75\sin(200\pi f)$ mA, determine (a) the frequency and (b) the time taken for it to reach 35 mA, for the first time, after passing through zero.

$$i = 75\sin(200\pi t) = I_m \sin(2\pi f t)$$

(a) $$\text{so, } 2\pi f = 200\pi$$

$$\text{and } f = \frac{200\pi}{2\pi} = 100\,\text{Hz}$$

$$35 = 75\sin(200\pi t)$$

$$\frac{35}{75} = \sin(200\pi t) = 0.4667$$

(b) $$\text{therefore, } 200\pi t = \sin^{-1} 0.4667^{*} = 0.4855\text{ rad}$$

$$\text{so, } t = \frac{0.4855}{200\pi} = 0.773\,\text{ms}$$

*Remember, use *radian* mode on your calculator.

8.4 AVERAGE AND R.M.S. VALUE AND PEAK AND FORM FACTOR

Figure 8.4 shows one cycle of a sinusoidal current.

From this it is apparent that the area under the curve in the positive half is exactly the same as that for the negative half. Thus, the average value over one complete cycle must be zero. For this reason, the average value is taken to be the average over one half cycle. This average may be obtained in a number of ways. These include the mid-ordinate rule,

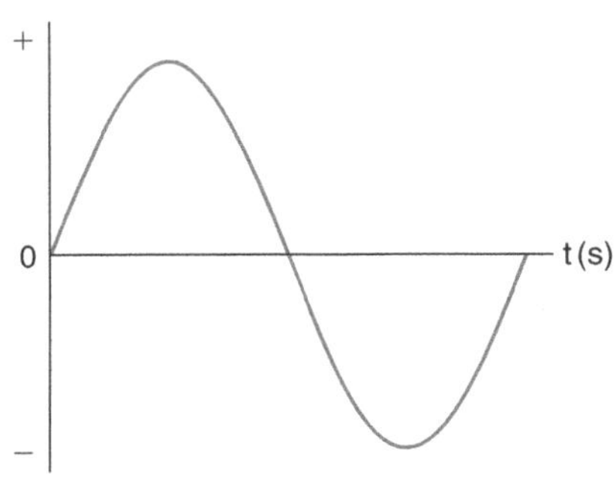

Figure 8.4 One cycle of a sinusoidal current

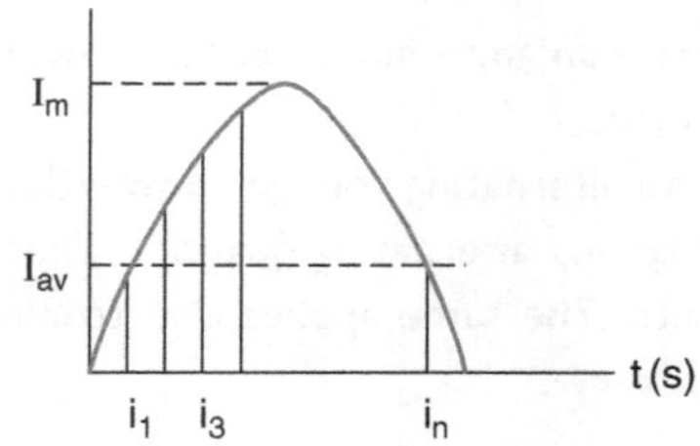

Figure 8.5 The average over one half cycle

the trapezoidal rule, Simpson's rule and integral calculus. The simplest of these is the mid-ordinate rule, and this will be used here to illustrate the average value in Figure 8.5.

A number of equally spaced intervals are selected, along the time axis of the graph. At each of these intervals, the instantaneous value is determined. This results in values for a number of ordinates, i_1, i_2, …, i_n, where n is the number of ordinates chosen. The larger the number of ordinates chosen, the more accurate will be the final average value obtained. The average is simply found by adding together all the ordinate values, and then dividing this figure by the number of ordinates chosen, thus

$$I_{av} = \frac{i_1 + i_2 + i_3 + \cdots + i_n}{n}$$

The average value will of course depend upon the shape of the waveform, and *for a sinewave only* it is

$$I_{av} = \frac{2}{\pi} I_m = 0.637 I_m \tag{8.8}$$

WORKED EXAMPLE 8.3

Q A sinusoidal alternating voltage has an average value of 3.5 V and a period of 6.67 ms. Write down the standard (trigonometrical) expression for this voltage. $V_{av} = 3.5\,\text{V}$; $T = 6.67 \times 10^{-3}$ s

The standard expression is of the form $v = V_m \sin(2\pi ft)$

$$V_{av} = 0.637\,V_m$$

$$\text{so, } V_m = \frac{V_{av}}{0.637} = \frac{3.5}{0.637} = 5.5\,V$$

$$f = \frac{1}{T} = \frac{1}{6.67 \times 10^{-3}} \text{ and, } f = 150\,\text{Hz}$$

$$v = 5.5 \sin(2\pi \times 150 \times t)$$

$$\text{so, } v = 5.5 \sin(300\pi t)$$

WORKED EXAMPLE 8.4

Q For the waveform specified in Worked Example 8.3, after the waveform passes through zero, going positive, determine its instantaneous value (a) 0.5 ms later, (b) 4.5 ms later and (c) the time taken for the voltage to reach 3 V for the first time.

(a) $t = 0.5 \times 10^{-3}$ s; (b) $t = 4.5 \times 10^{-3}$ s; (c) $v = 3$ V

(a) $v = 5.5\sin\left(300\pi \times 0.5 \times 10^{-3}\right) = 5.5\sin 0.4712 = 5.5 \times 0.454 = 2.5\,\text{V}$

(b) $v = 5.5\sin\left(300\pi \times 4.5 \times 10^{-3}\right) = 5.5\sin 4.241 = 5.5 \times (-0.891) = -4.9\,\text{V}$

Note: Remember that the expression inside the brackets is an angle in *radian*.

(c)
$$3 = 5.5\sin(300\pi t)$$
$$\text{so, } \sin(300\pi t) = \frac{3}{5.5} = 0.5455$$
$$300\pi t = \sin^{-1} 0.5455 = 0.5769 \text{ rad}$$
$$t = \frac{0.5769}{300\pi} = 6.12 \times 10^{-4} = 0.612\,\text{ms}$$

A sketch graph illustrating these answers is shown in Figure 8.6.

The r.m.s. value of an alternating current is equivalent to that value of direct current, which when passed through an identical circuit will dissipate exactly the same amount of power. The r.m.s. value of an a.c. thus provides a means of making a comparison between a.c. and d.c. systems.

The term r.m.s. is an abbreviation of the square Root of the Means Squared. The technique for finding the r.m.s. value may be based on the same ways as were used to find the average value. However, the r.m.s. value applies to the complete cycle of the waveform. For simplicity, we will again consider the use of the mid-ordinate rule technique.

Considering Figure 8.5, the ordinates would be selected and measured in the same way as before. The value of each ordinate is then squared. The resulting values are then summed, and the average found. Finally, the square root of this average (or mean) value is determined. This is illustrated below:

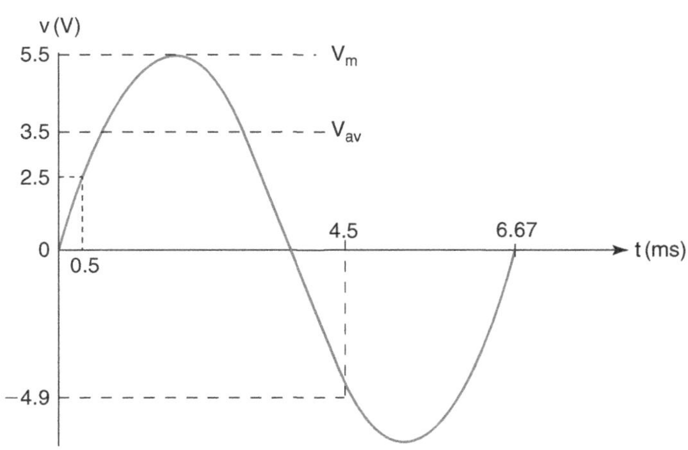

Figure 8.6 A sketch graph for Worked Example 8.4

$$I_{rms} = \sqrt{\frac{i_1^2 + i_2^2 + i_3^2 + \cdots + i_n^2}{n}}$$

and, for a sine wave only, $I_{rms} = \frac{1}{\sqrt{2}} I_m = 0.707 I_m$ (8.9)

Other waveforms will have a different ratio between r.m.s. and peak values.

Note: The r.m.s. value of an a.c. is the value normally used and quoted. For example, if reference is made to a 230 V a.c. supply, then 230 V is the r.m.s. value. In general therefore, if an unqualified value for an a.c. is given, the assumption is made that this is the r.m.s. value. Since r.m.s. values are those commonly used, the subscript letters r.m.s. are not normally included. I_{rms} has been used above, simply for emphasis. The following convention is used:

i, *v*, *e*, represent instantaneous values
I_{av}, V_{av}, E_{av}, represent average values
I_m, V_m, E_m, represent maximum or peak values, or *amplitude*
I, *V*, *E*, represent r.m.s. values

In many parts of the world, a voltage (nominally) of 230 V and frequency of 50 Hz is distributed via the power grid. In North America, however, the most common combination is 120 V, with a frequency of 60 Hz. Also other combinations exist, like 230 V at 60 Hz. Plugs and sockets are mostly non-interchangeable to provide protection from accidental use of appliances with incompatible voltage and/or frequency requirements.

The peak factor is defined as the ratio of the peak or maximum value, to the r.m.s. value, of a waveform. Thus, *for a sinewave only*

$$\text{peak factor} = \frac{\text{maximum value}}{\text{r.m.s. value}}$$

$$= \frac{V_m}{0.707 V_m} = \sqrt{2} \text{ or } 1.414$$

WORKED EXAMPLE 8.5

Q Calculate the amplitude of the household 230 V supply.

Since this supply is sinusoidal, the peak factor will be $\sqrt{2}$ so $V_m = \sqrt{2} \times V = \sqrt{2} \times 230 = 325.3\,\text{V}$

WORKED EXAMPLE 8.6

Q A non-sinusoidal waveform has a peak factor of 2.5, and an r.m.s. value of 230 V. It is proposed to use a capacitor in a circuit connected to this supply. Determine the minimum safe working voltage rating required for the capacitor.

peak factor = 2.5; $V = 230$

$V_m = 2.5 \times V = 2.5 \times 230 = 575\,\text{V}$

Thus the absolute minimum working voltage must be 575 V.

In practice, a capacitor having a higher working voltage would be selected. This would then allow a factor of safety.

The form factor gives an indication of the form or shape of the waveform, as the name implies. It is defined as the ratio of the r.m.s. value to the average value.

Thus, *for a sinewave*,

$$\text{form factor} = \frac{\text{r.m.s. value}}{\text{average value}} = \frac{0.707}{0.637}$$

$$\text{so, form factor} = 1.11$$

For a rectangular waveform (a squarewave), form factor = 1, since the r.m.s. value, the peak value and the average value are all the same.

WORKED EXAMPLE 8.7

Q A rectangular coil, measuring 25 cm by 20 cm, has 80 turns. The coil is rotated about an axis parallel with its longer sides, in a magnetic field of density 75 mT. If the speed of rotation is 3000 rev/min, calculate, from first principles, (a) the amplitude, r.m.s. and average values of the emf, (b) the frequency and period of the generated waveform and (c) the instantaneous value, 2 ms after it is zero.

$$l = 0.25\ \text{m};\ d = 0.2\ \text{m};\ N = 80;\ B = 0.075\ \text{T};\ n = \frac{3000}{60}\ \text{rev/s};\ t = 2 \times 10^{-3}\ \text{s}$$

(a) $e = 2 \times NB\ell v \sin(2\pi ft)$

Now, we know the rotational speed *n*, but the above equation requires the tangential velocity, *v*, in metres per second. This may be found as follows.

Consider Figure 8.7, which shows the path travelled by the coil sides. The circumference of rotation = πd metre = 0.2π metre. The coil sides travel this distance in one revolution. The rotational speed n = 3000/60 = 50 rev/s. Hence the coil sides have a velocity, $v = 50 \times 0.2\pi$ m/s.

Therefore, $e = 2 \times 80 \times 0.075 \times 0.25 \times 50 \times 0.2\pi \sin(2\pi ft)$ and emf is a maximum value when $\sin(2\pi ft) = 1$

$$\text{so, } E_m = 2 \times 80 \times 0.075 \times 0.25 \times 50 \times 0.2\pi = 94.25\,\text{V}$$

Assuming a sinusoidal waveform:

$$E = 0.707\,E_m = 0.707 \times 94.25 = 66.64\,\text{V}$$

$$E_{av} = 0.637\,E_m = 0.637 \times 94.25$$

$$\text{so, } E_{av} = 60.04\,\text{V}$$

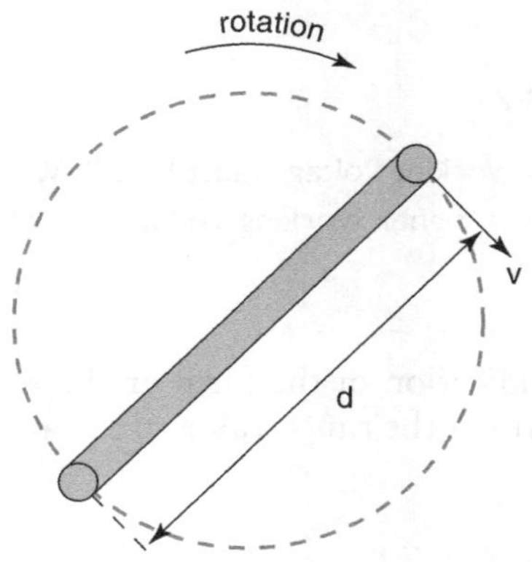

Figure 8.7 A sketch graph for Worked Example 8.7

Assuming a 2-pole field system, then $f = n$ therefore $f = 50$ Hz

(b) $T = \frac{1}{f} = \frac{1}{50} = 20\,\text{ms}$

(c) $e = E_m \sin(2\pi ft) = 94.25 \sin(2\pi \times 50 \times 2 \times 10^{-3}) = 94.25 \times 0.5878 = 55.4\,\text{V}$

8.5 RECTIFIERS

A rectifier is a circuit which converts a.c. to d.c. The essential component of any rectifier circuit is a diode, allowing current to flow through it in one direction only. Two different rectifier circuits are discussed: the half-wave rectifier and the full-wave bridge rectifier.

The half-wave rectifier is the simplest form of rectifier circuit. It consists of a single diode, placed between an a.c. supply and the load, for which d.c. is required. The arrangement is shown in Figure 8.8, where the resistor R represents the load.

Let us assume that, in the first half cycle of the applied voltage, the instantaneous polarities at the input terminals are as shown in Figure 8.9. Under this condition, the diode is forward biased. A half sinewave of current will therefore flow through the load resistor, in the direction shown.

In the next half cycle of the input waveform, the instantaneous polarities will be reversed. The diode is therefore reverse biased, and no current will flow. This is illustrated in Figure 8.10.

The graphs of the applied a.c. voltage, and the corresponding load current, are shown in Figure 8.11. The load p.d. will be of exactly the same waveshape as the load current. Both of these quantities are unidirectional, and so by definition, are d.c. quantities. The 'quality' of the d.c. so produced is very poor, since it exists only in pulses of current. The average value of this current is determined over the time period, 0 to t_2. The average value from 0 to time t_1 will be $0.637I_m$. From t_1 to t_2 it will be zero. The average value of the d.c. will therefore be $I_{av} = 0.318I_m$.

Both the 'quality' and average value of the d.c. need to be improved. This may be achieved by utilising the other half cycle of the a.c. supply. The circuit of a full-wave bridge rectifier consists of four diodes, connected in a 'bridge' configuration, as shown in Figure 8.12.

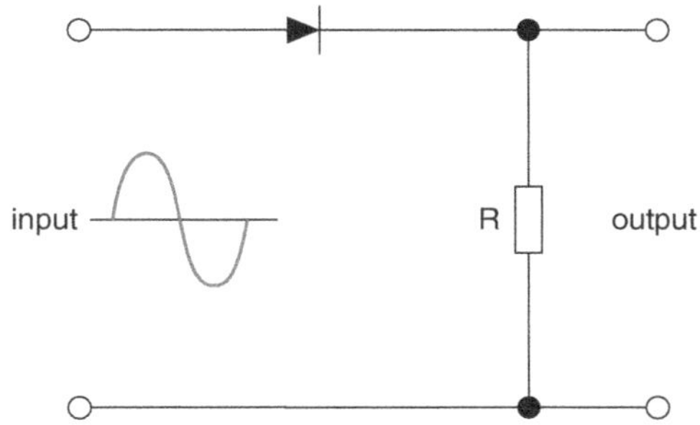

Figure 8.8 A half-wave rectifier

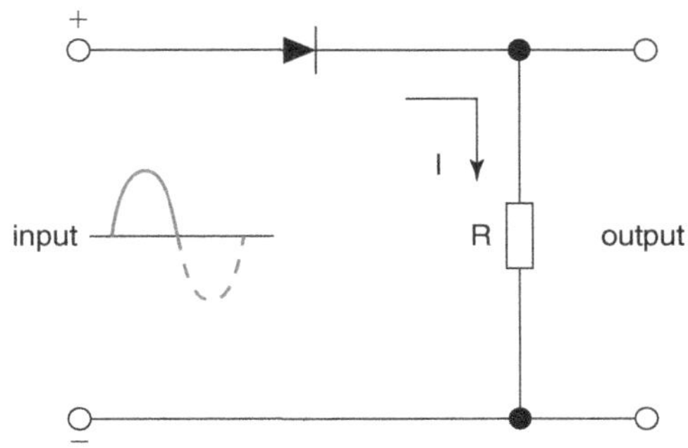

Figure 8.9 The first half cycle of a half-wave rectifier

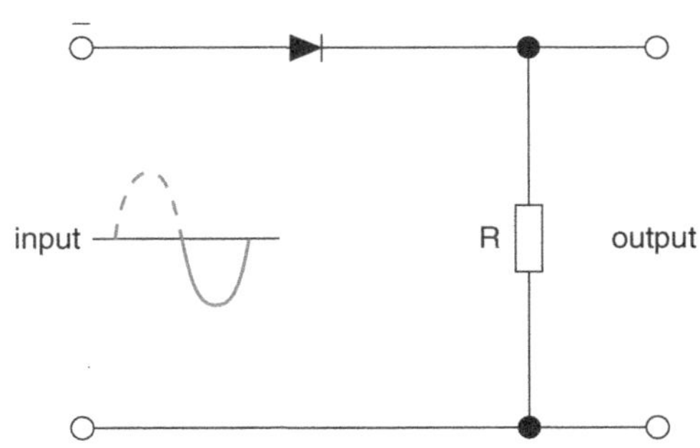

Figure 8.10 The second half cycle of a half-wave rectifier

We will again assume the instantaneous polarities for the first half cycle as shown. In this case, diodes D_1 and D_4 will be forward biased. Diodes D_2 and D_3 will be reverse biased. Thus, D_1 and D_4 allow current to flow, as shown. In the next half cycle, the polarities are reversed. Hence, D_2 and D_3 will conduct, whilst D_1 and D_4 are reverse biased. Current will therefore flow as shown in Figure 8.13. Notice that the current through the load resistor is in the same direction for the whole cycle of the a.c. supply.

The relevant waveforms are shown in Figure 8.14. It should be apparent that the average value of the a.c. will now be twice that in the previous circuit. That is, $I_{av} = 0.637I_m$.

Further improving the 'quality' and average value of the d.c. can be done by adding in parallel a capacitor filter across the output of the bridge rectifier. The capacitor resists voltage

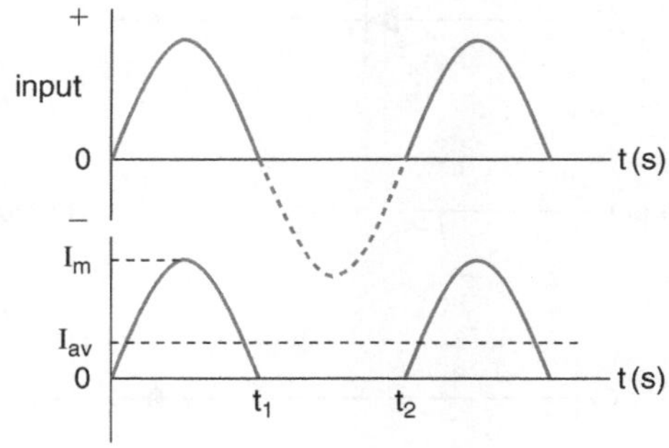

Figure 8.11 The graphs of the applied a.c. voltage and the load current,

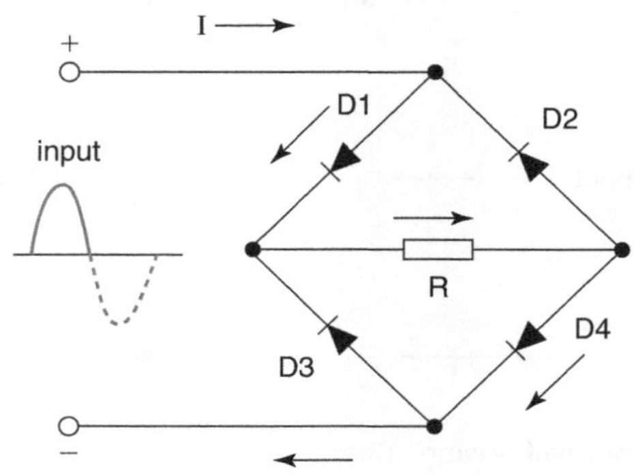

Figure 8.12 The first half of a full-wave bridge rectifier

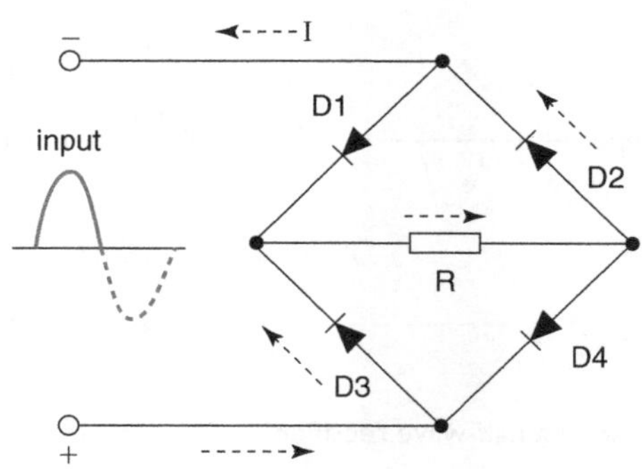

Figure 8.13 The second half of a full-wave bridge rectifier

changes and hence it attenuates the ripple of the rectified d.c. current. When the voltage across the capacitor increases, the charge on the plates of that capacitor also rises. However when the bridge rectifier voltage decreases, the voltage across that capacitor also decreases, but noticeably slower than the bridge rectifier voltage goes down. The voltage is equalised and the existing difference between the minimum and the maximum output voltage is called the ripple. It is often expressed as a percentage of the average voltage. The larger the capacitor value, the lower this ripple.

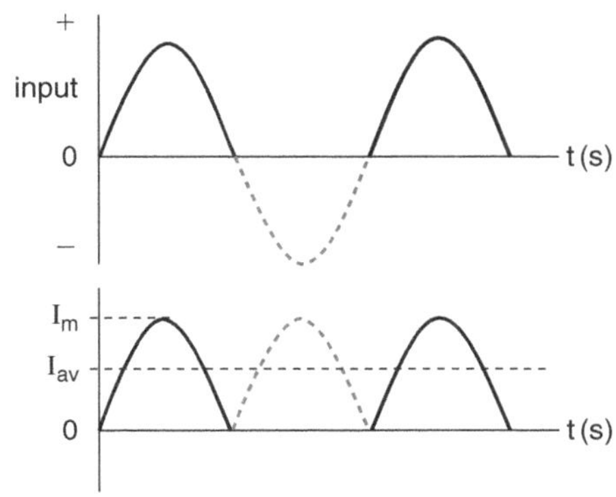

Figure 8.14 The graphs of the applied a.c. voltage and the load current

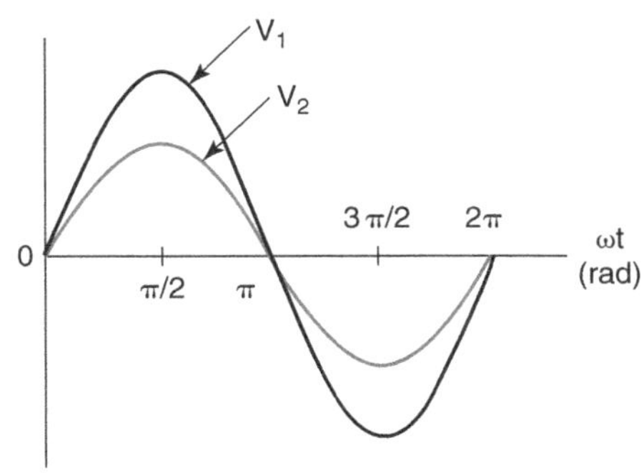

Figure 8.15 Two a.c. voltages with the same frequency

A full-wave rectifier in combination with a capacitor is used in switching power supplies (or Switched Mode Power Supply or SMPS). The mains voltage is rectified and filtered with a large capacitor. The resulting d.c. voltage is then again converted into an a.c. voltage with a much higher frequency (typically 10 kHz – 1 MHz). This is done by a chopper, an electronic switching circuit that very quickly (hence high frequency) switches the input voltage on and off. It is followed by a small transformer to lower the mains voltage and by a rectification and filtering step. The advantage is that the transformer can be made much more compact and lighter in weight. Such switching power supplies are used in household electronics where the voltage of the public electricity network must be efficiently converted to d.c. voltages.

8.6 PHASE AND PHASE ANGLE

Consider two a.c. voltages, of the same frequency, as shown in Figure 8.15. Both voltages pass through the zero on the horizontal axis at the same time. They also reach their positive and negative peaks at the same time. Thus, the two voltages are exactly synchronised, and are said to be *in phase* with each other.

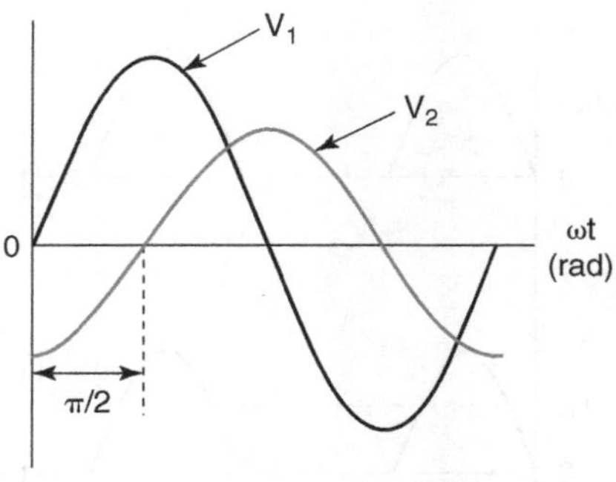

Figure 8.16 Two a.c. voltages with the same frequency and one lagging behind.

Figure 8.16 shows the same two voltages, but in this case let v_2 reach its maximum value $\pi/2$ radian (90°) after v_1. It is necessary to consider one of the waveforms as the *reference* waveform. It is normal practice to consider the waveform that passes through zero, going positive, at the beginning of the cycle, as the reference waveform. So, for the two waveforms shown, v_1 is taken as the reference. In this case, v_2 is said to *lag* v_1 by $\pi/2$ radian or by 90°. The standard expressions for the two voltages would therefore be written as follows:

$$v_1 = V_{m1}\sin(2\pi ft), \text{or } V_{m1}\sin\theta$$

$$v_2 = V_{m2}\sin\left(2\pi ft - \frac{\pi}{2}\right), \text{or } V_{m2}\left(\sin\theta - 90^\circ\right)$$

The minus signs, in the brackets of the above expressions, indicate that v_2 lags the reference by the angle quoted. This angle is known as the *phase angle*, or *phase difference*, between the two waveforms.

In general, the standard expression for an a.c. voltage is:

$$\begin{aligned} v &= V_m \sin(2\pi ft \pm \phi), \\ \text{or } v &= V_m \sin(\omega t \pm \phi) \end{aligned} \tag{8.10}$$

Although it would be usual to take v_1 as the reference in the above example, it is not mandatory. Thus, if for some good reason v_2 was chosen as the reference, v_1 is said to *lead* v_2 by $\pi/2$ radian or by 90°. The expressions would be written as:

$$v_2 = V_{m2}\sin(\omega t), \text{or } V_{m2}\sin\theta$$

$$v_1 = V_{m1}\sin\left(\omega t + \frac{\pi}{2}\right), \text{or } V_{m1}\sin(\theta + 90^\circ)$$

Note: When the relevant phase angle, ϕ, is quoted in the standard expression, do *not* mix degrees with radians. Thus, if the initial angular data is in radian (ωt or $2\pi ft$), then ϕ must also be expressed in radian. Similarly, if the angular data is initially in degrees (θ), the ϕ must also be quoted in degrees.

WORKED EXAMPLE 8.8

Q Three alternating currents are specified below. Determine the frequency, and for each current, determine its phase angle and amplitude.

$$i_1 = 5\sin\left(80\pi t + \frac{\pi}{6}\right)$$

$$i_2 = 3\sin 80\pi t$$

$$i_3 = 6\sin(80\pi t - \pi/4)$$

All three waveforms have the same value of ω, namely 80π rad/s. Thus all three have the same frequency:

$$\omega = 2\pi f = 80\pi \text{ rad/s}$$

$$\text{therefore, } f = \frac{80\pi}{2\pi} = 40\,\text{Hz}$$

Since zero phase angle is quoted for i_2, this is the reference waveform, of amplitude 3 A.

$I_{m1} = 5$ A, and leads i_2 by $\pi/6$ rad (30°)

$I_{m3} = 6$ A, and lags i_2 by $\pi/4$ rad (45°)

The majority of people can appreciate the relative magnitudes of angles when they are expressed in degrees. Angles expressed in radians are more difficult to appreciate. Some of the principal angles encountered are listed below. This should help you to gain a better 'feel' for radian measure.

Degrees	*Radians*	*Radians*	*Degrees*
360	$2\pi \approx 6.28$	0.1	5.73
270	$3\pi/2 \approx 4.71$	0.2	11.46
180	$\pi \approx 3.14$	0.3	17.19
120	$2\pi/3 \approx 2.09$	0.4	22.92
90	$\pi/2 \approx 1.57$	0.5	28.65
60	$\pi/3 \approx 1.05$	1.0	57.30
45	$\pi/4 \approx 0.79$	1.5	85.94
30	$\pi/6 \approx 0.52$	2.0	114.60

8.7 PHASOR REPRESENTATION

A phasor is a rotating vector. Apart from the fact that a phasor rotates at a constant velocity, it has exactly the same properties as any other vector. Thus its length corresponds to the magnitude of a quantity. It has one end arrowed, to show the direction of action of the quantity.

Consider two such rotating vectors, v_1 and v_2, rotating at the same angular velocity, ω rad/s. Let them rotate in a counterclockwise direction, with v_2 lagging behind v_1 by $\pi/6$ radian (30°). This situation is illustrated in Figure 8.17.

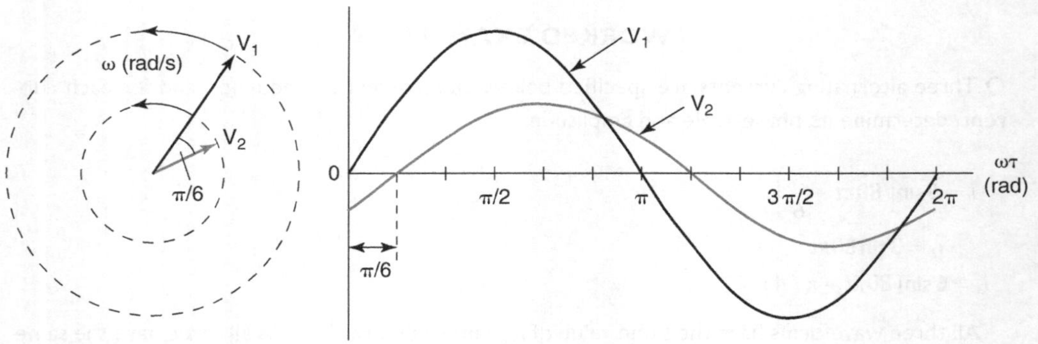

Figure 8.17 Phasor representation with one lagging behind

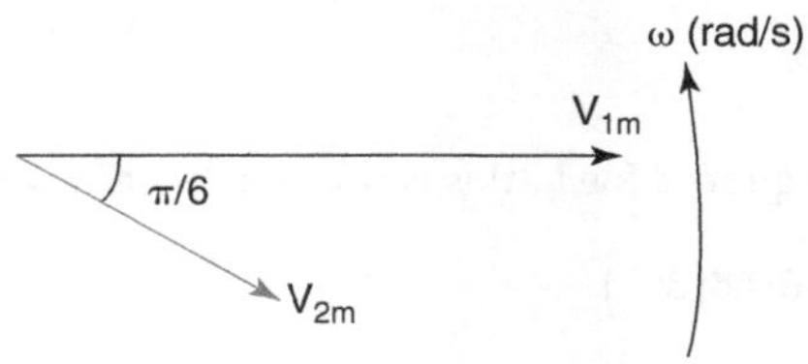

Figure 8.18 Referenced phasor representation with one lagging behind

The instantaneous vertical height of each vector is then plotted for one complete revolution. The result will be the two sinewaves shown. Notice that the angular difference between v_1 and v_2 is also maintained throughout the waveform diagram. Also note that the peaks of the two waveforms correspond to the magnitudes, or amplitudes, of the two vectors. In this case, these two waveforms could equally well represent either two a.c. voltages, or currents. If this were the case, then the two a.c. quantities would be of the *same frequency*. This is because the value of ω is the same for both. The angular difference, of π/6 radian, would then be described as the phase difference between them.

We can therefore represent an alternating quantity by means of a phasor. The length of the phasor represents the amplitude. Its angle, with respect to some reference axis, will represent its phase angle. Considering the two waveforms in Figure 8.17, the plot has been started with v_1 in the horizontal position (vertical component of $v_1 = 0$). This horizontal axis is therefore taken as being the reference axis. Thus, if these waveforms represent two voltages, v_1 and v_2, the standard expressions would be:

$$v_1 = V_{m1}\sin(\omega t)$$
$$v_2 = V_{m2}\sin(\omega t - \pi/6)$$

The inconvenience of representing a.c. quantities in graphical form was pointed out earlier, in Section 8.3. This section introduced the concept of using a standard mathematical expression for an a.c. However, a visual representation is also desirable. We now have a much simpler means of providing a visual representation. It is called a phasor diagram. Thus the two voltages we have been considering above may be represented as in Figure 8.18.

Notice that v_1 has been chosen as the reference phasor. This is because the standard expression for this voltage has a phase angle of zero (there is no $\pm\phi$ term in the bracket). Also, since the phasors are rotating counterclockwise, and v_2 is lagging v_1 by π/6 radian, then v_2 is shown at this angle *below* the reference axis.

Notes

1 Any a.c. quantity can be represented by a phasor, *provided that it is a sinewave.*
2 Any number of a.c. voltages and/or currents may be shown on the same phasor diagram, *provided that they are all of the same frequency.*
3 Figure 8.18 shows a *counterclockwise* arrow, with ω rad/s. This has been shown here to emphasise the point that phasors *must* rotate in this direction only. It is normal practice to omit this from the diagram.
4 When dealing with a.c. circuits, r.ms. values are used almost exclusively. In this case, it is normal to draw the phasors to lengths that correspond to r.m.s. values.

WORKED EXAMPLE 8.9

Q Four currents are as shown below. Draw to scale the corresponding phasor diagram.

$$i_1 = 2.5\sin(\omega t + \pi/4) \quad i_2 = 4\sin(\omega t - \pi/3)$$
$$i_3 = 6\sin\omega t \quad i_4 = 3\cos\omega t$$

Before the diagram is drawn, we need to select a reference waveform (if one exists). The currents i_1 and i_2 do not meet this criterion, since they both have an associated phase angle.

This leaves the other two currents. Neither of these has a phase angle shown. However, i_3 is a sinewave, whilst i_4 is a *cosine* waveform. Now, a cosine wave *leads* a sinewave by 90°, or $\pi/2$ radian.

Therefore, i_4 may also be expressed as $i_4 = 3\sin(\omega t + \pi/2)$. Thus i_3 is chosen as the reference waveform, and will therefore be drawn along the horizontal axis.
The resulting phasor diagram is shown in Figure 8.19.

WORKED EXAMPLE 8.10

Q The phasor diagram representing four alternating currents is shown in Figure 8.20, where the length of each phasor represents the amplitude of that waveform. Write down the standard expression for each waveform.

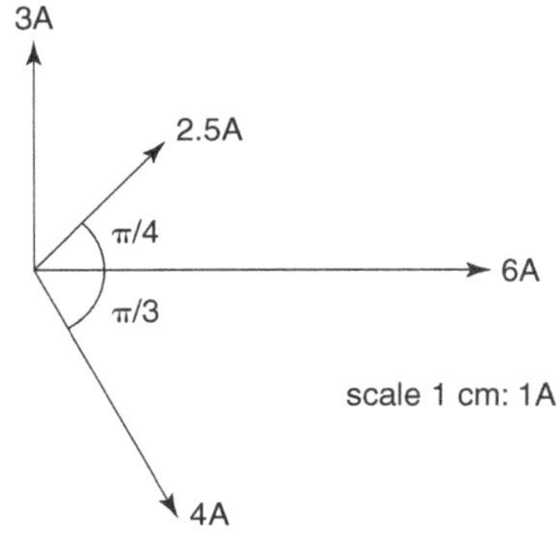

Figure 8.19 The resulting phasor diagram for Worked Example 8.9

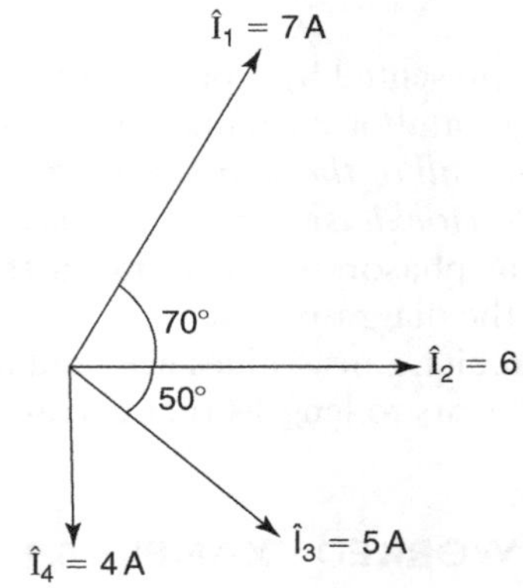

Figure 8.20 The phasor diagram for Worked Example 8.10

$$I_{m1} = 7\text{ A}; \phi_1 = +70° = \frac{70\pi}{180°} = 1.22\text{ rad}$$

$$I_{m2} = 6\text{ A}; \phi_2 = 0° = 0\text{ rad}$$

$$I_{m3} = 5\text{ A}; \phi_3 = -50° = \frac{-50\pi}{180°} = -0.873\text{ rad}$$

$$I_{m4} = 4\text{ A}; \phi_4 = -90° = \frac{-90\pi}{180°} = -1.571\text{ rad}$$

hence,

$$i_1 = 7\sin(\omega t + 1.22)$$
$$i_2 = 6\sin\omega t$$
$$i_3 = 5\sin(\omega t - 0.873)$$
$$i_4 = 4\sin(\omega t - 1.571)$$

8.8 ADDITION OF ALTERNATING QUANTITIES

Consider two alternating currents, $i_1 = I_{m1} \sin \omega t$ and $i_2 = I_{m2} \sin (\omega t - \pi/4)$, that are to be added together. There are three methods of doing this, as listed below.

(a) Plotting them on graph paper. Their ordinates are then added together, and the resultant waveform plotted. This is illustrated in Figure 8.21. The amplitude I_m and the phase angle ϕ of the resultant current are then measured from the two axes.

Thus, $i = I_m \sin(\omega t - \phi)$

Note: Although $i = i_1 + i_2$, the *amplitude* of the resultant is *not* $I_{m1} + I_{m2}$ amp. This would only be the case if i_1 and i_2 were in phase with each other.

(b) Drawing a scaled phasor diagram, as illustrated in Figure 8.22. The resultant is found by completing the parallelogram of vectors. The amplitude and phase angle are then measured on the diagram.

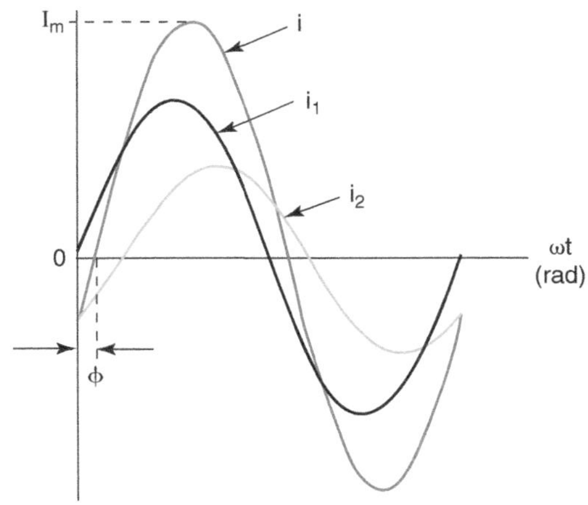

Figure 8.21 Two alternating currents added together on graph paper

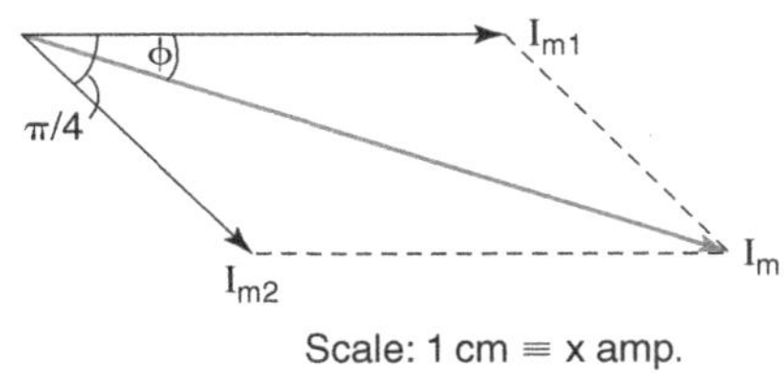

Figure 8.22 Two alternating currents added together with a phasor diagram

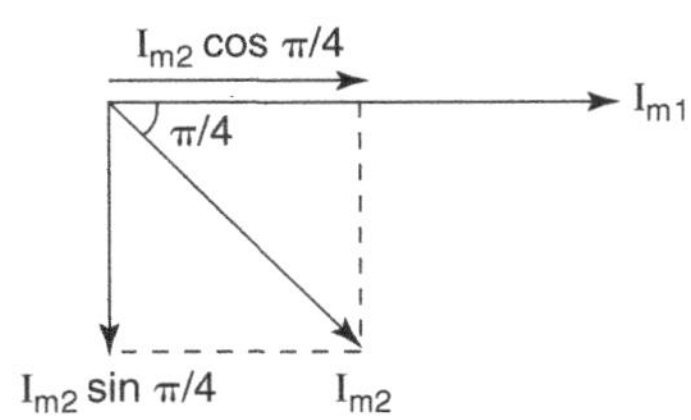

Figure 8.23 The horizontal and vertical component of the currents

(c) Resolving the two currents, into horizontal and vertical components, and applying Pythagoras' theorem. This method involves using a *sketch* of the phasor diagram, followed by a purely mathematical process. This phasor diagram, including the identification of the horizontal and vertical components, is shown in Figure 8.23.

Horizontal Component (H.C.):

$$\begin{aligned}\text{H.C.} &= I_{m1}\cos 0 + I_{m2}\cos \pi/4 \\ &= I_{m1} + (0.707 \times I_{m2})\end{aligned}$$

Vertical Component (V.C.):

$$\begin{aligned}\text{V.C.} &= I_{m1}\sin 0 + I_{m2}\sin \pi/4 \\ &= 0 + (0.707 \times I_{m2})\end{aligned}$$

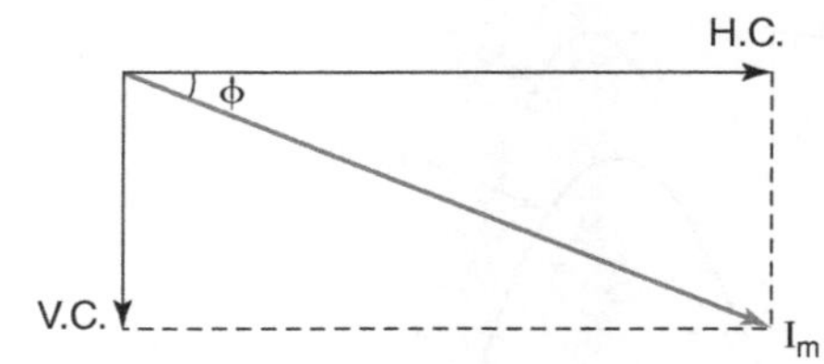

Figure 8.24 The triangle of horizontal and vertical component, and the resultant current

The triangle of H.C., V.C., and the resultant current, is shown in Figure 8.24. From this, we can apply Pythagoras' theorem to determine the amplitude and phase angle, thus:

$$I_m = \sqrt{\text{H.C.}^2 + \text{V.C.}^2}$$

$$\tan\phi = \frac{\text{V.C.}}{\text{H.C.}}, \text{ so } \phi = \tan^{-1}\frac{\text{V.C.}}{\text{H.C.}}$$

The final answer, regardless of the method used, would then be expressed in the form $i = I_m \sin(\omega t \pm \phi)$.

Let us now compare the three methods, for calculation speed, convenience and accuracy.

The graphical technique is very time-consuming (even for the addition of only two quantities). The accuracy also leaves much to be desired; in particular, determining the exact point for the maximum value of the resultant. The determination of the precise phase angle is also very difficult. This method is therefore not recommended.

A phasor diagram, drawn to scale, can be the quickest method of solution. However, it does require considerable care, in order to ensure a reasonable degree of accuracy. Even so, the precision with which the length – and (even more so) the angle – can be measured, leaves a lot to be desired. This is particularly true when three or more phasors are involved. This method is therefore recommended only for a rapid estimate of the answer.

The use of the resolution of phasors is, with practice, a rapid technique, and yields a high degree of accuracy. Unless specified otherwise it is the technique you should use. Although, at first acquaintance, it may seem to be rather a complicated method, this is not the case. With a little practice, the technique will be found to be relatively simple and quick. Some worked examples now follow.

WORKED EXAMPLE 8.11

Q Determine the phasor sum of the two voltages specified below.

$$v_1 = 25\sin(314t + \pi/3)\text{ V and } v_2 = 15\sin(314t - \pi/6)\text{ V}$$

Figure 8.25 shows the sketch of the phasor diagram.

Note: Always sketch a phasor diagram.

$$\text{H.C.} = 25\cos\pi/3 + 15\cos(-\pi/6) = (25\times0.5) + (15\times0.866) = 12.5 + 12.99 = 25.49\text{ V}$$

$$\text{V.C.} = 25\sin\pi/3 + 15\sin(-\pi/6) = (25\times0.866) + (15\times(-0.5)) = 21.65 - 7.5 = 14.15\text{ V}$$

Figure 8.26 shows the phasor diagram for H.C., V.C. and V_m.

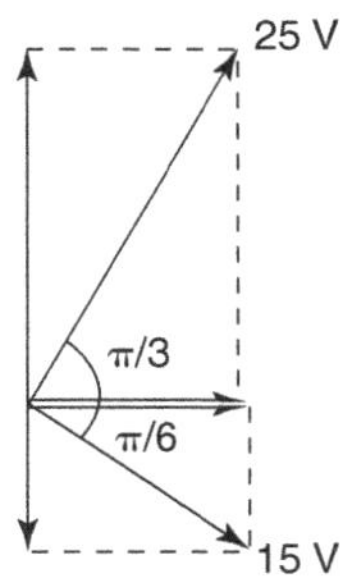

Figure 8.25 The phasor diagram for Worked Example 8.11

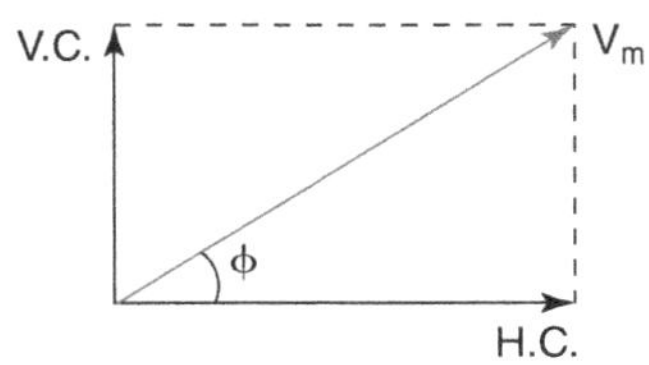

Figure 8.26 The phasor diagram of H.C, V.C. and *Vm* for Worked Example 8.11

$$V_m = \sqrt{\text{H.C.}^2 + \text{V.C.}^2} = \sqrt{25.49^2 + 14.15^2} = 29.15\,\text{V}$$

$$\tan\phi = \frac{\text{V.C.}}{\text{H.C.}} = \frac{14.15}{25.49} = 0.555\,*$$

$$\phi = \tan^{-1} 0.555 = 0.507\,\text{rad}$$

$$\text{therefore, } v = 29.15\sin(314t + 0.507)\,\text{V}$$

*radian mode required

WORKED EXAMPLE 8.12

Q Calculate the phasor sum of the three currents listed below.

$$i_1 = 6\sin\omega t\ \text{A}$$

$$i_2 = 8\sin\left(\omega t - \frac{\pi}{2}\right)\text{A}$$

$$i_3 = 4\sin\left(\omega t + \frac{\pi}{6}\right)\text{A}$$

The relevant phasor diagrams are shown in Figures 8.27 and 8.28.

$$\text{H.C.} = 6\cos 0 + 8\cos(-\pi/2) + 4\cos\pi/6 = (6\times 1) + (8\times 0) + (4\times 0.866) = 6 + 3.46 = 9.46\,\text{A}$$

$$\text{V.C.} = 6\sin 0 + 8\sin(-\pi/2) + 4\sin\pi/6 = (6\times 0) + (8\times[-1]) + (4\times 0.5) = -8 + 2 = -6\,\text{A}$$

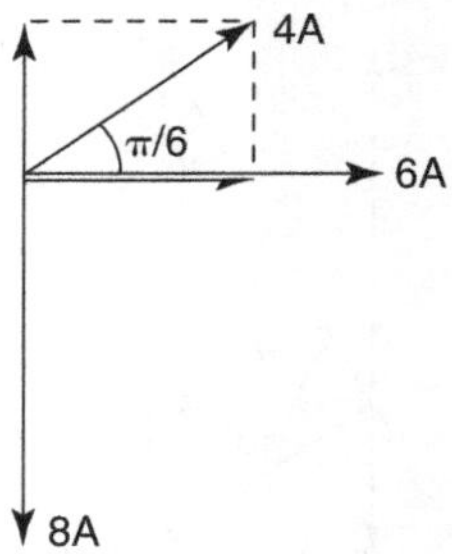

Figure 8.27 The phasor diagram for Worked Example 8.12

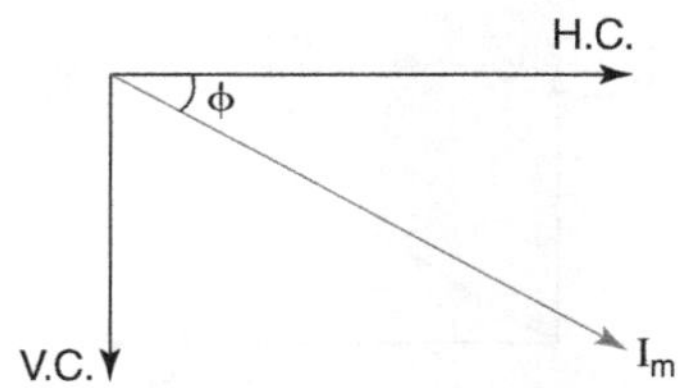

Figure 8.28 The phasor diagram of H.C, V.C. and *Vm* for Worked Example 8.12

$$I_m = \sqrt{\text{H.C.}^2 + \text{V.C.}^2} = \sqrt{9.46^2 + (-6)^2} = 11.2\,\text{A}$$

$$\phi = \tan^{-1}\frac{\text{V.C.}}{\text{H.C.}} = \tan^{-1}\frac{-6}{9.46} = \tan^{-1}(-0.6342) = -0.565\,\text{rad}$$

$$\text{therefore, } i = 11.2\sin(\omega t - 0.565)\,\text{A}$$

WORKED EXAMPLE 8.13

Q Three alternating voltages and one current are as specified in the expressions below.

$V_1 = 10\sin(628t - \pi/6)\,\text{V}$

$V_2 = 8\sin(628t + \pi/3)\,\text{V}$

$V_3 = 12\sin(628t + \pi/4)\,\text{V}$

$i = 6\sin(628t)\,\text{A}$

(a) For each voltage determine the frequency, phase angle and amplitude.

(b) Determine the phasor sum of the three voltages.

(a) All four waveforms have the same value of ω = 628 rad/s, so they are all of the same frequency, hence

$$\omega = 2\pi f = 628\,\text{rad/s}$$

$$f = \frac{628}{2\pi} = 100\,\text{Hz}$$

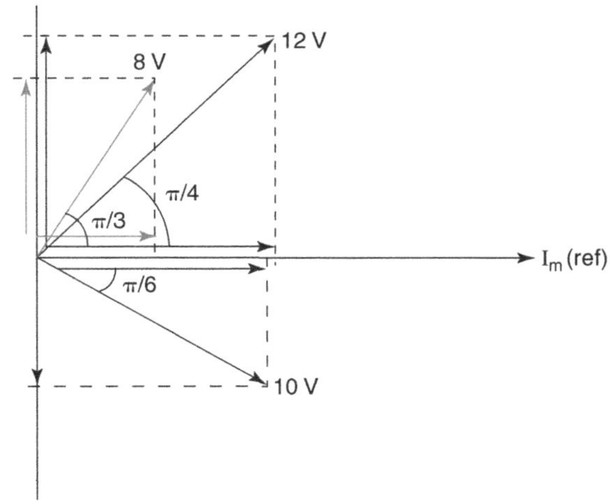

Figure 8.29 The phasor diagram for Worked Example 8.13

$$\text{for } V_1, \phi_1 = -\frac{\pi}{6}\text{ rad or } -30^\circ;\text{ and } V_m = 10\text{ V}$$

$$\text{for } V_2, \phi_2 = \frac{\pi}{3}\text{ rad or } 60^\circ;\text{ and } V_m = 8\text{ V}$$

$$\text{for } V_3, \phi_3 = \frac{\pi}{4}\text{ rad or } 45^\circ;\text{ and } V_m = 12\text{ V}$$

(b) Firstly the phasor diagram (Figure 8.29) is sketched, very roughly to scale. In order to do this a reference waveform needs to be selected, and since the current has a zero phase angle, this is chosen as the reference. However, if the current waveform had not been specified, the horizontal axis would still be taken as the reference from which all phase angles are measured. Since v_2 and v_3 have positive phase angles, and phasors rotate anticlockwise, these two phasors will appear above the reference axis. The voltage v_1, having a negative phase angle will appear below the reference axis. Also shown on the phasor diagram are the horizontal and vertical components of each voltage.

$$\text{H.C.} = 12\cos\pi/4 + 8\cos\pi/3 + 10\cos\pi/6 = (12\times 0.707) + (8\times 0.5) + (10\times 0.866)$$

$$= 8.48 + 4 + 8.66 = 21.44\text{ V}$$

$$\text{V.C.} = 12\sin\pi/4 + 8\sin\pi/3 + 10\sin\pi/6 = (12\times 0.707) + (8\times 0.866) + (10\times 0.5)$$

$$= 8.48 + 6.928 - 5 = 10.412\text{ V}$$

The phasor diagram for H.C. and V.C., and the resultant phasor sum, is Figure 8.30.

$$V_m = \sqrt{\text{H.C.}^2 + \text{V.C.}^2} = \sqrt{21.44^2 + 10.412^2} = 23.83\text{ V}$$

$$\phi = \tan^{-1}\frac{\text{V.C.}}{\text{H.C.}} = \tan^{-1}\frac{10.412}{21.44} = \tan^{-1}(0.4856)$$

$$\phi = 0.452\text{ rad}$$

Hence, the phasor sum, $V = 23.83\sin(628t + 0.452)\text{ V}$

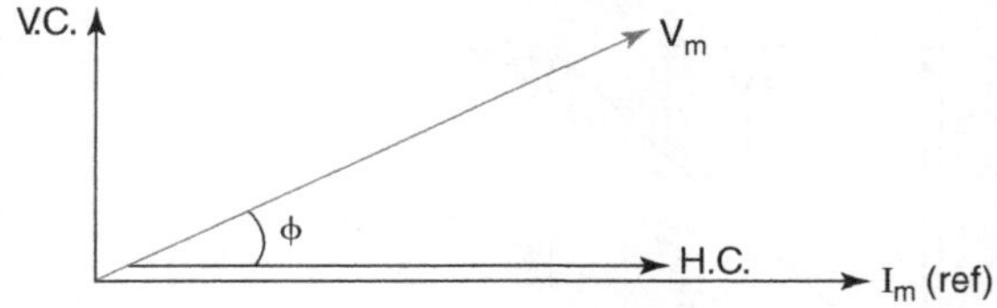

Figure 8.30 The phasor diagram of H.C, V.C. and *Vm* for Worked Example 8.13

8.9 OSCILLOSCOPE

An oscilloscope (more often abbreviated to scope) is a very versatile instrument that may be used to measure both a.c. and d.c. voltages and can display electrical voltages in a graphical way on a two-dimensional plot as a function of time. For d.c. measurements, a voltmeter is usually more convenient to use. The principal advantages of the oscilloscope when used to measure a.c. quantities are:

1. A visual indication of the waveform is produced.
2. The frequency, period and phase angle of the waveform(s) can be determined.
3. It can be used to measure very high frequency waveforms.
4. Any waveshape can be displayed, and measured with equal accuracy.
5. The input resistance (impedance) is of the same order as a voltmeter. It therefore applies minimal loading effect to a circuit to which it is connected.
6. Some oscilloscopes can display two or more waveforms simultaneously.

Very often two traces can be visualised, having the advantage that two waveforms can be displayed simultaneously. This enables waveforms to be compared, in terms of their amplitudes, shape, phase angle or frequency.

Oscilloscopes are used in all kinds of disciplines in engineering, sciences and medicine. The visualisation of the heartbeat on an electrocardiogram, for instance, is done by such a oscilloscope. But also the troubleshooting of malfunctioning electronic equipment can be performed by using a scope, because it can reveal that the circuit is oscillating or that the shape or timing is not as expected.

WORKED EXAMPLE 8.14

Q Figure 8.31 shows two traces obtained on an oscilloscope. The graticule is marked in 1 cm squares. The channel 1 input is displayed by the upper trace. If settings of the controls for the two channels are as follows, determine the amplitude, r.m.s. value and frequency of each input.

Channel 1: timebase of 0.1 ms/cm; Y-amp setting of 5 V/cm
Channel 2: timebase of 10 μs/cm; Y-amp setting of 0.5 V/cm

Channel 1: peak to peak occupies 3 cm, so

$$V_{p-p} = 3 \times 5 = 15\,\text{V}$$

$$V_m = \frac{V_{p-p}}{2} = 7.5\,\text{V}$$

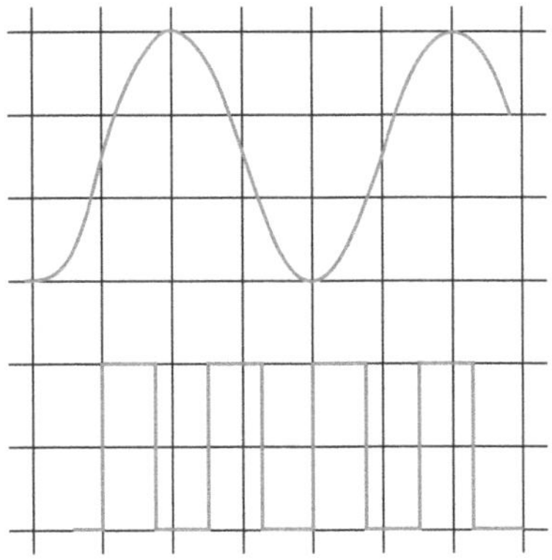

Figure 8.31 Two traces obtained on a oscilloscope for Worked Example 8.14

As the waveform is a sinewave, then r.m.s. value $V = V_m / \sqrt{2}$

$$V = \frac{7.5}{\sqrt{2}} = 5.3\,\text{V}$$

One cycle occupies 4 cm, so T = 4 cm × 0.1 ms/cm = 0.4 ms

$$f = \frac{1}{T} = \frac{1}{0.4 \times 10^{-3}\,\text{s}} = 2.5\,\text{kHz}$$

Channel 2: peak to peak occupies 2 cm, so

$$V_{p-p} = 2\,\text{cm} \times 0.5\frac{\text{V}}{\text{cm}} = 1\text{V, and } V_m = 0.5\,\text{V}$$

Since it is a squarewave, then r.m.s. value = amplitude,

hence $V = 0.5\,\text{V}$

Two cycles occur in 3 cm, so one cycle occurs in 2/3 cm. Therefore, $T = 0.6667 \times 10 = 6.667\,\mu\text{s}$

$$f = \frac{1}{T} = \frac{1}{6.667 \times 10^{-6}\,\text{s}} = 150\,\text{kHz}$$

SUMMARY OF EQUATIONS

Frequency generated: $f = np$

Periodic time: $T = \frac{1}{f}$

Angular velocity: $\omega = 2\pi f$

Standard expression for a sinewave: $e = E_m \sin(\theta \pm \phi) = E_m \sin(\omega t \pm \phi) = E_m \sin(2\pi f t \pm \phi)$

Average value for a sinewave: $I_{av} = \frac{2I_m}{\pi} = 0.637\,I_m$

R.m.s. value for a sinewave: $I = \frac{I_m}{\sqrt{2}} = 0.707\,I_m$

Peak factor for a sinewave: $\frac{\text{max. value}}{\text{r.m.s. value}} = 1.414$

Form factor for a sinewave: $\frac{\text{r.m.s. value}}{\text{ave value}} = 1.11$

ASSIGNMENT QUESTIONS

1 A coil is rotated between a pair of poles. Calculate the frequency of the generated emf if the rotational speed is (a) 150 rev/s, (b) 900 rev/minute and (c) 200 rad/s.

2 An alternator has eight poles. If the motor winding is rotated at 1500 rev/min, determine (a) the frequency of the generated emf and (b) the speed of rotation required to produce frequency of 50 Hz.

3 A frequency of 240 Hz is to be generated by a coil, rotating at 1200 rev/min. Calculate the number of poles required.

4 A sinewave is shown in Figure 8.32. Determine its amplitude, periodic time and frequency.

5 A sinusoidal current has a peak-to-peak value of 15 mA and a frequency of 100 Hz. (a) Plot this waveform, to a base of time and (b) write down the standard expression for the waveform.

6. A sinusoidal voltage is generated by an 85-turn coil, of dimensions 20 cm by 16 cm. The coil is rotated at 3000 rev/min, with its longer sides parallel to the faces of a pair of poles. If the flux density produced by the poles is 0.5 T, calculate (a) the amplitude of the generated emf, (b) the frequency and (c) the r.m.s. and average values.

7 Write down the standard expression for a voltage of r.m.s. value 45 V and frequency 1.5 kHz. Hence, calculate the instantaneous value 38 μs after the waveform passes through its zero value.

8 For each of the following alternating quantities, determine (a) the amplitude and r.m.s. value and (b) the frequency and period.

i $e = 250 \sin 50\pi t$ V

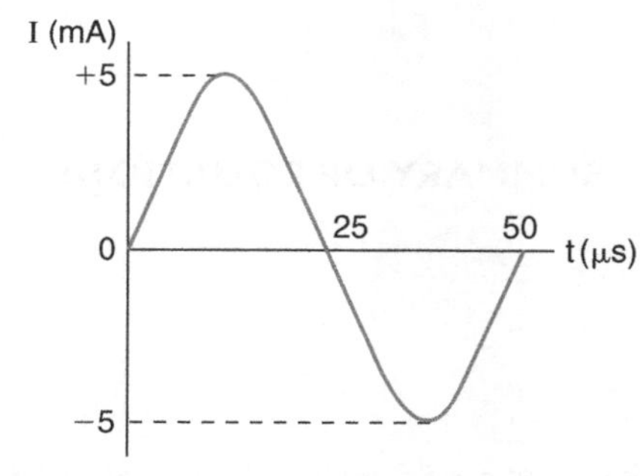

Figure 8.32 The sinewave current for Assignment Question 4

ii $i = 75 \sin 628.3t$ mA
iii $\varphi = 20 \sin 100\pi t$ mWb
iv $v = 6.8 \sin (942.8t + \varphi)$ V.

9 For a current of r.m.s. value 5 A, and frequency 2kHz, write down the standard expression. Hence, calculate (a) the instantaneous value 150 μs after it passes through zero and (b) the time taken for it to reach 4 A, after passing through zero for the first time.

10 Calculate the peak and average values for a 250 V sinusoidal supply.

11 A sinusoidal current has an average value of 3.8 mA. Calculate its r.m.s. and peak values.

12 An alternating voltage has an amplitude of 500 V and an r.m.s. value of 350 V. Calculate the peak factor.

13 A waveform has a form factor of 1.6 and an average value of 10 V. Calculate its r.m.s. value.

14 A moving coil voltmeter, calibrated for sinewaves, is used to measure a voltage waveform having a form factor of 1.25. Determine the true r.m.s. value of this voltage if the meter indicates 25 V. Explain why the meter does not indicate the true value.

15 Explain why only sinusoidal waveforms can be represented by phasors.

16 Sketch the phasor diagram for the two waveforms shown in Figure 8.33.

17 Sketch the phasor diagram for the two voltages represented by the following expressions:

$$v_1 = 12 \sin 314t \text{ V}$$
$$v_2 = 8 \sin(314t + \pi/3) \text{ V}$$

18 Determine the phasor sum of the two voltages specified in Question 17 above.

19 Three currents in an a.c. circuit meet at a junction. Calculate the phasor sum, if the currents are:

$$i_1 = 10 \sin \omega t \text{ A}$$
$$i_2 = 5 \sin(\omega t + \pi/4) \text{ A}$$
$$i_3 = 14 \sin(\omega t + \pi/3) \text{ A}$$

20 Determine the phasor sum of the following voltages, all of which are sinewaves of the same frequency:

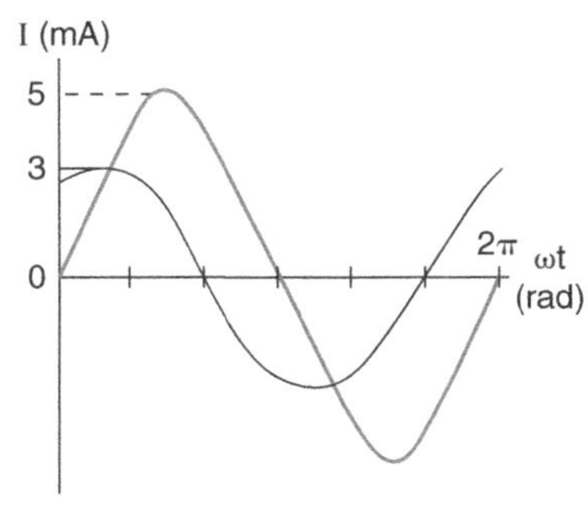

Figure 8.33 Two current waveforms for Assignment Question 16

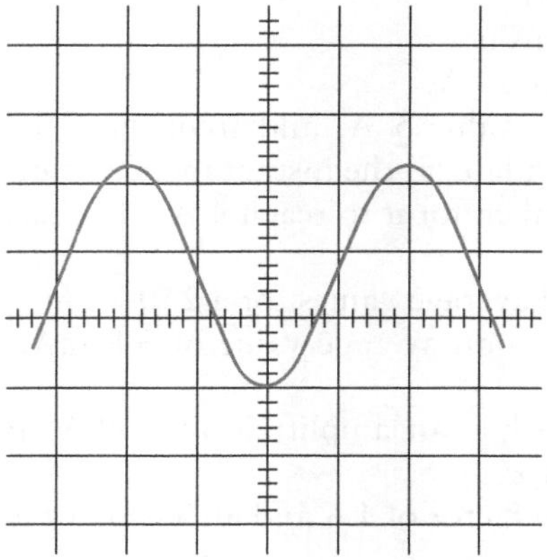

Figure 8.34 The trace obtained on a oscilloscope for Assignment Question 23

v_1 has an amplitude of 25 V, and a phase angle of zero.
v_2 has an amplitude of 13.5 V, and lags v_1 by 25°.
v_3 has an amplitude of 7.6 V, and leads v_2 by 40°.

21 By means of a phasor diagram, drawn to scale, check your answer to Question 19 above.

22 Plot, on the same axes, the graphs of the following two voltages. By adding ordinates, determine the sum of these voltages. Express the result in the form

$$V = V_m \sin(\omega t \pm \phi)$$

$$v_1 = 12 \sin t \text{ and } v_2 = 8 \sin(\omega t - \pi/6) \text{ V}.$$

23 The waveform displayed on an oscilloscope is as shown in Figure 8.34. The timebase is set to 100 μs/cm, and the Y-amp is set to 2 V/cm. Determine the amplitude, r.m.s. value, periodic time and frequency of this waveform.

Chapter 9

D.C. Machines

LEARNING OUTCOMES

This chapter covers the operating principles of d.c. generators and motors, their characteristics and applications. On completion you should be able to:

1 Understand and explain generator/motor duality.
2 Appreciate the need for a commutator.
3 Identify the different types of d.c. generator, and describe their characteristics. Carry out practical tests to compare the practical and theoretical characteristics.

9.1 MOTOR/GENERATOR DUALITY

An electric motor is a rotating machine which converts an electrical input power into a mechanical power output. A generator converts a mechanical power input into an electrical power output. Since one process is the converse of the other, a motor may be made to operate as a generator, and vice versa. This duality of function is not confined to d.c. machines. An alternator can be made to operate as a synchronous a.c. motor, and vice versa.

To demonstrate the conversion process involved, let us reconsider two simple cases that were met when dealing with electromagnetic induction.

Consider a conductor being moved at constant velocity, through a magnetic field of density B tesla, by some externally applied force F newton. This situation is illustrated in Figure 9.1. Work W, expressed in Nm or newton-metre, done in moving the conductor,

$$W = Fd$$

mechanical power input P_1, expressed in watt,

$$P_1 = \frac{W}{t} = \frac{Fd}{t}$$

and since d/t is the velocity, v at which the conductor is moved, then

$$P_1 = Fv \qquad [1]$$

However, when the conductor is moved, an emf will be induced into it. Provided that the conductor forms part of a closed circuit, then the resulting current flow will be as shown in Figure 9.2. This induced current, i, produces its own magnetic field, which reacts with the main field, producing a reaction force, F_r, in direct opposition to the applied force, F.

DOI: 10.1201/9781003308294-9

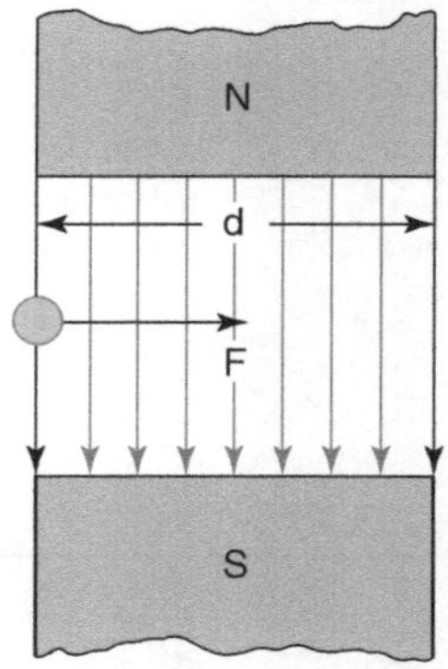

Figure 9.1 A conductor moved at constant velocity, through a magnetic field

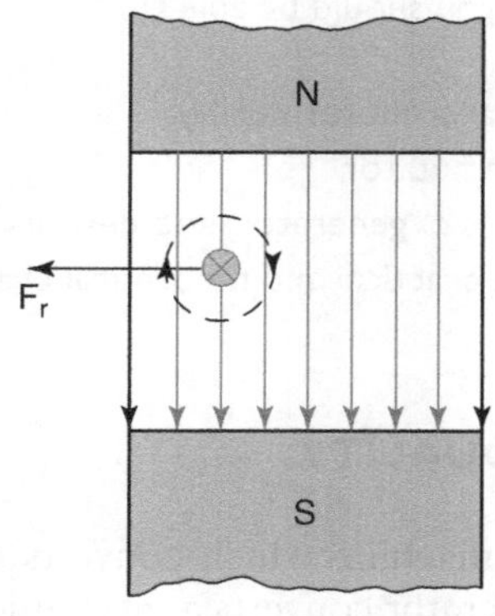

Figure 9.2 The resulting current flow for moving conductor in a magnetic field

$$F_r = Bi\ell$$

Assuming no frictional or other losses, then the applied force F has only to overcome the reaction force F_r, such that:

$$F = F_r = Bi\ell$$

$$P_1 = Bi\ell v \qquad [2]$$

$$e = B\ell v$$

$$P_2 = ei \qquad [3]$$

$$P_2 = Bi\ell v$$

Since [3] = [2], then the electrical power generated P_2 is equal to the mechanical power input P_1 (assuming no losses). Now consider the conductor returned to its original starting position. Let an external source of emf, *e* volt pass a current of *i* ampere through the conductor. Provided that the direction of this current is opposite to that shown in Figure 9.2, then the conductor will experience a force that will propel it across the field. In this case, the same basic arrangement exhibits the motor effect, since the electrical input power is converted into mechanical power.

Although the above examples involve linear movement of the conductor, exactly the same principles apply to a rotating machine.

9.2 THE GENERATION OF D.C. VOLTAGE

We have seen in Chapter 8 already that if a single-loop coil is rotated between a pair of magnetic poles, then an *alternating* emf is induced into it. This is the principle of a simple form of alternator. Of course, this a.c. output could be converted to d.c. by employing a rectifier circuit. Indeed, that is exactly what is done with vehicle electrical systems. However, in order to have a truly d.c. machine, this rectification process needs to be automatically accomplished within the machine itself. This process is achieved by means of a commutator, the principle and action of which will now be described.

Consider a simple loop coil and the two ends of which are connected to a single 'split' slip-ring, as illustrated in Figure 9.3. Each half of this slip-ring is insulated from the other half, and also from the shaft on which it is mounted. This arrangement forms a simple commutator, where the connections to the external circuit are via a pair of carbon brushes. The rectifying action is demonstrated in the series of diagrams of Figure 9.4. In these diagrams, one side of the coil and its associated commutator segment are identified by a thickened line edge. For the sake of clarity, the physical connection of each end of the coil to its associated commutator segment is not shown. Figure 9.4(a) shows the instant when maximum emf is induced in the coil. The current directions have been determined by applying Fleming's right-hand rule. At this instant current will be fed out from the coil, through the external circuit from right to left, and back into the other side of the coil. As the coil continues to rotate from this position, the value of induced emf and current will decrease. Figure 9.4(b) shows the instant when the brushes short-circuit the two commutator segments. However, the induced emf is also zero at this instant, so no current flows through the external circuit. Further rotation of the coil results in an increasing emf, but of the opposite polarity to that induced before. Figure 9.4(c) shows the instant when the emf has reached its next maximum. Although the generated emf is now reversed, the current through the external circuit will be in the same direction as before. The load current will therefore be a series of half-sinewave pulses, of the same polarity. Thus the commutator is providing a d.c. output to the load, whereas the armature-generated emf is alternating.

A single-turn coil will generate only a very small emf. An increased amplitude of the emf may be achieved by using a multi-turn coil.

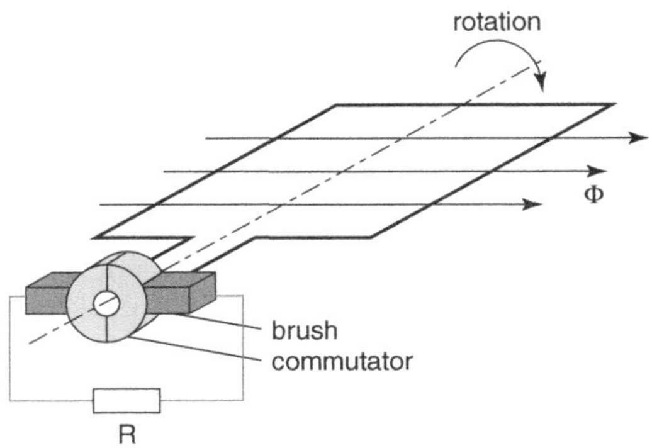

Figure 9.3 A simple loop coil connected to a single 'split' slipring

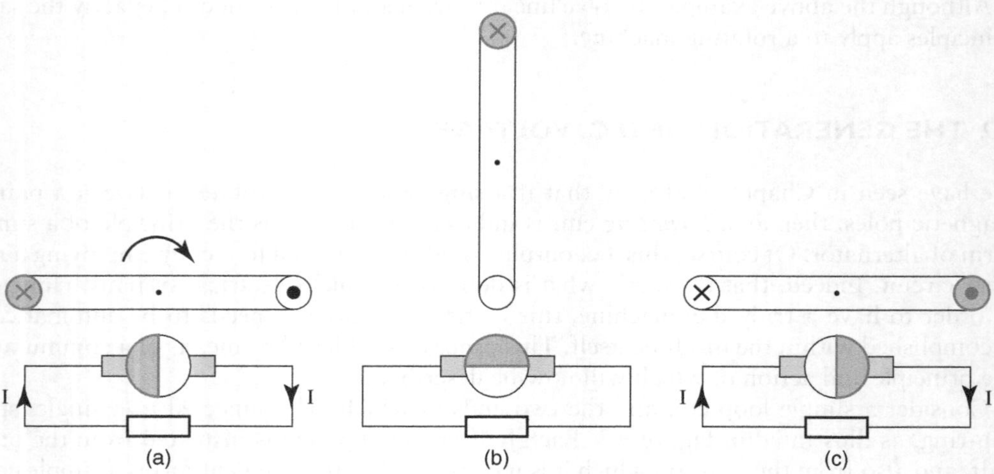

Figure 9.4 The corresponding rectifying action

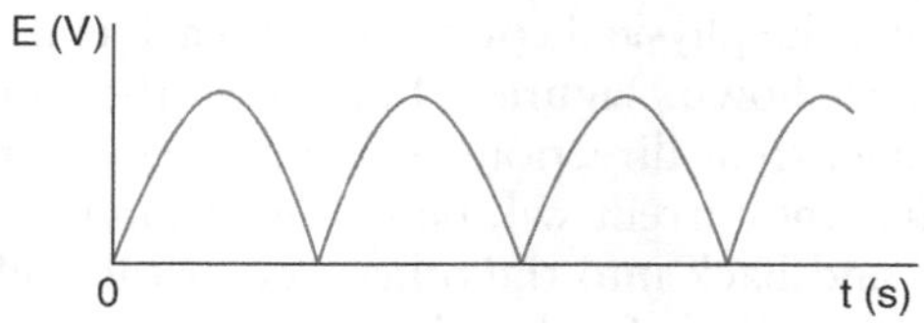

Figure 9.5 The resulting output voltage waveform

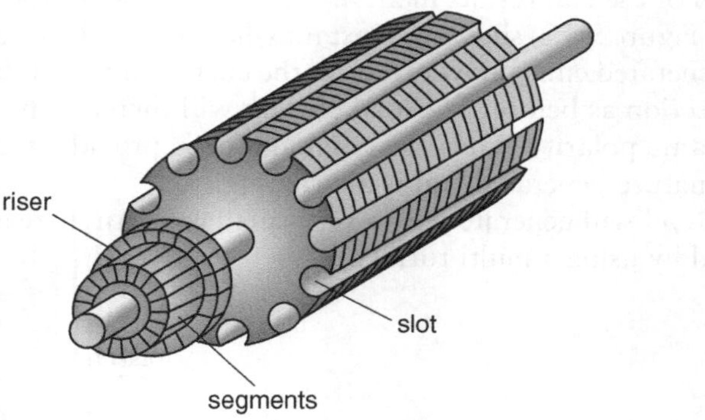

Figure 9.6 The armature construction (before the coils have been inserted)

The resulting output voltage waveform is shown in Figure 9.5. Although this emf is unidirectional, and may have a satisfactory amplitude, it is not a satisfactory d.c. waveform. The problem is that we have a concentrated winding. In a practical machine the armature has a number of multi-turn coils. These are distributed evenly in slots around the periphery of a laminated steel core. Each multi-turn coil has its own pair of slots, and the two ends are connected to its own pair of commutator segments. Figure 9.6 shows the armature construction (before the coils have been inserted). The riser is the section of the commutator to which the ends of the coils are soldered. Due to the distribution of the coils around the

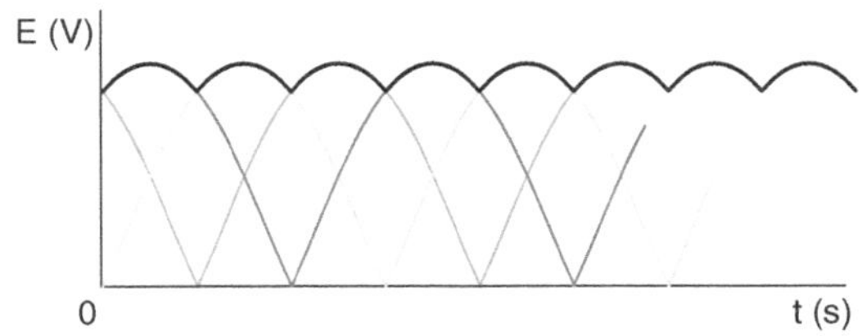

Figure 9.7 The maximum induced emfs occurred one after the other

Figure 9.8 The various parts of a small d.c. machine

armature, their maximum induced emfs will occur one after the other, i.e. they will be out of phase with each other. Figure 9.7 illustrates this, but for simplicity, only three coils have been considered.

Nevertheless, the effect on the resultant machine output voltage is apparent, and is shown by the thick line along the peaks of the waveform. With a large number of armature coils the ripple on the resultant waveform will be negligible, and a smooth d.c. output is produced.

The various parts of a small d.c. machine are shown separately in Figure 9.8, with the exception that neither the field nor armature windings have been included. The frame shell (bottom left) contains the pole pieces, around which the field winding would be wound. One end frame (top left) would simply contain a bearing for the armature shaft.

The other end frame (bottom right) contains the brushgear assembly in addition to the other armature shaft bearing. The armature (top right) construction has already been described. The slots are skewed to provide a smooth starting and slow-speed torque.

All d.c. generators are classified according to whether the field winding is electrically connected to the armature winding, and if it is, whether it is connected in parallel with or in series with the armature. The field current may also be referred to as the excitation current. If this current is supplied internally, by the armature, the machine is said to be self-excited. When the field current is supplied from an external d.c. source, the machine is said to be separately excited. The circuit symbol used for the field winding of a d.c. machine is simply the same as that used to represent any other form of winding. The armature is represented by a circle and two 'brushes'. The armature conductors, as such, are not shown.

9.3 SEPARATELY EXCITED GENERATOR

The circuit diagram of a separately excited generator is shown in Figure 9.9. The variable resistor, R_1, is included so that the field excitation current, I_f, can be varied. This diagram also shows the armature being driven at constant speed by some prime mover. Since the armature of any generator must be driven, this drive is not normally shown. The load, R_L, being supplied by the generator may be connected or disconnected by switch S_2. The resistance of the armature circuit is represented by R_a.

Consider the generator being driven, with switches S_1 and S_2 both open. Despite the fact that there will be zero field current, a small emf would be measured. This emf is due to the small amount of residual magnetism retained in the poles. With switch S_1 now closed, the field current may be increased in discrete steps, and the corresponding values of generated emf noted. A graph of generated emf versus field current will be as shown in Figure 9.10, and is known as the open-circuit characteristic of the machine.

It will be seen that the shape of this graph is similar to the magnetisation curve for a magnetic material. This is to be expected, since the emf will be directly proportional to the

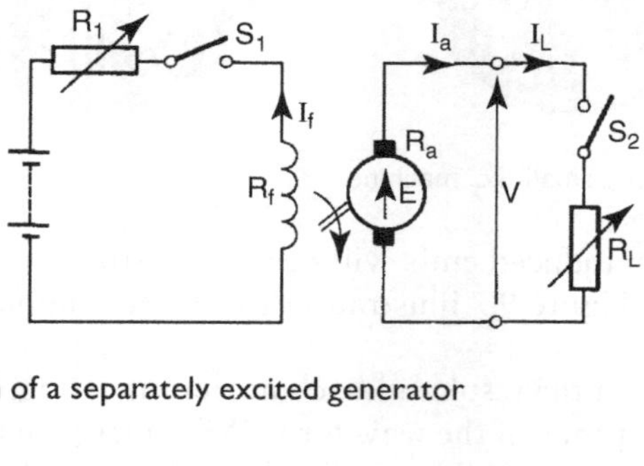

Figure 9.9 The circuit diagram of a separately excited generator

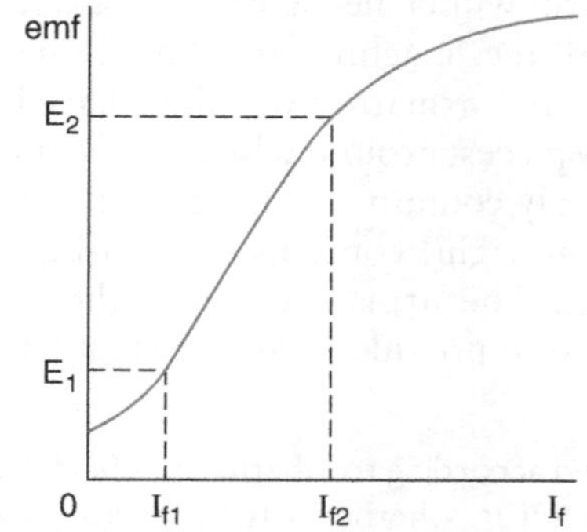

Figure 9.10 The generated emf as function of the field current

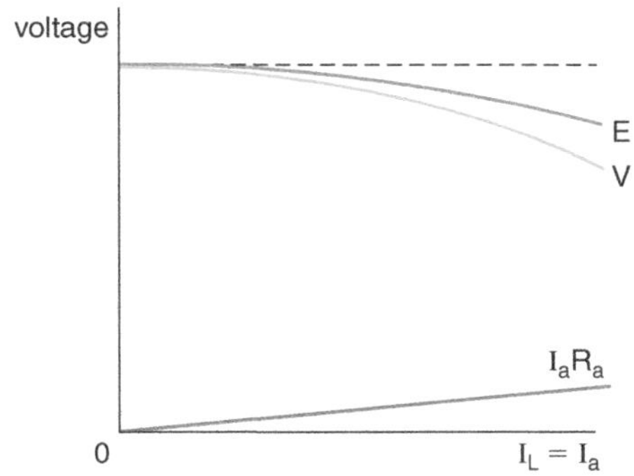

Figure 9.11 The output characteristic of the generator

pole flux. The 'flattening' of the emf graph indicates the onset of saturation of the machine's magnetic circuit. When the machine is used in practice, the field current would normally be set to some value within the range indicated by I_{f1} and I_{f2} on the graph. This means that the facility exists to vary the emf between the limits E_1 and E_2, simply by adjusting the variable resistor R_1.

Let the emf be set to some value E, within the range specified above. If the load is now varied, the corresponding values of terminal voltage V and load current I_L may be measured. Note that with this machine the armature current is the same as the load current. The graph of V versus I_L is known as the output characteristic of the generator, and is shown in Figure 9.11. The terminal p.d. of the machine will be less than the generated emf, by the amount of internal voltage drop due to R_a, such that:

$$V = E - I_a R_a \qquad (9.1)$$

Ideally, the graph of E versus I_L would be a horizontal line. However, an effect known as armature reaction causes this graph to 'droop' at the higher values of current. The main advantage of this type of generator is that there is some scope for increasing the generated emf in order to offset the internal voltage drop $I_a R_a$ as the load is increased.

The big disadvantage is the necessity for a separate d.c. supply for the field excitation. Therefore, in practice such a separately excited generator is rarely used.

9.4 SHUNT GENERATOR

This is a self-excited machine, where the field winding is connected in parallel (shunt) with the armature winding. The circuit diagram is shown in Figure 9.12, and from this it may be seen that the armature has to supply current to both the load and the field, such that:

$$I_a = I_L + I_f \qquad (9.2)$$

This self-excitation process can take place only if there is some residual flux in the poles, and if the resistance of the field circuit is less than some critical value. The open-circuit characteristic is illustrated in Figure 9.13.

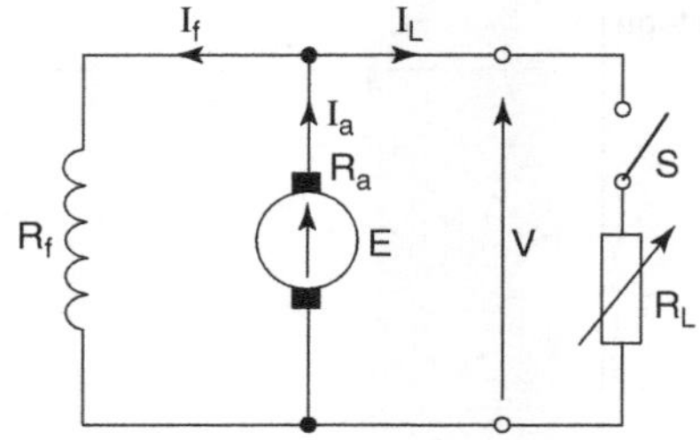

Figure 9.12 The circuit diagram of self-excited machine with shunt generator

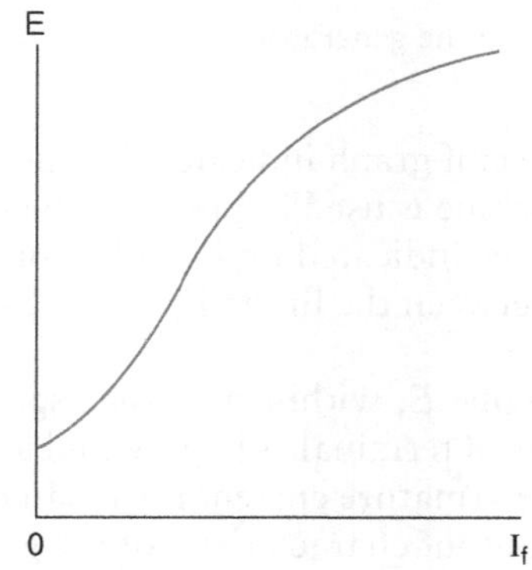

Figure 9.13 The open-circuit characteristic

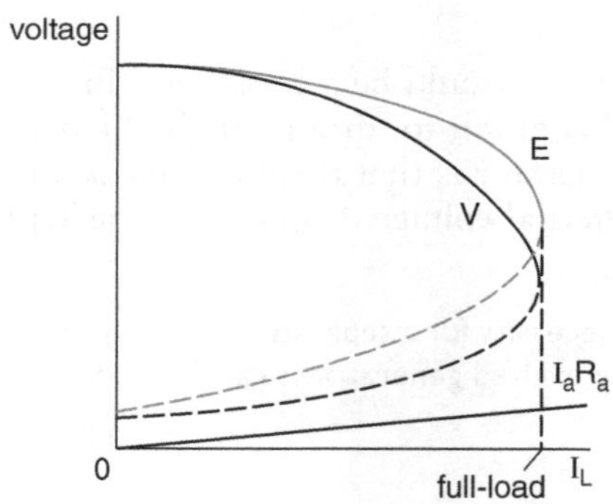

Figure 9.14 The output characteristic

The resistance of the field winding, R_f, is constant and of a relatively high value compared with R_a. Typically, I_f will be in the order of 1 A to 10 A, and will remain reasonably constant. The shunt machine is therefore considered to be a constant-flux machine. When switch S is closed, the armature current will increase in order to supply the demanded load current, I_L. Thus $I_a \infty I_L$, and as the load current is increased, so the terminal voltage will fall, according to the equation, $V = E - I_aR_a$. The output characteristic will therefore follow much the same shape as that for the separately excited generator and is shown in Figure 9.14. This condition applies until the machine is providing its rated full-load output. If the load should now demand even more current, i.e. the machine is overloaded, the result is that the generator simply stops generating. This effect is shown by the dotted lines in the output characteristics.

The shunt generator is the most commonly used d.c. generator, since it provides a reasonably constant output voltage over its normal operating range. Its other obvious advantage is the fact that it is self-exciting, and therefore requires only some mechanical means of driving the armature.

An example of a series generator is the dynamo on a bicycle, transforming the mechanical bike movement into energy to power running lights and other equipment. Bottle dynamos engage the bicycle's tyre, and hub dynamos are permanently attached to the bicycle's drive train.

9.5 SERIES GENERATOR

In this machine the field winding is connected in series with the armature winding and the load, as shown in Figure 9.15. In this case, $I_L = I_a = I_f$, so this is a variable-flux machine. Since the field winding must be capable of carrying the full-load current (which could be in hundreds of amps for a large machine), it is usually made from a few turns of heavy gauge wire or even copper strip. This also has the advantage of offering a very low resistance. This generator is a self-excited machine, provided that it is connected to a load when started. Note that a *shunt* generator will self-excite only when *disconnected* from its load.

When the load on a series generator is increased, the flux produced will increase, in almost direct proportion. The generated emf will therefore increase with the demanded load. The increase of flux, and hence voltage, will continue until the onset of magnetic saturation, as shown in the output characteristic of Figure 9.16. The terminal voltage is related to the emf by the equation:

$$V = E - I_a\left(R_a + R_f\right) \tag{9.3}$$

The variation of terminal voltage with load is not normally a requirement for a generator, so this form of machine is seldom used. However, the rising voltage characteristic of a series-connected field winding is put to good use in the compound machine, which is described in *Further Electrical and Electronic Principles*.

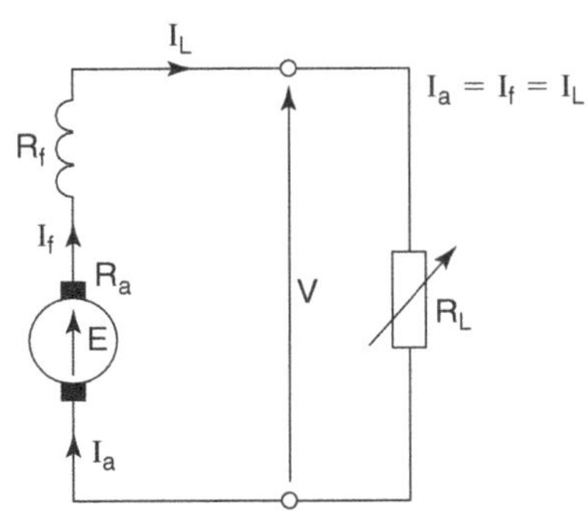

Figure 9.15 The circuit diagram of the series generator

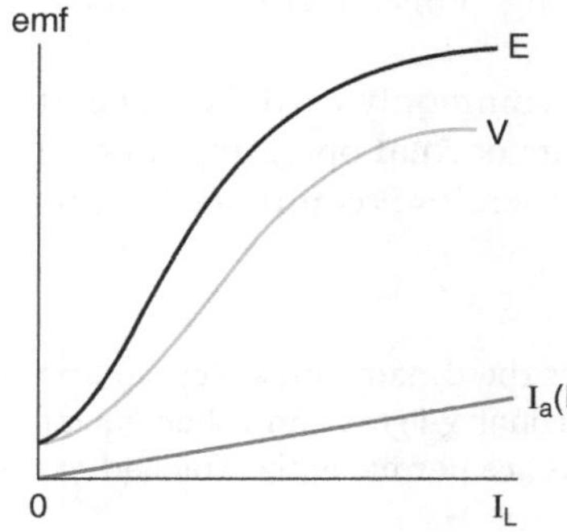

Figure 9.16 The output characteristic

WORKED EXAMPLE 9.1

Q The resistance of the field winding of a shunt generator is 200 Ω. When the machine is delivering 80 kW the generated emf and terminal voltage are 475 V and 450 V respectively. Calculate (a) the armature resistance and (b) the value of generated emf when the output is 50 kW, the terminal voltage then being 460 V.

$R_f = 200\ \Omega;\ P_o = 80 \times 10^3$ W; $V = 450$ V; $E = 475$ V

The circuit diagram is shown in Figure 9.17. It is always good practice to sketch the appropriate circuit diagram when solving machine problems.

(a)

$$P_o = VI_L$$

$$I_L = \frac{P_o}{V} = \frac{80 \times 10^3}{450} = 177.8 \text{ A}$$

$$I_f = \frac{V}{R_f} = \frac{450}{200} = 2.25 \text{ A}$$

$$I_a = I_L + I_f = 180.05 \text{ A}$$

$$I_a R_a = E - V = 475 - 450 = 25 \text{ V}$$

$$R_a = \frac{25}{180.05} = 0.139\ \Omega$$

(b)

$$P_o = 50 \times 10^3 \text{ W},\ V = 460 \text{ V}$$

$$I_L = \frac{50 \times 10^3}{460} = 108.7 \text{ A}$$

$$I_f = \frac{V}{R_f} = \frac{460}{200} = 2.3 \text{ A}$$

$$I_a = 108.7 + 2.3 = 111 \text{ A}$$

$$E = V + I_a R_a = 460 + (111 \times 0.139) = 475.4 \text{ V}$$

Note: Although the load had changed by about 60%, the field current has changed by only about 2.2%. This justifies the statement that a shunt generator is considered to be a constant-flux machine.

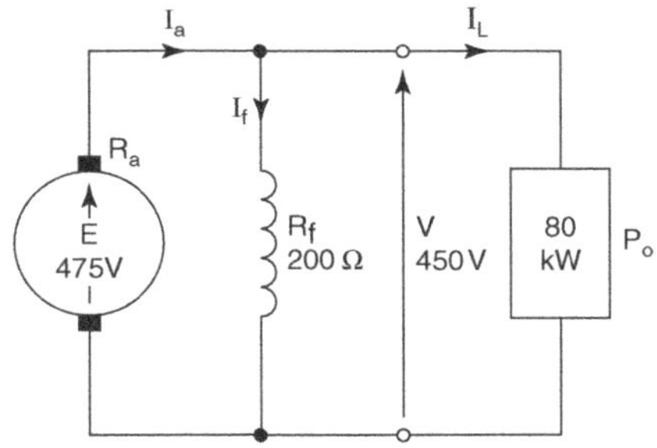

Figure 9.17 The circuit diagram for Worked Example 9.1

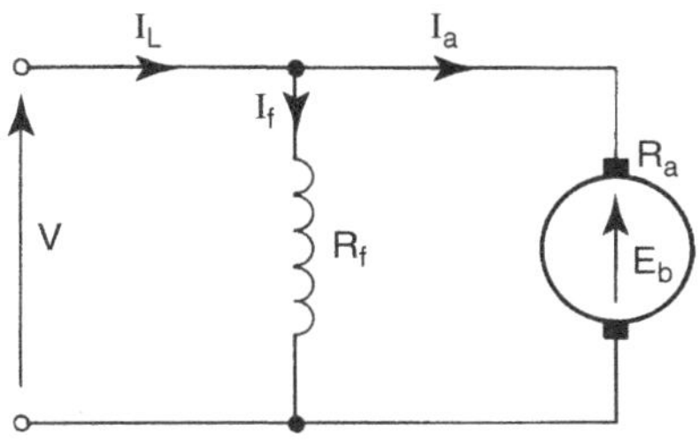

Figure 9.18 The circuit diagram of the shunt motor

9.6 SHUNT MOTOR

All of the d.c. generators so far described could be operated as motors, provided that they were connected to an appropriate d.c. supply. When the machine is used as a motor, the armature-generated emf is referred to as the back-emf, E_b, which is directly proportional to the speed of rotation. However, the speed is *inversely* proportional to the field flux Φ. In addition, the torque produced by the machine is proportional to both the flux and the armature current. Bearing these points in mind, we can say that:

$$\text{Speed } \omega \infty \frac{E_b}{\Phi} \tag{9.4}$$

$$\text{and torque} T \infty \Phi I_a \tag{9.5}$$

When the machine reaches its normal operating temperature, R_f will remain constant. Since the field winding is connected directly to a fixed supply voltage V, then I_f will be fixed. Thus, the shunt motor (Figure 9.18) is a constant-flux machine.

As the back-emf will have the same shape graph as that for the generator emf, and using Equations (9.4) and (9.5) the graphs of speed and torque versus current will be as in Figure 9.19. Note that when the machine is used as a motor, the supply current is identified as I_L. In this case, the subscript 'L' represents the word 'line'. Thus I_L identifies the line current drawn from the supply, and I_a is directly proportional to I_L.

Shunt motors are used for applications where a reasonably constant speed is required, between no-load and full-load conditions. A shunt motor can never run wild when the load drops (unless

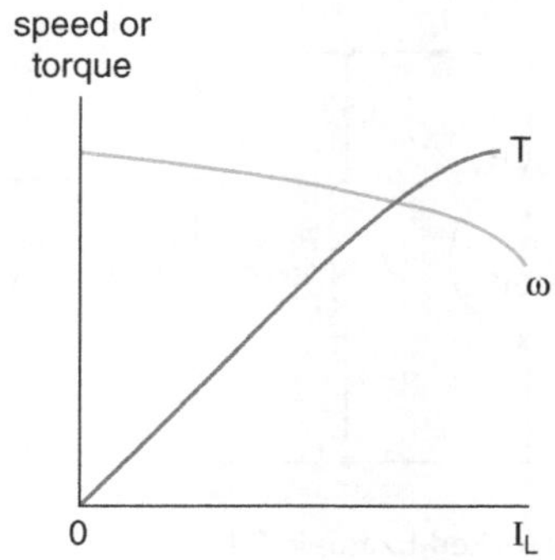

Figure 9.19 The graphs of speed and torque as function of the current

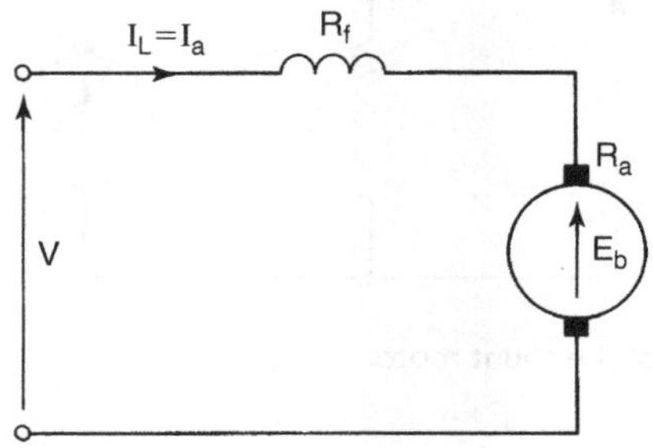

Figure 9.20 The circuit diagram of the series motor

the flux drops). The speed of shunt motors can be controlled within wide limits and very precisely, without much loss. The shunt motor is very often combined with a conveyor belt and is mainly used for machines that operate at a constant speed with varying loads, such as cranes and lifts, as well as for drives where the speed must be controlled.

9.7 SERIES MOTOR

Like the series generator, this machine is a variable-flux machine. Despite this, the back-emf of this motor remains almost constant, from light-load to full-load conditions. This fact is best illustrated by considering the circuit diagram (Figure 9.20), with some typical values.

$$E_b = V - I_a\left(R_a + R_f\right) \tag{9.6}$$

Let us assume the following: V = 200 V; R_a = 0.15 Ω; R_f = 0.03 Ω; I_a = 5 A on light load; I_a = 50 A on full load

$$\text{Lightload: } E_b = 200 - 5(0.15 + 0.03) = 199.1 \text{ V}$$
$$\text{Fullload: } E_b = 200 - 50(0.15 + 0.03) = 191 \text{ V}$$

From the above figures, it may be seen that although the armature current has increased tenfold, the back-emf has decreased by only 4%. Hence, E_b remains sensibly constant. Since $\omega \infty \frac{E_b}{\Phi}$, and E_b is constant, then:

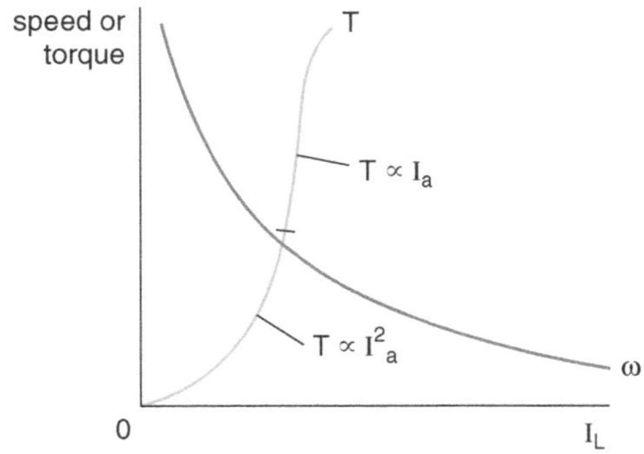

Figure 9.21 The graphs of speed and torque as function of the current

$$\omega \propto \frac{1}{\Phi} \quad [1]$$

Similarly, $T \propto \Phi I_a$ and since $\Phi \propto I_a$ until the onset of magnetic saturation,

$$\text{then } T \propto I_a^2 \quad [2]$$

$$\text{and after saturation, } T \propto I_a \quad [3]$$

Using [1] to [3] above, the speed and torque characteristics shown in Figure 9.21 may be deduced.

Note: From the speed characteristic it is clear that, on very light loads, the motor speed would be excessive. *Theoretically*, the no-load speed would be infinite! For this reason a series motor must *never* be started unless it is connected to a mechanical load sufficient to prevent a dangerously high speed. Similarly, a series motor must not be used to operate belt-driven machinery, lifting cranes, etc., due to the possibility of the load being suddenly disconnected. If a series motor is allowed to run on a very light load, its speed builds up very quickly. The probable outcome of this is the disintegration of the machine, with the consequent dangers to personnel and plant.

The series motor has a high starting torque due to the 'square-law' response of the torque characteristic, which means that it is powerful and accelerates quickly. For this reason, it tends to be used mainly for traction purposes.

Examples of the series motor can be found where a load is always present, for example through direct coupling to the machine to be driven. The series motor is used where high starting torques are required, such as a traction motor in electric trains and trams, in order to overcome the massive inertia of a stationary train. The starter motor in the car is also a series motor.

SUMMARY OF EQUATIONS

Generators:

Shunt generator:
$$I_a = I_L + I_f$$
$$V = E - I_a R_a$$

Series generator: $I_a = I_L = I_f$
$V = E - I_a(R_a + R_f)$

Motors:

Shunt motor: $E_b = V - I_aR_a$

Series motor: $E_b = V - I_a(R_a + R_f)$

Speed equation: $n \propto \frac{E_b}{\Phi}$; or $\omega \propto \frac{E_b}{\Phi}$

Torque equation: $T \propto \Phi I_a$

ASSIGNMENT QUESTIONS

1 A shunt generator supplies a current of 85 A at a terminal p.d. of 380 V. Calculate the generated emf if the armature and field resistances are 0.4 Ω and 95 Ω respectively.
2 A generator produces an armature current of 50 A when generating an emf of 400 V. If the terminal p.d. is 390 V, calculate (a) the value of the armature resistance and (b) the power loss in the armature circuit.
3 A d.c. shunt generator supplies a 50 kW load at a terminal voltage of 250 V. The armature and field circuit resistances are 0.15 Ω and 50 Ω respectively. Calculate the generated emf.

Chapter 10

D.C. Transients

LEARNING OUTCOMES

This chapter explains the response of capacitor-resistor (*C-R*), and inductor-resistor (*L-R*) circuits, when they are connected to and disconnected from a d.c. supply.

On completion of this chapter you should be able to:

1. Show how the current and capacitor voltage in a series *C-R* circuit varies with time, when connected to/disconnected from a d.c. supply.
2. Show how the current through, and p.d. across an inductor in a series *L-R* circuit varies with time, when connected to/disconnected from a d.c. supply.
3. Define the term 'time constant' for both types of above circuits.

10.1 CAPACITOR-RESISTOR SERIES CIRCUIT: CHARGING

Before dealing with the charging process for a *C-R* circuit, let us firstly consider an analogous situation. Imagine that you need to inflate a 'flat' tyre with a foot pump. Initially it is fairly easy to pump air into the tyre. However, as the air pressure inside the tyre builds up, it becomes progressively more difficult to force more air in. Also, as the internal pressure builds up, the rate at which air can be pumped in decreases. Comparing the two situations, the capacitor (which is to be charged) is analogous to the tyre; the d.c. supply behaves like the pump; the charging current compares to the air flow rate; and the p.d. developed between the plates of the capacitor has the same effect as the tyre pressure. From these comparisons we can conclude that as the capacitor voltage builds up, it reacts against the emf of the supply, so slowing down the charging rate. Thus, the capacitor will charge at a non-uniform rate, and will continue to charge until the p.d. between its plates is equal to the supply emf. This last point would also apply to tyre inflation, when the tyre pressure reaches the maximum pressure available from the pump. At this point the air flow into the tyre would cease. Similarly, when the capacitor has been fully charged, the charging current will cease.

Let us now consider the *C-R* charging circuit in more detail. Such a circuit is shown in Figure 10.1. Let us assume that the capacitor is initially fully discharged, i.e. the p.d. between its plates (v_C) is zero, as will be the charge, q. Note that the lowercase letters v and q are used because, during the charging sequence, they will have continuously changing values, as will the p.d. across the resistor (v_R) and the charging current, i. Thus these quantities are said to have *transient* values.

At some time $t = 0$, let the switch in Figure 10.1 be moved from position 'A' to position 'B'. At this instant the charging current will start to flow. Since there will be no opposition

DOI: 10.1201/9781003308294-10

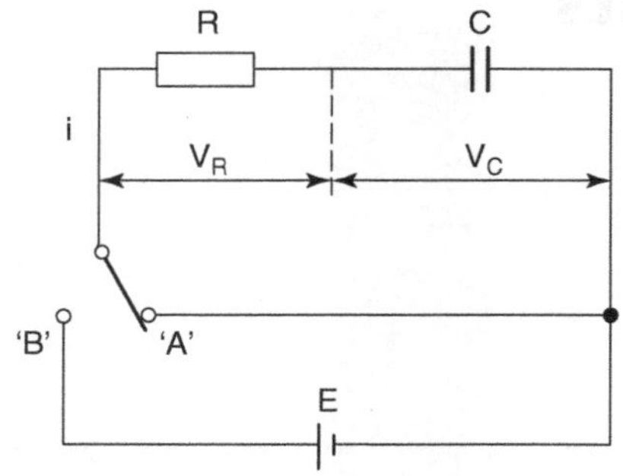

Figure 10.1 The charging of the capacitor-resistor series circuit

offered by capacitor p.d. ($v_C = 0$), only the resistor R will offer any opposition. Consequently, the initial charging current (I_0) will have the maximum possible value for the circuit. This initial charging current is therefore given by:

$$I_0 = \frac{E}{R} \tag{10.1}$$

Since we are dealing with a series d.c. circuit, the following equation must apply *at all times*:

$$E = v_R + v_C \qquad [1]$$

thus, at time $t = 0$

$$E = v_R + 0$$

i.e. the full emf of E volt is developed across the resistor at the instant the supply is connected to the circuit. Since $v_R = iR$, and at time $t = 0$, $i = I_0$, this confirms Equation (10.1) above.

Let us now consider the situation when the capacitor has reached its fully charged state. In this case, it will have a p.d. of E volt, a charge of Q coulomb and the charging current, $i = 0$. If there is no current flow then the p.d. across the resistor, $v_R = 0$, and Equation [1] is:

$$E = 0 + v_C$$

Having confirmed the initial and final values for the transients, we now need to consider how they vary, with time, between these limits. It has already been stated that the variations will be non-linear (i.e. not a straight line graph). In fact the variations follow an *exponential* law. Any quantity that varies in an exponential fashion will have a graph like that shown in Figure 10.2(a) if it increases with time, and as in Figure 10.2(b) for a decreasing function.

In Figure 10.2(a), X represents the final steady-state value of the variable x, and in Figure 10.2(b), X_0 represents the initial value of x. In each case the straight line (tangent to the curve at time $t = 0$) indicates the initial rate of change of x. The time interval shown as τ on both graphs is known as the time constant, which is defined as follows: The time constant is the time that it would take the variable to reach its final steady state if it continued to change at its initial rate. From the above figures it can be seen that for an increasing exponential function, the variable will reach 63.2% of its final value after one time constant, and for a decreasing function it will fall to 36.8% of its initial value after τ seconds.

Note: Considering *any* point on the graph, it would take one time constant for the variable to reach its final steady value if it continued to change at the same rate as at that point. Thus an exponential graph may be considered as being formed from an infinite number of tangents, each of which represents the slope at a particular instant in time. This is illustrated

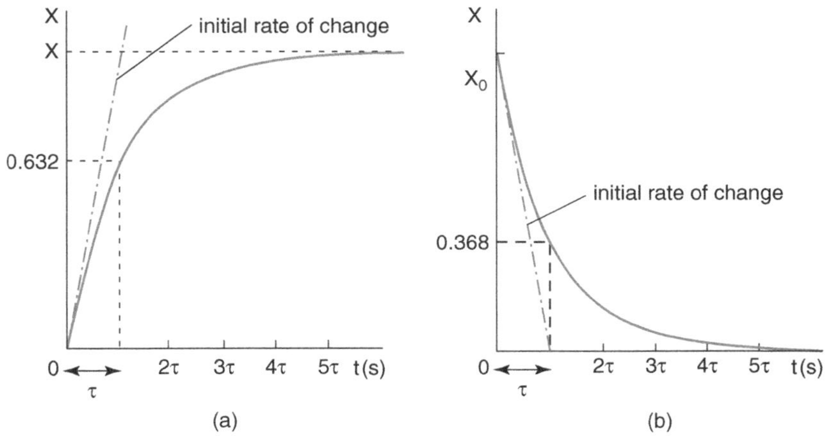

Figure 10.2 The variations following an exponential law.

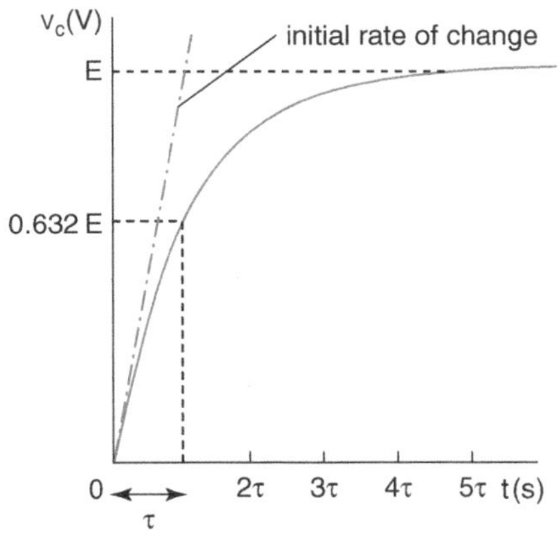

Figure 10.3 The variations of capacitor voltage as function of the time

in Figure 10.4. Also, *theoretically*, an exponential function can never actually reach its final steady state. However, for *practical* purposes it is assumed that the final steady state is achieved after five time constants. This is justifiable since the variable will be within 0.67% of the final value after 5τ seconds. So for Figure 10.2(a), after 5τ seconds, $x = 0.9973\ X$.

Considering the circuit of Figure 10.1, assuming that the capacitor is fully discharged, let the switch be moved to position 'B'. The capacitor will now charge via resistor R until the p.d. between its plates, $v_C = E$ volts. Once fully charged, the circuit current will be zero. The variations of capacitor voltage and charge, p.d. across the resistance and charging current are shown in Figures 10.3–10.6.

For such a C-R circuit the time constant, τ (Greek letter tau), is CR seconds. It may appear strange that the product of capacitance and resistance yields a result having units of time. This may be justified by considering a simple dimensional analysis, as follows:

$$C = \frac{Q}{V} = \frac{It}{V} \text{ and } R = \frac{V}{I}$$

$$\text{so, } CR = \frac{It}{V} \times \frac{V}{I} = t \text{ and is expressed in seconds}$$

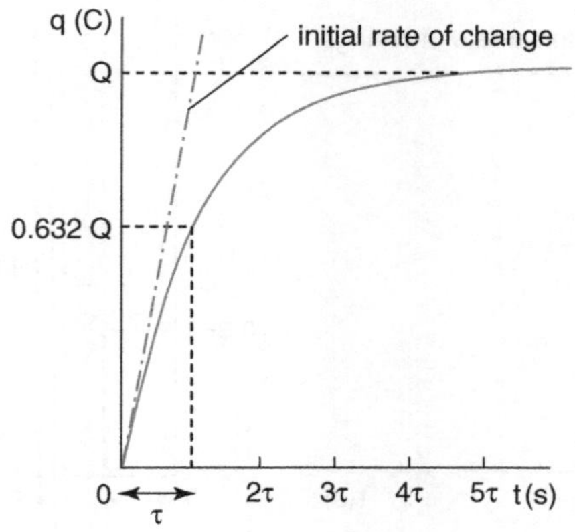

Figure 10.4 The variations of charge as function of the time

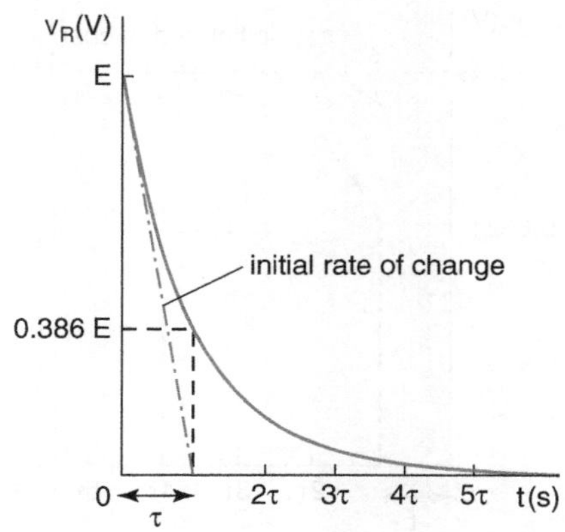

Figure 10.5 The variations of p.d. across the resistance as function of time

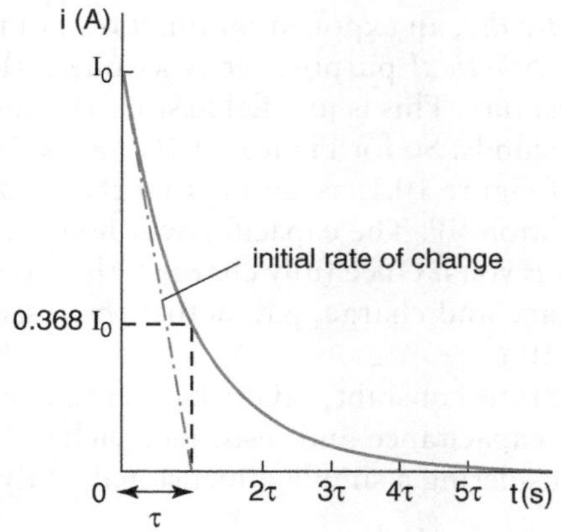

Figure 10.6 The variations of charging current as function of time

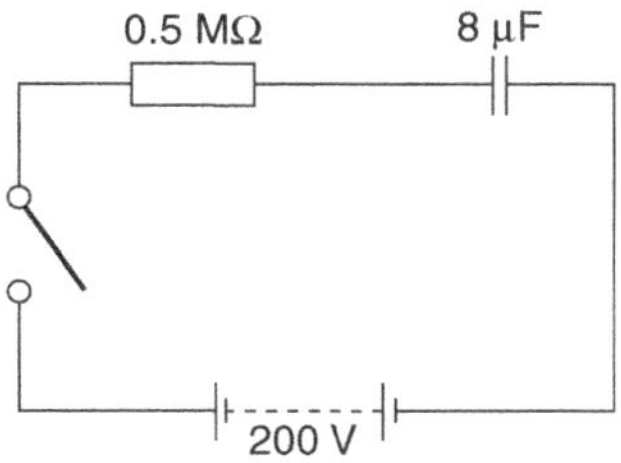

Figure 10.7 The circuit diagram for Worked Example 10.1

WORKED EXAMPLE 10.1

Q An 8 μF capacitor is connected in series with a 0.5 MΩ resistor, across a 200 V d.c. supply, as is shown in Figure 10.7. Calculate (a) the circuit time constant, (b) the initial charging current and (c) the p.d.s across the capacitor and resistor 4 seconds after the supply is connected. You may assume that the capacitor is initially fully discharged

$C = 8 \times 10^{-6}$ F; $R = 0.5 \times 10^{6}$ Ω; $E = 200$ V

(a) $\tau = CR = 8 \times 10^{-6} \times 0.5 \times 10^{6} = 4$ s

(b) $I_0 = \frac{E}{R} = \frac{200}{0.5 \times 10^{6}} = 400\ \mu\text{A}$

(c) After τ seconds, $v_C = 0.632\,E = 0.632 \times 200 = 126.4$ V
$v_R = E - v_C = 200 - 126.4 = 73.6$ V

10.2 CAPACITOR-RESISTOR SERIES CIRCUIT: DISCHARGING

Consider the circuit of Figure 10.1, where the switch has been in position 'B' for sufficient time to allow the charging process to be completed. Thus the charging current will be zero, the p.d. across the resistor will be zero, the p.d. across the capacitor will be E volt and it will have stored a charge of Q coulomb.

At some time $t = 0$, let the switch be moved back to position 'A'. The capacitor will now be able to discharge through resistor R. The general equation for the voltages in the circuit will still apply. In other words, $E = v_R + v_C$. But, at the instant the switch is moved to position 'A', the source of emf is removed. Applying this condition to the general equation above yields:

$$0 = v_R + v_C;\ \text{where } v_C = E \text{ and } v_R = I_0 R$$

$$0 = I_0 R + E$$

$$\text{hence } I_0 = -\frac{E}{R} \qquad (10.2)$$

This means that the initial discharge current has the same value as the initial charging current, but (as you would expect) it flows in the opposite direction.

Since the capacitor is discharging, then its voltage will decay from E volt to zero, its charge will decay from Q coulomb to zero and the discharge current will also decay from I_0

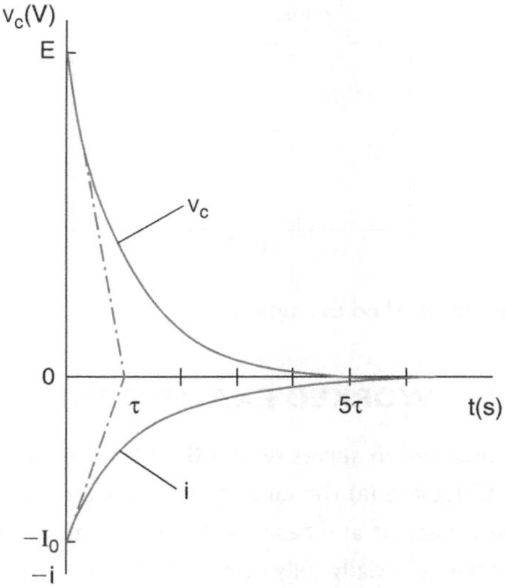

Figure 10.8 The discharging of the capacitor-resistor series circuit

to zero. The circuit time constant will be the same as before, i.e. $\tau = CR$ seconds. The graphs for v_C and i are shown in Figure 10.8.

Note: The time constant for the *C-R* circuit was defined previously in terms of the capacitor charging. However, a time constant also applies to the discharge conditions. It is therefore better to define the time constant in a more general manner, as follows: The time constant of a circuit is the time that it would have taken for any transient variable to change, from one steady state to a new steady state, if it had maintained its rate of change existing at the time of the first steady state.

When such an *C-R* circuit is switched on and off very quickly, the capacitor is charged and discharged at high speed and a filter characteristic is obtained. Measured across the capacitor of this *C-R* circuit, this is hence called a low-pass filter. At low frequencies, charging and discharging is slow. There is sufficient time to fully charge and also discharge the capacitor. At high frequencies, the changes are too fast and the capacitor voltage hardly changes. This is equivalent to saying that low-frequency signals are passed through and high-frequency signals are blocked. This can be used for filtering audio signals, for instance. When measured across the resistor, the opposite behaviour is obtained: a high-pass filter passes the high frequencies and blocks the low frequencies. This a.c. transient behaviour is discussed more in *Further Electrical and Electronic Principles*.

WORKED EXAMPLE 10.2

Q A *C-R* charge/discharge circuit is shown in Figure 10.9. The switch has been in position 'A' for a sufficient time to allow the capacitor to become fully discharged.

(a) If the switch is now moved to position 'B', calculate the time constant and initial charging current.

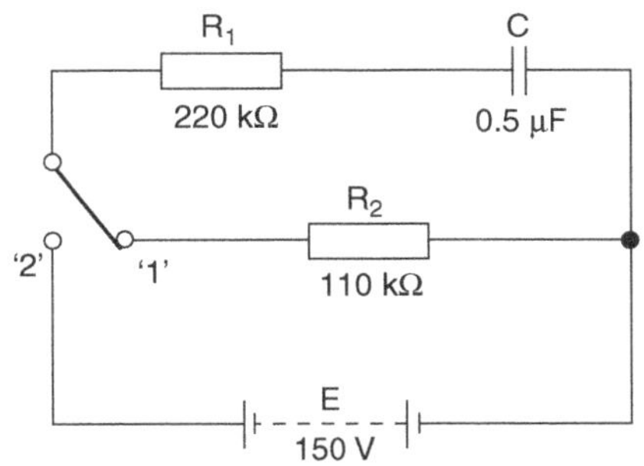

Figure 10.9 A C-R charge/discharge circuit for Worked Example 10.2

$C = 0.5\ \mu F;\ R_1 = 220\ k\Omega;\ R_2 = 110\ k\Omega;\ E = 150\ V$

(b) After the capacitor has completely charged, the switch is moved back to position 'A'. Calculate the time constant and the p.d across R_2 at this time.

(a) When charging, only resistor R_1 is connected in series with the capacitor, so R_2 may be ignored.

$$\tau = CR_1 = 0.5 \times 10^{-6} \times 220 \times 10^{3} = 0.11\ s$$

$$I_0 = \frac{E}{R} = \frac{150}{220 \times 10^{3}} = 682\ \mu A$$

(b) When discharging, both R_1 and R_2 are connected in series with the capacitor, so their combined resistance $R = R_1 + R_2$, will determine the discharge time constant.

$$\tau = CR = 0.5 \times 10^{-6} \times 330 \times 10^{3} = 0.16\ s$$

After one time constant the discharge current will have fallen to 0.368 I_0

$$I_0 = \frac{E}{R} = \frac{150}{330 \times 10^{3}} = 454.5\ \mu A$$

$$i = 0.368 \times 454 \times 10^{-6} = 167.26\ \mu A$$

$$v_{R2} = iR_2 = 167.26 \times 10^{-6} \times 110 \times 10^{3} = 18.4\ V$$

10.3 INDUCTOR-RESISTOR SERIES CIRCUIT: CHARGING

Consider the circuit of Figure 10.10, where an inductor is connected in series with a resistor. At some time $t = 0$, the switch is moved from position 'A' to position 'B'. The connection to the supply is now complete, and current will start to flow, increasing towards its final steady value.

However, whilst the current is *changing* it will induce a back-emf across the inductor, of e volt. From electromagnetic induction theory we know that this induced emf will have a value given by:

$$e = -L\frac{di}{dt}$$

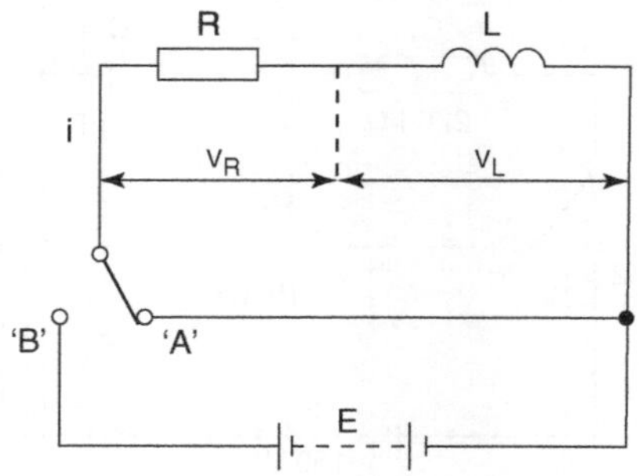

Figure 10.10 The charging of the inductor-resistor series circuit

Being a simple series circuit, Kirchhoff's voltage law will apply, such that the sum of the p.d.s equals the applied emf. Also, since we are considering a perfect inductor (the resistor shown may be considered as the coil's resistance), the p.d. across the inductor will be exactly equal but opposite in polarity to the induced emf.

$$\text{Therefore, } v_L = -e = L\frac{\mathrm{d}i}{\mathrm{d}t}$$

$$\text{hence, } E = v_R + v_L \qquad [1]$$

$$\text{or, } E = iR + L\frac{\mathrm{d}i}{\mathrm{d}t}$$

Comparing this equation with that for the *C-R* circuit, it may be seen that they are both of the same form. Using the analogy technique, we can conclude that both systems will respond in a similar manner. In the case of the *L-R* circuit, the *current* will increase from zero to its final steady value, following an exponential law.

At the instant that the switch is moved from 'A' to 'B' (t = 0), the current will have an instantaneous value of zero, but it *will* have a certain rate of change, d*i*/d*t* amp/s. From Equation [1] above, this initial rate of change can be obtained, thus:

$$E = 0 + L\frac{\mathrm{d}i}{\mathrm{d}t}$$

$$\text{so, initial } \frac{\mathrm{d}i}{\mathrm{d}t} = \frac{E}{L} \qquad (10.3)$$

When the current reaches its final steady value, there will be no back-emf across the inductor, and hence no p.d. across it. Thus the only limiting factor on the current will then be the resistance of the circuit. The final steady current is therefore given by:

$$I = \frac{E}{R} \qquad (10.4)$$

The time constant of the circuit, expressed in seconds, is obtained by dividing the inductance by the resistance.

$$\tau = \frac{L}{R} \qquad (10.5)$$

The above equation may be confirmed by using a simple form of dimensional analysis, as follows.

$$\text{In general, } V = \frac{LI}{t}; \text{ so } L = \frac{Vt}{I} \text{ and } R = \frac{V}{I}$$

$$\text{therefore, } \frac{L}{R} = \frac{Vt}{I} \times \frac{I}{V} = t \text{ and is expressed in seconds}$$

The time constant of the circuit may be defined in the general terms given in the '*Note*', in the previous section, dealing with the *C-R* circuit.

The rate of change of current will be at its maximum value at time $t = 0$, so the p.d. across the inductor will be at its maximum value at this time. This p.d. therefore decays exponentially from E volt to zero. The graphs for i, v_R and v_L are shown in Figures 10.11–10.13 respectively.

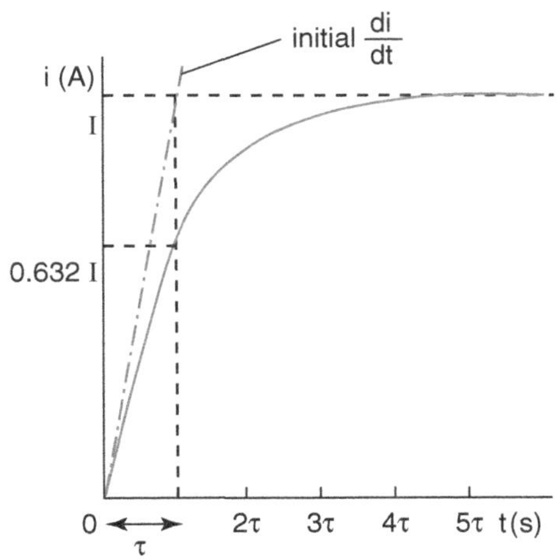

Figure 10.11 The variations of discharging current as function of time

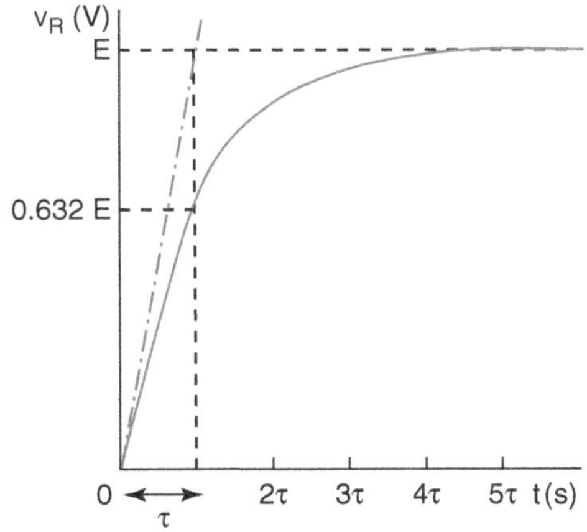

Figure 10.12 The variations of p.d. across the resistance as function of time

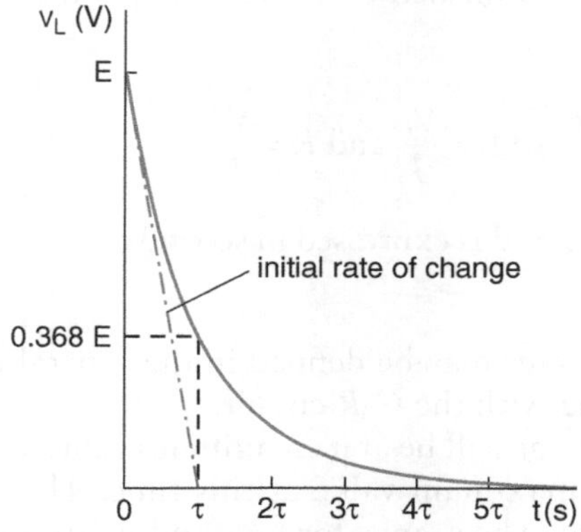

Figure 10.13 The variations of inductor voltage as function of the time

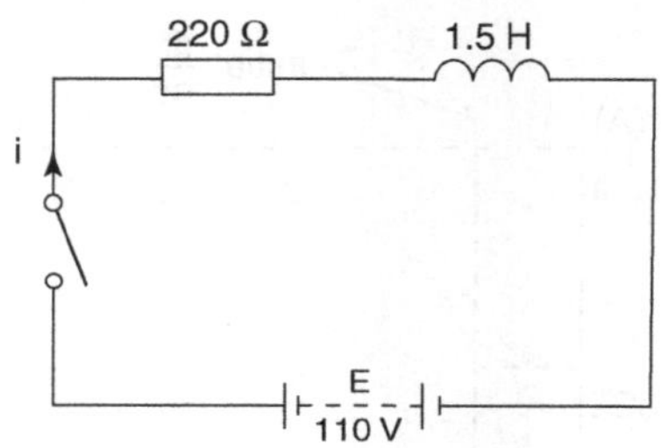

Figure 10.14 The circuit diagram for Worked Example 10.3

WORKED EXAMPLE 10.3

Q The field winding of a 110 V d.c. motor has an inductance of 1.5 H, and a resistance of 220 Ω. From the instant that the machine is connected to a 110 V supply, calculate (a) the initial rate of change of current, (b) the final steady current and (c) the time taken for the current to reach its final steady value.

$E = 110$ V; $L = 1.5$ H; $R = 220\ \Omega$

The circuit diagram is shown in Figure 10.14.

(a) initial $\frac{di}{dt} = \frac{E}{L} = \frac{110}{1.5} = 73.33$ A / s

(b) final current $I = \frac{E}{R} = \frac{110}{220} = 0.5$ A

(c) $\tau = \frac{L}{R} = \frac{1.5}{220} = 6.82$ ms

Since the system takes approximately 5τ seconds to reach its new steady state, the current will reach its final steady value in a time:

$t = 5 \times 6.82 = 34.1$ ms

10.4 INDUCTOR-RESISTOR SERIES CIRCUIT: DISCHARGING

Figure 10.15 shows such a circuit, connected to a d.c. supply. Assume that the current has reached its final steady value of I amps. Let the switch now be returned to position 'A' (at time $t = 0$). The current will now decay to zero in an exponential manner. However, the decaying current will induce a back-emf across the coil. This emf must oppose the change of current. Therefore, the decaying current will flow *in the same direction* as the original steady current. In other words, the back-emf will try to maintain the original current flow. The graph of the decaying current, with respect to time, will therefore be as shown in Figure 10.16. The time constant of the circuit will, of course, still be L/R second, and the current will decay from a value of $I = E/R$ amp. The initial rate of decay will also be E/L amp/s.

Just like fast charging and discharging an C-R circuit, this can also be performed with an R-L circuit. When the output is measured across the inductor, a high-pass filter is obtained. It results in a low-pass filter when the output is measured across the resistor. Although this R-L circuit behaves as a low-pass or high-pass filter, it is less common in practice. An inductor is usually bigger than a capacitor and also has more resistive losses. When both a C-R circuit and a L-R circuit are combined to a R-L-C circuit, a band-pass filter (passing through some central frequencies) and a band-stop filter (blocking those frequencies) are possible. This is described more in *Further Electrical and Electronic Principles*.

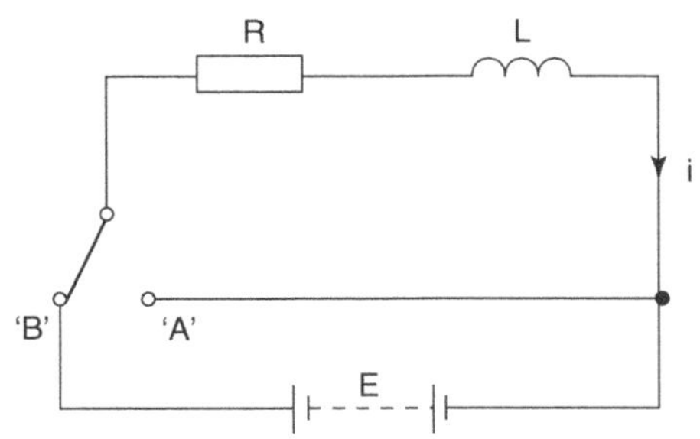

Figure 10.15 The discharging of the inductor-resistor series circuit

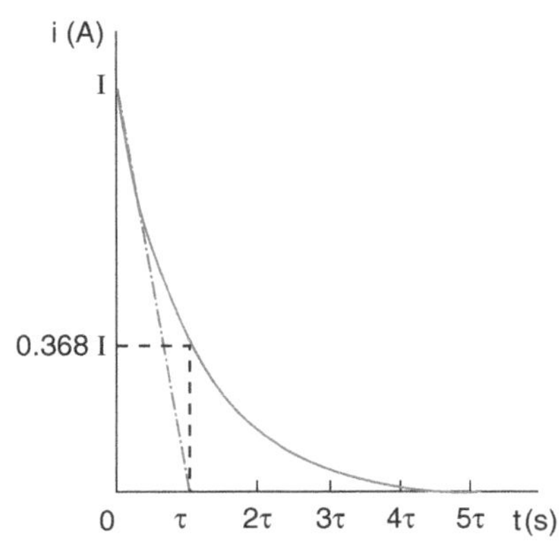

Figure 10.16 The variations of discharging current as function of time

SUMMARY OF EQUATIONS

C-R circuit:

Time constant: $\tau = CR$

Initial current: $I_0 = \dfrac{E}{R}$

Steady-state conditions after approx. 5τ seconds

after, τ second: $v_C = 0.632\ E$; and $i = 0.368\ I_0$

L-R circuit:

Time constant: $\tau = \dfrac{L}{R}$

Initial rate of change of current: $\dfrac{di}{dt} = \dfrac{E}{L}$

Final current flowing: $I = \dfrac{E}{R}$

Steady-state conditions after approx. 5τ seconds

after τ second: $V_L = 0.368\ E$; and $i = 0.632\ I$

ASSIGNMENT QUESTIONS

1 A 47 μF capacitor is connected in series with a 39 kΩ resistor, across a 24 V d.c. supply. Calculate (a) the circuit time constant, (b) the values for initial and final charging current and (c) the time taken for the capacitor to become fully charged.

2 A 150 mH inductor of resistance 50 Ω is connected to a 50 V d.c. supply. Determine (a) the initial rate of change of current, (b) the final steady current and (c) the time taken for the current to change from zero to its final steady value.

3 An inductor of negligible resistance and of inductance 0.25 H, is connected in series with a 1.5 kΩ resistor, across a 24 V d.c. supply. Calculate the current flowing after one time constant.

4 A 5 H inductor has a resistance *R* ohm. This inductor is connected in series with a 10 Ω resistor, across a 140 V d.c. supply. If the resulting circuit time constant is 0.4 s, determine (a) the value of the coil resistance and (b) the final steady current.

5 Define the *time constant* of a capacitor-resistor series circuit. Such a circuit comprises a 50 μF capacitor and a resistor, connected to a 100 V d.c. supply via a switch. If the circuit time constant is to be 5 s, determine (a) the resistor value and (b) the initial charging current.

6 The dielectric of a 20 μF capacitor has a resistance of 65 MΩ. This capacitor is fully charged from a 120 V d.c. supply. Calculate the time taken, after disconnection from the supply, for the capacitor to become fully discharged.

SUGGESTED PRACTICAL ASSIGNMENTS

Assignment 1

To investigate the variation of capacitor voltage and current during charge and discharge cycles.

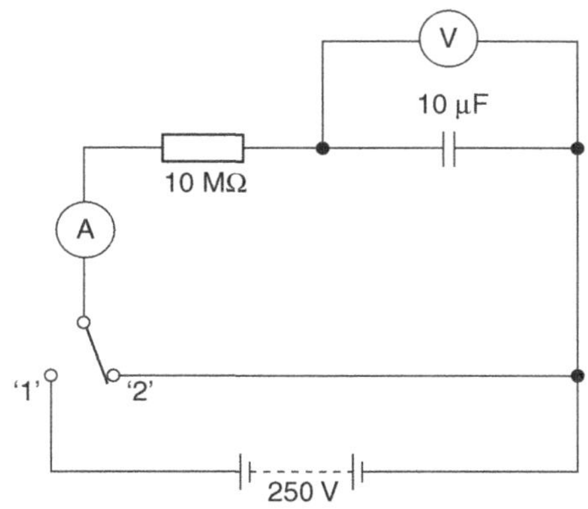

Figure 10.17 The circuit diagram for Practical Assignment 1

Apparatus

1 × 10 μF capacitor
1 × 10 MΩ resistor
1 × 2-pole switch
1 × d.c. power supply
2 × multimeter: one for the current and one for the voltage
1 × stopwatch

Method

1 Connect the circuit of Figure 10.17, and adjust the power supply output to 250 V.
2 Simultaneously move the switch to position 'B' and start the stopwatch.
3 Record the circuit current and capacitor p.d. at 10 s intervals, for the first 60 s.
4 Continue recording the current and voltage readings, at 20 s intervals, for a further 4 minutes. Reset the stopwatch to zero. Reverse the connections to the ammeter.
5 Move the switch back to position 'A', and repeat the procedures of paragraphs (3) and (4) above.
6 Plot graphs of current and capacitor p.d., versus time, for both the charging and discharging cycles.
7 Submit a complete assignment report, which should include the following:
   i The comparison of the actual time constant (determined from the plotted graphs) to the theoretical value. Explain any discrepancy found.
   ii Explain why both the charging and discharging currents tend to 'level off' at some small value, rather than continuing to decrease to zero.

5. Move the switch back to position 'A' and record the discharge values as for steps (3) and (4) above.

Appendix A: Physical Quantities with SI and other preferred units

General quantities	*Symbol*	*Units*
Acceleration, linear	a	m/s^2 (metre/second/second)
Area	A	m^2 (square metre)
Energy or work	W	J (joule)
Force	F	N (newton)
Length	l	m (metre)
Mass	m	kg (kilogram)
Power	P	W (watt)
Pressure	p	Pa (pascal)
Temperature value	θ	K or °C (Kelvin or degree Celsius)
Time	t	s (second)
Torque	T	Nm (newton metre)
Velocity, angular	ω	rad/s (radian/second)
Velocity, linear	v or u	m/s (metre/second)
Volume	V	m^3 (cubic metre)
Wavelength	λ	m metre

Electrical quantities	*Symbol*	*Units*
Admittance	Y	Ω (ohm)
Charge (quantity)	Q	C (coulomb)
Conductance	G	S (Siemens)
Current	I	A (ampere)
Current density	$\mathbf{J}$	A/m^2 (ampere/square metre)
Electromotive force (emf)	E	V (volts)
Frequency	f	Hz (hertz)
Impedance	Z	Ω (ohm)
Period	T	s (second)
Potential difference (p.d.)	V	V (volt)
Power, active	P	W (watt)
Power, apparent	S	VA (volt ampere)
Power, reactive	Q	VAr (volt ampere reactive)
Reactance	X	Ω (ohm)
Resistance	R	Ω (ohm)
Resistivity	ρ	Ωm (ohm metre)
Time constant	τ	s (second)

Electrostatic quantities	*Symbol*	*Unit*
Capacitance	C	F (farad)
Field strength	$\mathbf{E}$	V/m (volt/metre)
Flux	ψ	C (coulomb)
Flux density	D	C/m^2 (coulomb/square metre)
Permittivity, absolute	$\in$	F/m (farad/metre)

General quantities	*Symbol*	*Unit*
Permittivity, relative	ε_r	no units
Permittivity, of free space	ε_0	F/m (farad/metre)

Electromagnetic quantities	*Symbol*	*Unit*
Field strength	H	A/m (ampere/metre)[a]
Flux	Φ	Wb (weber)
Flux density	B	T (tesla)
Inductance, mutual	M	H (henry)
Inductance, self	L	H (henry)
Magnetomotive force (mmf)	F	A (ampere)[b]
Permeability, absolute	μ	H/m (henry/metre)
Permeability, relative	μ_r	no units
Permeability, of free space	μ_0	H/m (henry/metre)
Reluctance	S	A/Wb (ampere/weber)[c]

[a] At/m (ampere turn/metre) in this book
[b] (ampere turn) in this book
[c] At/Wb (ampere turn/weber) in this book

Appendix B: Answers to Assignment Questions

CHAPTER 1

1.1
- a 4.563×10^2
- b 9.023×10^5
- c 2.85×10^{-4}
- d 8×10^3
- e 4.712×10^{-2}
- f 1.8×10^{-4}A
- g 3.8×10^{-2}V
- h 8×10^{10}N
- i 2×10^{-3}F

1.2
- a 1.5 kΩ
- b 3.3 mΩ
- c 25 μA
- d 750 V or 0.75 kV
- e 800 kV
- f 47 nF

1.3 750 C
1.4 0.104 s
1.5 0.917 A
1.6
- a 2.25 kV
- b 18.75 V

1.7
- a 2.273 A
- b 61 mA
- c 18.52 μA
- d 0.512 mA

1.8 54.14 MJ or 15.04 kWh
1.9 £55.88
1.10
- a 0.5 A
- b 280 V
- c 140 W
- d 42 kJ

1.11
- a 49.64 V
- b 27.58 Ω

1.12
- a 0.15 Ω
- b 2.25 Ω

1.13 225 W; 67.5 kJ
1.14
- a 2 A
- b 8 V
- c 2.4 kC

1.15 2 m/s^2; 2.67 m/s^2; 0.5 m/s^2
1.16 8 μA
1.17 19.2 Ω; 12.5 A
1.18 15.65 kW
1.19 71.72 °C
1.20 33.6 Ω

CHAPTER 2

2.1 0.375 A
2.2
- a 29.1 Ω
- b 24.75 A

2.3 0.45 A
2.4 14 Ω
2.5
- a 15.67 Ω
- b 0.766 A
- c 0.426 A

2.6
- a 2.371 A; 1.779 A; 1.804 A; 1.443 A; 0.902 A
- b 35.37 V (15 Ω and 20 Ω); 14.43 V (8 Ω, 10 Ω, 16 Ω)
- c 63.26 W

2.7 2.5 V
2.8
a 12.63 V
b 2.4 A
c 0.947 A
2.9
a 3.68 Ω
b 5.44 A
c 9.32 V
d 2.33 A
2.10 12.8 Ω
2.11
a 1.57 Ω
b 127.33 Ω
2.12 40 Ω
2.13 80 Ω
2.14
a 1.304 A
b 5.217 V
c 0.87 A
2.15
a 6.67 A
b 26.67 A
c 13.33 A
2.16
a 2 V
b 19.33 V
2.17 0.024 Ω
2.18
a 9.6 A
b 43.2 V
c 4.32 A; 2.16 A; 1.08 A
2.19 3.094 V
2.20 6.82 V
2.21
a 20 Ω
b 10 A
c 60 V
d 2 kW
e 300 W
2.22
a 1 A
b 30 V; 36 V
c 54 kJ
d 180 C
2.23 2.368 A; – 0.263 A; 2.105 A; 10.526 V
2.24 1 A (discharge); 0.5 A (discharge); 1.5 A; 9 W
2.25
a –0.183 A (charge); 5.725 A (discharge); 5.542 A (discharge)
b 108.55 V
2.26
a 2.273 A (discharge); –0.455 A (charge);
b 1.32 A
c 11.363 V
2.27 0.376 A; 0.388 A; 0.764 A; 61.14 V
2.28 1.194 V

CHAPTER 3

3.1
a 0.2 μC
b 2.29 μC/m^2
3.2 16.94 kV/m
3.3 50 kV/m
3.4 60 mC/m^2
3.5 125 kV/m
3.6 40 mC
3.7 165.96 V
3.8 300 V
3.9 500 pF or 0.5 nF
3.10 240 pF or 0.24 nF
3.11 3
3.12 5.53 nF
3.13
a 442.7 pF
b 0.177 μC
c 400 kV/m
3.14 0.089 mm
3.15 188 nF
3.16 1.36
3.17 25
3.18
a 4.8 mm
b 1.213 × 10^{-3}m^2
3.19
a 14 μF
b 2.86 μF
3.20
i 1.463 μF; 17 μF
ii 0.013 μF; 0.29 μF
iii 19.18 pF; 490 pF
iv 215 pF; 10.2 nF
3.21 8 nF

3.22
- a 5 μF
- b 200 V
- c 3 mC

3.23
- a 24 μF
- b 480 μC
- c 1.44 mC

3.24
- a 13.85 V (4 nF); 6.15 V
- b 18.46 nC

3.25 C_2 = 4.57 μF; C_3 = 3.56 μF
3.26 200 V; 200 V; 1.2 mC; 2 mC; 3.2 mC
3.27 V_1 = 360 V; V_2 = 240 V; C_3 = 40 μF
3.28 200 V
3.29 80.67 cm²
3.30
- a 48 pF
- b 267 pC
- c 40 V

3.31 625 mJ
3.32 200 V
3.33 5 μF
3.34
- a 40 nF
- b 0.8 mJ
- c 400 k V/m

3.35
- a 1.6 mC; 0.32 J
- b 266.7 V; 0.213 J

3.36
- a 0.6 mC; 150 V; 100 V
- b 120 V; 0.48 mC; 0.72 mC

3.37
- a 1.5 mm
- b 52.94 cm²
- c 75 nC; 28 μJ
- d 14.2 μC/m²

3.38 0.5 μm

CHAPTER 4

4.1 0.417 T
4.2 1.98 mWb
4.3 40 cm²
4.4 21.25 At
4.5 0.8 A
4.6 270
4.7 5633 At/m
4.8 112.5 At
4.9 2A
4.10
- a 900 At
- b 1.11 T
- c 5000 At/m

4.11 64
4.12 278
4.13
- a 1200 At
- b 5457 At/m
- c 1.37 T
- d 549 μWb

4.14
- a 0.206 A
- b 1777

4.15 2.95 A
4.16
- a 360 At/m; 1830 At/m
- b 582

4.17 176.7
4.18 5.14 A
4.19 1.975 A
4.20 1.54 T
4.21 11.57 μWb

CHAPTER 5

5.1 352.9 V
5.2 0.571 ms
5.3 37.5 V
5.4
- a 32 V
- b 5.33 V

5.7 1 Wb/s
5.8 0.05 T
5.9
- a 0.5 V
- b 0.433 V
- c 0.354 V

5.11 1.125 N
5.12 2.5 A
5.13 0.143 m
5.14 3 N
5.15 12.5 mN
5.16 1750 A

5.17 1.12 H
5.18 4 A
5.19 15 625
5.20 5 H; 800 V
5.21 120
5.22 22.5 A/s
5.23 0.75 mH
5.24
a 1364
b 33.6 mH
c 1.344 V
5.25 0.12 H
5.26 30 V
5.27
a 628.3 mH
b 56.5 mH
c 5.09 V
5.28
a 0.3 mH
b 1.125 mH
c 0.3 V; 1.125 V
5.29 58.67 mH
5.30
a 150 V
b 5.625 mWb
5.31 12.5 kV
5.33 12 V
5.34 27.5 V; 400 mA; 11 W
5.35
a 42.67 V
b 3.56 A
5.36 51.2 mJ
5.37 60.4 mH
5.38 2.45 A

CHAPTER 6

6.1
a 67.5 Ω [NPV 68 Ω, 0.5 W]
b 50.7 mA

6.2
a 108 Ω [NPV 120 Ω, 2 W]
b 117 mA
c 12 mV
6.3 At least 12

CHAPTER 7

7.1 50; 2.04 mA
7.2 630 Ω
7.3 636 Ω

CHAPTER 8

8.1
a 150 Hz
b 15 Hz
c 31.83 Hz
8.2
a 100 Hz
b 12.5 rev/s
8.3 24
8.4 5 mA; 50 ±s; 20 kHz
8.5 (b) $i = 7.5 \sin(200\pi t)$milliamp
8.6
a 427.3 V
b 50 Hz
c 302 V; 272.2 V
8.7 $\upsilon = 63.64 \sin(3000\pi t)$ volt; 22.31 V
8.8
a 250 V; 176.8 V; 75 mA; 53 mA; 20 mWb; 14.14 mWb; 6.8 V; 4.81 V
b 25 Hz; 100 Hz; 50 Hz; 1.5 kHz
8.9 $i = 7.07 \sin(4000\pi t)$ amp
a 6.724 A
b 47.86 μs
8.10 353.6 V; 159.25 V
8.11 4.22 mA; 5.97 mA
8.12 1.429
8.13 16 V
8.14 22.2 V
8.18 $\upsilon = 17.44 \sin(314t + 0.409)$ volt
8.19 $i = 22.26 \sin(\omega t — 0.396)$ amp
8.20 $\upsilon = 43.06 \sin(\omega t — 0.019)$ volt
8.23 3.2 V; 2.26 V; 400 μs; 2.5 kHz

CHAPTER 9

9.1 415.6 V
9.2
a 0.2 Ω
b 500 W
9.3 280.75 V

CHAPTER 10

10.1
- a 0.183 s
- b 6.15 mA; 0 A
- c 0.915 s

10.2
- a 333.3 A/s
- b 1 A
- c 15 ms

10.3
- a 10.1 mA

10.4
- a 2.5 Ω

10.5
- a 100 Ω
- b 1 mA

10.6 1.8 h

Index

For Product Safety Concerns and Information please contact our
EU representative GPSR@taylorandfrancis.com Taylor & Francis
Verlag GmbH, Kaufingerstraße 24, 80331 München, Germany